Springer Proceedings in Physics

Volume 404

Indexed by Scopus

The series Springer Proceedings in Physics, founded in 1984, is devoted to timely reports of state-of-the-art developments in physics and related sciences. Typically based on material presented at conferences, workshops and similar scientific meetings, volumes published in this series will constitute a comprehensive up to date source of reference on a field or subfield of relevance in contemporary physics. Proposals must include the following:

- Name, place and date of the scientific meeting
- A link to the committees (local organization, international advisors etc.)
- Scientific description of the meeting
- List of invited/plenary speakers
- An estimate of the planned proceedings book parameters (number of pages/articles, requested number of bulk copies, submission deadline).

Please contact:

For Americas and Europe: Dr. Zachary Evenson; zachary.evenson@springer.com
For Asia, Australia and New Zealand: Dr. Loyola DSilva; loyola.dsilva@springer.com

Ramis Örlü · Alessandro Talamelli ·
Joachim Peinke · Martin Oberlack
Editors

Progress in Turbulence X

Proceedings of the iTi Conference on Turbulence 2023

Springer

Editors
Ramis Örlü
Department of Engineering Mechanics
KTH Royal Institute of Technology
Stockholm, Sweden

Department of Mechanical, Electrical
and Chemical Engineering
OsloMet–Oslo Metropolitan University
Oslo, Norway

Alessandro Talamelli
Department of Industrial Engineering
Università di Bologna
Forlì, Italy

Martin Oberlack
Department of Mechanical Engineering
TU Darmstadt
Darmstadt, Germany

Joachim Peinke
ForWind–Center for Wind Energy
Research, Institute of Physics
AG TWiSt—Turbulenz
Wind Energy and Stochastics
Carl-von-Ossietzky University Oldenburg
Oldenburg, Niedersachsen, Germany

ISSN 0930-8989 ISSN 1867-4941 (electronic)
Springer Proceedings in Physics
ISBN 978-3-031-55930-3 ISBN 978-3-031-55924-2 (eBook)
https://doi.org/10.1007/978-3-031-55924-2

This Springer imprint is published by the registered company Springer Nature Switzerland AG
The registered company address is: Gewerbestrasse 11, 6330 Cham, Switzerland

Paper in this product is recyclable.

Preface

The iTi has become an established biannual conference on turbulence research taking place in the years between the ETC—(European Turbulence Conference) and TSFP—(Turbulence and Shear Flow Phenomena) conferences. With 80–100 participants and a single session only, the iTi conference places value on creating an environment to stimulate discussions and personal contacts in a unique location like the beautiful town of Bertinoro in Northern Italy close to Bologna, continuing a tradition that has started in Bad Zwischenahn/Germany with the first edition of the conference in 2003. The size of the conference allows to have no parallel sessions and gives time to special topics to be stressed. The content-related focus areas of the conference are the interdisciplinary aspects of turbulence, defining the abbreviation iTi—interdisciplinary Turbulence initiative. iTi attracts scientist from the engineering, physics and mathematics communities.

The 10th iTi conference in 2023 (Fig. 1) hosted 105 scientists from 22 different countries. Due to the very large number of abstracts received, it had been very hard for the organizing committee to select and limit the number of accepted contributions. We were therefore forced to reject nearly half of the submissions and could only accept 1/3 of the submissions for oral contributions. In total there were 87 accepted contributions comprising 5 invited talks, 48 regular oral presentations, and 34 poster presentations covering a wide range of aspects of current turbulence research. Advances in the basics of understanding and modelling turbulence were addressed as well as practical implications such as the control of turbulence.

The content of the 10th iTi conference is documented in this volume comprising 46 contributions. All contributions were thoroughly reviewed by external reviewers, to whom we want to express our thanks for their valuable and important contribution.

Based on the successful previous conferences, we will continue with this initiative for subsequent years with the 11th iTi Conference planned for July 2025.

Oslo/Stockholm, Norway/Sweden — Ramis Örlü
Forlì, Italy — Alessandro Talamelli
Darmstadt, Germany — Martin Oberlack
Oldenburg, Germany — Joachim Peinke

Fig. 1 Group photo of the tenth edition of iTi

Looking Back at 20 Years of iTi

The original idea of the interdisciplinary turbulence initiative goes back to Dr. Lachenmeier, who was responsible for funding the fluid mechanics sciences at the German Research Foundation (DFG) at the turn of the millennium. At the end of the 90s, the DFG strove to become more interdisciplinary. Dr. Lachenmeier had the impression that in particular turbulence research is suitable for this objective and suggested to conduct several dual joint research projects between fluid dynamicists and physicists. For this, he approached Joachim Peinke and Martin Oberlack about the organization. This type of research funding was organized in similarity to the well-known 6-year DFG Priority Programmes. To enhance the interdisciplinary exchange, a central point was the holding of a turbulence conference with the participation of international scientists at the halfway point after three years and after the end of the programme after six years. This led to the first two iTi conferences in Bad Zwischenahn in the years 2003 and 2005. Already at the first meeting around 100 scientists from four continents and ten countries participated. In this meeting we also started the tradition of a conference dinner with a dinner speech, which was held in 2003 by Robert (Bob) Moser.

At the last of these two iTi founding conferences, and after the end of DFG funding, the idea was born to continue these conferences in an international setting. Enthusiastic about the first two conferences, Hassan Nagib and Alessandro Talamelli suggested to organize the third conference at the *University Residential Centre at Bertinoro* (Ce.U.B.). The idea to organize the next iTi in Bertinoro was further stimulated by the presence of CICLoPE (Centre of International Cooperation in Long Pipe Experiment) in the close town of Predappio, which was in the process of being set up by Alessandro Talamelli at the time. The CICLoPE, inspired by a group of different universities and research centres, run by University of Bologna, is located beside the old factory of the Caproni Industry, in a tunnel complex excavated in between the two world wars. In 2005, the tunnels were given by the Air Force to the University of Bologna specifically for turbulence research. The complex comprises two 130 m long tunnels with a diameter of about 9 m each. The stability of the ambient conditions and the complete absence of any background noise (vibration, electrical noise, etc.) are the characteristics that make Predappio an ideal site for this laboratory.

In 2007, time of the first visit of Ce.U.B., CICLoPE was in its early phase of construction. The tunnels were at the beginning of the renovation phase, being used before as a site for growing mushrooms (see Fig. 2). The visit by Martin and Joachim was organized by Alessandro and Henrik Alfredsson who was Guest Professor at the University of Bologna at that time. The first approach was not encouraging. During the way to a relaxing dinner in a restaurant on a hill, we were hit by a car, spending the entire evening waiting and talking to the police. The day after was much better. We finally realized that Ce.U.B. in Bertinoro could have been a very good candidate to host the conference, fully matching the spirit of its participants. The third conference was a success. The participants liked the site, a small centre, ideal for a single session meeting. All the researchers enjoyed the pleasant time in Bertinoro, discussing about their research topics, proposing new ideas in an interdisciplinary way, and setting up new collaborations.

Fig. 2 Photo of the tunnel in 2008

"Alea Iacta Est", Caesar said crossing the Rubicone, a small river not far from Bertinoro. With these words Caesar meant that the point of no return had been reached. And 2008 was the "Rubicone" for iTi, as from 2008 iTi has not moved from Bertinoro.

In each edition of the iTi a visit to the CICLoPE was organized. It was very nice to show to the scientific community the progress in the installation of the infrastructure (Fig. 3). The visits were always a chance to discuss possible future experiments or cooperation. The historical interest of the site was anyway a reason to organize a visit even when the laboratory was not completed (Fig. 4).

In 2015, the CICLoPE has been inaugurated and we took the chance to organize in 2016, just before the seventh edition of the conference, a special workshop with the title "Eu-HIT Workshop on High Reynolds number turbulent flows—A large-scale infrastructure perspective" (Fig. 5).

The conference dinner has been always a key point in the iTi organization. We must confess that it was not very complicated, given the area's offerings, to find places of interest for the participants. We visited places like the beautiful Borgo Condè with its cellar where "noble" fluids are studied, the restaurants of Cesenatico, the land that hosted one of the first researchers in turbulence, Leonardo da Vinci, for which he designed the beautiful harbour "Porto Canale", the Academy of food, "Casa Artusi" in Forlimpopoli, birthplace of Pellegrino Artusi, the author of the first monography

Fig. 3 Photo of the tunnel in 2010

Fig. 4 Photo of the tunnel in 2015

Fig. 5 Photo of the inauguration in 2015

"Science in the kitchen and the art of eating well". During this dinner we hosted the introductory lecture "Fluid mechanics in the kitchen" by Henrik Alfredsson.

The iTi organizing committee has been always ready to open the participation to colleagues. Starting from the diarchy of Martin and Joachim, in 2008 Alessandro was very welcomed to join, as a third member and local organizer. As conference organizers the committee has often received mails with compliments and very seldom complaints.

In 2008, an email was received: "*Dear iTi2008 organising committee. According to your homepage, the acceptance of abstracts should be announced June 30th, 2008, with a "short delay". As an engineer such a term means nothing to me without a reference. A fly would have died several times in this "short" time. I would appreciate a more realistic estimate of a date [...] I apologise for not being patient, but I really take deadlines seriously. Ramis Örlü*". Since similar complaints were received in 2010, 2012 and 2014, the only option to satisfy this participant was to make him a co-organizer/editor of iTi, and since then we have not had any complaints from him or from anyone else.

On the contrary, and certainly also because we now had Ramis Örlü on board, we have received a lot of very positive feedback over the years from participants from very different areas of turbulence research, which we are very proud to have been authorized to show:

Michael Wilczek: "Congratulations on running this successful format for twenty years and counting! [...] You have been bringing together researchers from various disciplines working on turbulence in a truly interdisciplinary atmosphere. [...] I literally grew up scientifically with this conference since I was a Ph.D. student. Many of my esteemed senior colleagues I first met at iTi meetings [...] Last but not least, I have to admit that I also came for the food and the beautiful view!"

Christos Vassilicos: "iTi has now a niche opportunity to be our long term single session conference on turbulence in Europe (and perhaps worldwide)".

Gregory Falkovich: "It was one of the few unforgettable conferences in my life, both science and place".

Dwight Barkley: "I hope that this significant, stimulating meeting on turbulence continues for many years to come".

Philipp Schlatter: "Most important has always been the people that attend the conference, as participants, as invited speakers, as organisers. There have been many interesting discussions in the last 20 years, with friends and colleagues at the iTi. I hope to come back many times more in the future!"

The magnificent castle of Ce.U.B. in Bertinoro has persuaded us not to move the location for the iTi ever since. In the following group photo from the first iTi conference in Bertinoro in 2008, all participants and the organizers are sitting in a small amphitheatre by the castle (Fig. 6).

At later iTi conferences, we then switched to taking the group photos at the entrance steps of the main entrance to the castle, as at the largest of all iTi conferences to date in 2023, see the picture in the Preface.

Fig. 6 Group photo of the third edition of iTi in 2008

However, there was one exception to the Bertinoro location and that was during the COVID pandemic in 2021, when we had to hold the iTi conference purely online. We took a group photo anyway (Fig. 7).

In addition to the iTi conference itself, from 2010 we also began to organize workshops, either before or after the respective conference, with a topical focus that was close to turbulence research:

iTi 2010 "The Impact of Turbulence on Wind Energy Converters" (co-organized by Luciano Castillo)
iTi 2012 "Workshop on symmetries and conservation laws"
iTi 2014 "Turbulence out of classical equilibrium in nature and engineering and multiscale-generated flows"
iTi 2016 "Eu-HIT Workshop on High Reynolds number turbulent flows—A large-scale infrastructure perspective"
iTi 2018 "Workshop on Turbulent aspects in wind-energy"

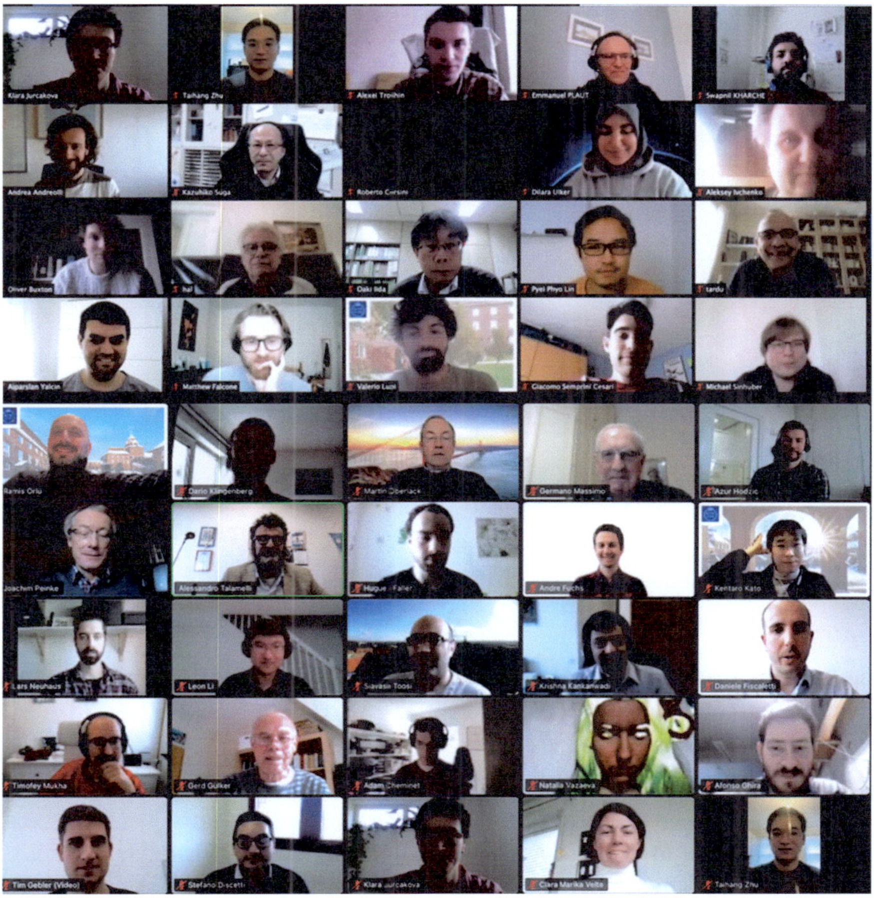

Fig. 7 Some of the participants of the online edition of iTi in 2021

With the iTi 2023, we say goodbye to Joachim Peinke as part of the iTi organizing committee and in recognition of his great achievements and as an organizer and source of ideas for the conference. Joachim was the main driving force behind the idea of the iTi from the very beginning and also had the extremely successful idea of organizing 1- or 2-day workshops in the context of the iTi. In this context, the first and at the same time most successful iTi workshop on wind energy in 2010

Fig. 8 Farewell cake for Joachim during iTi2023 during the dinner on the beach in Cervia

deserves special mention, for which the proceedings were downloaded more than 130 thousand times. We wish Joachim all the best and hope that he will continue to attend the next iTi conferences in Bertinoro as a participant (Fig. 8).

We are pleased to announce that Michael Wilczek from Uni Bayreuth will be part of the Organizing Committee starting from the next iTi conference. Welcome, Michael!

Stockholm/Oslo, Sweden/Norway — Ramis Örlü
Forlì, Italy — Alessandro Talamelli
Darmstadt, Germany — Martin Oberlack

Contents

Wall-Bounded Flows

Free Shear Flows

Contributors

H. Abitan Technical University of Denmark, Department of Civil and Mechanical Engineering, Kgs. Lyngby, Denmark

F. Ambrogi Queen's University, Kingston ON, Canada

E. Amico Department of Mechanical and Aerospace Engineering, Politecnico di Torino, Torino, Italy

M. Asadi Norwegian University of Science and Technology, Trondheim, Norway

C. Atkinson Laboratory for Turbulence Research in Aerospace and Combustion (LTRAC), Department of Mechanical and Aerospace Engineering, Monash University, Melbourne, Victoria, Australia

W. J. Baars Faculty of Aerospace Engineering, Delft University of Technology, Delft, The Netherlands

R. Barta Institute of Aerodynamics and Flow Technology, DLR, Göttingen, Germany

C. Bauer Department Ground Vehicles, Institute of Aerodynamics and Flow Technology, German Aerospace Center, Göttingen, Germany

V. Baxerres Illinois Institute of Technology, Chicago, IL, USA

G. Bellani Department of Industrial Engineering, University of Bologna, Forlì, Italy

N. Biswas Imperial College London, London, UK

C. Bogey Laboratoire de Mécanique des Fluides et d'Acoustique, Ecully, France

S. Brusco Faculty of Engineering, Western University, London, Canada

P. A. Bullee Norwegian University of Science and Technology, Trondheim, Norway

O. R. H. Buxton Imperial College of Science, Technology and Medicine, London, UK

G. Cafiero Department of Mechanical and Aerospace Engineering, Politecnico di Torino, Torino, Italy

J. Chen Imperial College London, London, UK

A. Cimarelli DIEF, University of Modena and Reggio Emilia, Modena, Italy

J. H. Conrad Chair of Fluid Dynamics (FDY), Darmstadt, Germany

G. Dacome Faculty of Aerospace Engineering, Delft University of Technology, Delft, The Netherlands

F. J. G. de Oliveira Imperial College of Science, Technology and Medicine, London, UK

T. Dokoza Technical University of Darmstadt, Chair of Fluid Dynamics, Department of Mechanical Engineering, Darmstadt, Germany

R. du Puits Aerodynamics Group, Technische Universitaet Ilmenau, Ilmenau, Germany

C. Egbers Department of Aerodynamics and Fluid Mechnaics, BTU Cottbus-Senftenberg, Cottbus, Germany

M. A. Ferreira Department of Process and Energy, Delft University of Technology, Delft, Netherlands

E. Foschi Department of Engineering Mechanics, KTH Royal Institute of Technology, Stockholm, Sweden

N. Furuichi National Institute of Advanced Industrial Science and Technology, Teragu, Tsukuba, Japan

B. Ganapathisubramani Department of Aeronautics and Astronautics, University of Southampton, Southampton, Hampshire, UK

D. F. Gayme Johns Hopkins University, Baltimore, MD, USA

S. R. Gomez California Institute of Technology, Pasadena, California, USA

S. Görtz Chair of Fluid Dynamics (FDY), Darmstadt, Germany

G. Hasanuzzaman Department of Aerodynamics and Fluid Mechnaics, BTU Cottbus-Senftenberg, Cottbus, Germany

R. J. Hearst Department of Energy and Processing Engineering, Norwegian University of Science and Technology, Trondheim, Norway

S. Heinz Department of Mathematics and Statistics, University of Wyoming, Laramie, WY, USA

A. Hemmati Department of Mechanical Engineering, University of Alberta, Edmonton, AB, Canada

K. Y.-N. Hinh University of Calgary, Calgary, Canada

A. Hodžić Technical University of Denmark, Department of Civil and Mechanical Engineering, Kgs. Lyngby, Denmark

S. Hoyas Instituto Universitario de Matemática Pura y Aplicada, Universitat Politècnica de València, València, Spain

N. Hutchins Department of Mechanical Engineering, University of Melbourne, Parkville, VIC, Australia

A. Hölling ForWind and Institute of Physics, University of Oldenburg, Oldenburg, Germany

M. Hölling Institute of Physics and For Wind, University of Oldenburg, Oldenburg, Germany

O. Iida Nagoya Institute of Technology, Nagoya, Japan

M. Iovieno DIMEAS, Politecnico di Torino, Torino, Italy

Y. Ito Graduate School of Engineering, Nagoya University, Nagoya, Japan

G. Iuso Department of Mechanical and Aerospace Engineering, Politecnico di Torino, Turin, Italy

K. Iwano Department of Mechanical Systems Engineering, Okayama University of Science, Okayama, Japan

C. T. Johansen University of Calgary, Calgary, Canada

C. Kaestner Production Technology Group, Technische Universitaet Ilmenau, Ilmenau, Germany

T. Kanda MHI Aerospace Systems Corp, Nagoya, Japan

E. Kannadasan Laboratory for Turbulence Research in Aerospace and Combustion (LTRAC), Department of Mechanical and Aerospace Engineering, Monash University, Melbourne, Victoria, Australia

R. Kapulla Paul Scherrer Institute, Villigen, Switzerland

Z. S. Khodaei Imperial College of Science, Technology and Medicine, London, UK

J. Klewicki Department of Mechanical Engineering, University of Melbourne, Parkville, Australia

J. Klinner Institute of Propulsion Technology, German Aerospace Center (DLR), Cologne, Germany

L. Lazzarini Department of Industrial Engineering, University of Bologna, Forlì, Italy

L. Li Department of Energy and Processing Engineering, Norwegian University of Science and Technology, Trondheim, Norway

G. Lunghi Dipartimento di Ingegneria Civile e Industriale, University of Pisa, Pisa, Italy

C. Y. MacDougall Queen's University, Kingston ON, Canada

A. Mariotti Dipartimento di Ingegneria Civile e Industriale, University of Pisa, Pisa, Italy

R. J. Martinuzzi Department of Mechanical and Manufacturing Engineering, University of Calgary, Calgary, Canada

Y. Masada Faculty of Science, Department of Applied Physics, Fukuoka University, Fukuoka, Japan

B. J. McKeon Stanford University, Stanford, California, USA

T. Medjnoun Department of Aeronautics and Astronautics, University of Southampton, Southampton, Hampshire, UK

B. A. Minnick Johns Hopkins University, Baltimore, MD, USA

M. Mommert Department Ground Vehicles, Institute of Aerodynamics and Flow Technology, German Aerospace Center, Göttingen, Germany

J. Monty Department of Mechanical Engineering, University of Melbourne, Parkville, VIC, Australia

S. Morita Institute of Fluid Science, Tohoku university, Sendai, Miyagi, Japan

C. Morton Department of Mechanical Engineering, McMaster University, Hamilton, Canada

T. Mukha Computer, Electrical and Mathematical Sciences and Engineering Division, King Abdullah University of Science and Technology, Thuwal, Saudi Arabia

H. Nagib Illinois Institute of Technology, Chicago, IL, USA

L. Neuhaus ForWind and Institute of Physics, University of Oldenburg, Oldenburg, Germany

I. Neunaber Norwegian University of Science and Technology, Trondheim, Norway

C. T. Nguyen Chair of Fluid Dynamics, TU Darmstadt, Darmstadt, Germany

B. Nugroho Department of Mechanical Engineering, University of Melbourne, Parkville, VIC, Australia

S. Obayashi Institute of Fluid Science, Tohoku university, Sendai, Miyagi, Japan

M. Oberlack Department of Mechanical Engineering, Technical University of Darmstadt, Chair of Fluid Dynamics, Darmstadt, Germany

M. Obligado University of Lille, CNRS, ONERA, Arts et Métiers ParisTech, Centrale Lille, UMR 9014 - LMFL - Laboratoire de Mécanique des fluides de Lille - Kampé de Feriet, Lille, France

P. J. Olesen Technical University of Denmark, Department of Civil and Mechanical Engineering, Kgs. Lyngby, Denmark

M. Ono National Institute of Advanced Industrial Science and Technology, Teragu, Tsukuba, Japan

R. Örlü Department of Mechanical, Electrical and Chemical Engineering, OsloMet–Oslo Metropolitan University, Oslo, Norway

J. Peinke ForWind and Institute of Physics, University of Oldenburg, Oldenburg, Germany

J. Philip Department of Mechanical Engineering, University of Melbourne, Parkville, Australia

G. Piccardo Dipartimento di Ingegneria Civile, Chimica e Ambientale, Polytechnic School, University of Genoa, Genoa, Italy

U. Piomelli Queen's University, Kingston ON, Canada

D. Plümacher Chair of Fluid Dynamics (FDY), Darmstadt, Germany

F. C. Putz Chair of Fluid Dynamics, Darmstadt, Germany

R. Reinartz Department of Applied Physics, Eindhoven University of Technology, Eindhoven, Netherlands

S. Rezaeiravesh Department of Fluids and Environment, The University of Manchester, Manchester, UK

S. L. Ribergaard Technical University of Denmark, Department of Civil and Mechanical Engineering, Kgs. Lyngby, Denmark

F. M. Ricci Department of Mechanical and Aerospace Engineering, Politecnico di Torino, Turin, Italy

S. Romero Department of Mechanical Engineering, University of Melbourne, Parkville, Australia

F. Romoli DIEF, University of Modena and Reggio Emilia, Modena, Italy

Y. Sakai Global Engagement Center, Nagoya University, Nagoya, Japan

M. V. Salvetti Dipartimento di Ingegneria Civile e Industriale, University of Pisa, Pisa, Italy

D. Schiepel Institute of Aerodynamics and Flow Technology, DLR, Göttingen, Germany

M. Schiødt Technical University of Denmark, Department of Civil and Mechanical Engineering, Kgs. Lyngby, Denmark

P. Schlatter Institute of Fluid Mechanics (LSTM), Friedrich–Alexander–Universität (FAU) Erlangen–Nürnberg, Erlangen, Germany

K. Seitz Department of Mechanical Engineering, University of Alberta, Edmonton, AB, Canada

J. Serpieri Department of Mechanical and Aerospace Engineering, Politecnico di Torino, Turin, Italy

J. Soria Laboratory for Turbulence Research in Aerospace and Combustion (LTRAC), Department of Mechanical and Aerospace Engineering, Monash University, Melbourne, Victoria, Australia

E. Stalio DIEF, University of Modena and Reggio Emilia, Modena, Italy

T. Takiwaki Division of Science, National Astronomical Observatory of Japan (NAOJ), Mitaka, Tokyo, Japan

A. Talamelli Department of Industrial Engineering, University of Bologna, Forlì, Italy

S. Tardu LEGI, Univ. Grenoble Alpes, CNRS, Grenoble-INP, Grenoble, France

Y. Tsuji Nagoya University, Furo-cho, Chikusa-ku, Nagoya, Japan

M. Uhlmann Institute for Water and Environment, Karlsruhe Institute of Technology, Karlsruhe, Germany

M. Umair LEGI, Univ. Grenoble Alpes, CNRS, Grenoble-INP, Grenoble, France

C. M. Velte Technical University of Denmark, Department of Civil and Mechanical Engineering, Kgs. Lyngby, Denmark

S. Verma Department of Mechanical Engineering, University of Alberta, Edmonton, AB, Canada

B. Viggiano Polytechnique Montréal, Montréal, QC, Canada

R. Vinuesa FLOW, Engineering Mechanics, KTH Royal Institute of Technology, Stockholm, Sweden

M. Vocke Department of Mechanical and Manufacturing Engineering, University of Calgary, Calgary, Canada

C. Wagner Department Ground Vehicles, Institute of Aerodynamics and Flow Technology, DLR, German Aerospace Center, Göttingen, Germany; Institute of Thermodynamics and Fluid Mechanics, Technische Universität Ilmenau, Ilmenau, Germany

M. Wang Graduate School of Engineering, Nagoya University, Nagoya, Japan

C. E. Willert Institute of Propulsion Technology, German Aerospace Center (DLR), Cologne, Germany

M. Wächter ForWind and Institute of Physics, University of Oldenburg, Oldenburg, Germany

A. Yakeno Institute of Fluid Science, Tohoku university, Sendai, Miyagi, Japan

N. Yokoi Institute of Industrial Science (IIS), University of Tokyo, Meguro, Tokyo, Japan

T. Yurikusa Graduate School of Engineering, Nagoya University, Nagoya, Japan

D. Zaza DIMEAS, Politecnico di Torino, Torino, Italy

Y. Zhang Technical University of Denmark, Department of Civil and Mechanical Engineering, Kgs. Lyngby, Denmark

S. Zimmerman Department of Mechanical Engineering, John Hopkins University, Baltimore, USA

Theory

Recent Lessons From Computations in Wall-Bounded Turbulence for Pipes and Channels

H. Nagib, V. Baxerres, R. Vinuesa, and S. Hoyas

Abstract Direct Numerical Simulations (DNSs) are one of the most powerful tools for studying turbulent flows. Even if the Reynolds numbers achievable with DNS are lower than those obtained with experimental means, there is a clear advantage since the entire velocity field is known, and any desired quantity can be evaluated. This also includes computation of derivatives of all relevant terms. One such derivative provides the indicator function, which may depend on mesh spacing and distribution, and the convergence of computation. The indicator function is a cornerstone to understanding inner and outer interactions in wall-bounded flows, and the description of the overlap region between them. We find dependence of the indicator function on the mesh distributions we examined, raising questions about the classical mesh and convergence requirements for DNS.

1 Introduction

Since the early work by Kim, Moin, and Moser [1], direct numerical simulation (DNS) has become a widely used approach to study fundamental aspects of wall-bounded turbulence [2–7]. One topic within the study of wall-bounded turbulence that

H. Nagib (✉) · V. Baxerres
Illinois Institute of Technology, 60616 Chicago, IL, USA
e-mail: nagib@iit.edu

V. Baxerres
e-mail: vbaxerres@hawk.iit.edu

R. Vinuesa
FLOW, Engineering Mechanics, KTH Royal Institute of Technology, SE-100 44 Stockholm, Sweden
e-mail: rvinuesa@mech.kth.se

S. Hoyas
Instituto Universitario de Matemática Pura y Aplicada, Universitat Politècnica de València, 46022 Valencia, Spain
e-mail: serhocal@mot.upv.es

R. Örlü et al. (eds.), *Progress in Turbulence X*, Springer Proceedings in Physics 404,
https://doi.org/10.1007/978-3-031-55924-2_1

has received extensive attention from the research community due to its significant implications on flow representation and modeling is the overlap region of the mean velocity profile. In this overlap region, the large structures that populate the outer layer co-exist with the smaller ones strongly influenced by the wall and viscous effects. According to the classical literature, the mean velocity profile in this overlap region follows the well-known logarithmic law, which contains the von Kármán coefficient, κ [8, 9]:

$$\overline{U_{x\text{in}}^{+}}(y^{+} \gg 1) = \frac{1}{\kappa} \ln y^{+} + B_0. \tag{1}$$

In this equation, $\overline{U_x}$ is the streamwise mean velocity. The "in" subindex indicates that we are using the wall characterization of this velocity. The independent variable y^+ is the inner-scaled wall-normal coordinate, $y^+ = yu_\tau/\nu$, where u_τ is the friction velocity and ν is the fluid kinematic viscosity.

Over the years, this has been a topic of active study, and universality of the log law and the von Kármán coefficient have been occasionally challenged or reaffirmed; e.g., see references [10–16]. Recently, Monkewitz and Nagib [17] (MN from now on) shed additional light on this topic, challenging the accumulated knowledge during the last century. MN's main point is to consider in the inner asymptotic expansion a term proportional to the wall-normal coordinate, $\mathcal{O}(Re_\tau^{-1})$. Here $Re_\tau = u_\tau\delta/\nu$ is the friction Reynolds number with δ a characteristic outer length. This δ can be thought of as the radius R in pipes, the semi-height in channels, h, or a length scale related to δ_{99}. Following MN, the pure log law is not observed in channels and pipes (as well as other flows with streamwise pressure gradients) even if Re_τ is not above $\mathcal{O}(10^4)$. MN's extended matched asymptotics reveals an additional term in the expansion of the velocity for this layer leading to an extra contribution in the overlap layer in the form $S_0 y^+ Re_\tau^{-1}$, where S_0 is a coefficient.

Interestingly, the overlap region represented by a combined logarithmic and linear terms exhibits values of the von Kármán coefficient κ consistent with those obtained from skin-friction relations, a fact that suggests that this approach can reveal high-Re trends even at moderate Reynolds numbers. The main point of this matched asymptotics theory is to use a new expression in the overlap region for $\overline{U}_x$. MN's formula reads:

$$\overline{U_{x\text{in}}^{+}}(y^{+} \gg 1) \sim \kappa^{-1} \ln y^{+} + B_0 + B_1/Re_\tau + S_0 y^{+}/Re_\tau. \tag{2}$$

To obtain accurate values of these coefficients, it is necessary to address two fundamental issues:

1. To find κ, the challenge is to identify the location and extent of the overlap region, which is weakly dependent on the Reynolds number [17].
2. To obtain the correct value of S_0, it is necessary to obtain with high accuracy the values of $\mathrm{d}\overline{\mathrm{U_x}}^{+}/\mathrm{d}y^{+}$.

Focusing on the second point above, it is possible to obtain an equation for the value of κ and S_0 through the indicator function:

$$\Xi(y^+) = y^+ \frac{\mathrm{d}\overline{U_x}^+}{\mathrm{d}y^+}. \tag{3}$$

From equations (2) and (3), one can obtain the equation for κ and S_0:

$$\Xi(y^+) = \kappa^{-1} + S_0 y^+ / Re_\tau.$$

Thus, to accurately determine S_0, it is necessary to compute the indicator function with high accuracy, an aspect which has not been sufficiently investigated in the literature. While a convergence criterion exists for fully developed flows [18], and the domain size of the problem has been examined in depth [19, 20], the necessary grid spacing has received much less attention. Examples can be found in some very large simulations [7, 21], where the authors only report their mesh size and compare their results with those in previous references, which basically do the same.

2 Direct Numerical Simulations

We have performed two different numerical experiments to examine this issue and its implications. The details can be found in Table 1. On the one hand, we have studied a smooth pipe of radius R and length $L_x = 10\pi R$ for $Re_\tau = 550$. We will use y to indicate the distance to the wall. We have used four different meshes, as seen in

Table 1 Summary of the simulations discussed in this work, where 'P' stands for pipe and 'C' for channel. The nomenclature 'LR' and 'HR' stands for low and high resolution in y, whereas 'C' and 'F' denote coarse and fine in x and z. Note that L_x and L_z are the periodic streamwise and spanwise dimensions while h is the channel half height. Δx^+ and Δz^+ are the inner-scaled resolutions in terms of dealiased Fourier modes. The wall-normal direction is indicated in both flows by y. N_x, N_z, and N_y are the numbers of collocation points in the three different directions. The time span of the simulation is given in terms of eddy turnovers $u_\tau T/\delta$, and ε is a measure of convergence, defined in [18]. Colors given in the first column are used throughout the paper, using dashed lines for the channel

Case	Re_τ	L_x/δ	L_z/δ	Δx^+	$\Delta(R\theta^+)$	$\max(\Delta y^+)$	$\min(\Delta y^+)$	N_x	N_z	N_y	ETT	ε
PLRF	549	10π	2π	5.62	3	3.2	0.018	3072	1152	256	93	1.2×10^{-4}
PHRF	549	10π	2π	5.62	3	1.6	0.005	3072	1152	512	87	2.6×10^{-4}
PLRC	549	10π	2π	11.2	4.5	3.2	0.018	1536	768	256	184	8.2×10^{-5}
PHRC	549	10π	2π	11.2	4.5	1.6	0.005	1536	768	512	56	3.8×10^{-4}
CLRC	546	8π	3π	9	5	5.85	0.8	1536	1152	251	150	7.6×10^{-5}
CHRC	546	8π	3π	9	5	1.68	0.05	1536	1152	901	150	3.8×10^{-5}

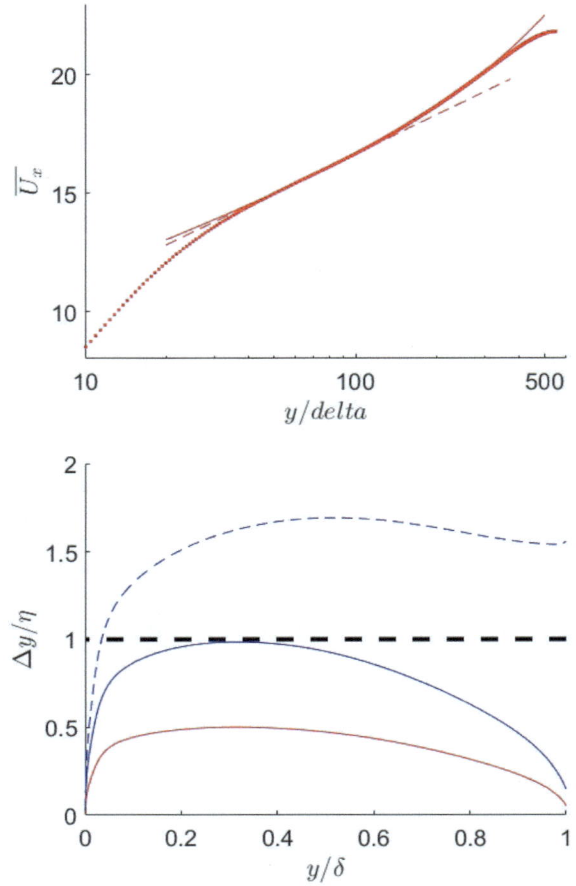

Fig. 1 (Top) Streamwise velocity for the PHRF case (thick line), together with the pure log (dashed line) and log-plus-linear (thin line) overlap layers. (Bottom) Grid distance in the wall-normal direction, scaled with the Kolmogorov length. Colors as in Table 1, where solid denotes pipe and dashed channel

Table 1. The base mesh is the case PLRC with the same resolution in x and Θ as a channel. On the other hand, the mesh in the radial direction has twice the number of points needed in a channel due to the skewness of the cells near the pipe center. The details of the mesh can be seen in right side of Fig. 1. Note that for the pipe cases, the wall-normal grid spacing is below the Kolmogorov scale througout the computational domain. Starting with the PLRC, we doubled the cells in the radial direction (PHRC), azimuthal, and streamwise directions (PLRF), and all of them (PHRF). All these simulations ran for at least 55 eddy-turnover times (ETT), defined as ETT $= Tu_\tau/R$. This is the primary motivation behind making our initial phase of DNS to Re_τ around 550.

On the other hand, we have run two channel-flow simulations. The first one (CLRC) is a simulation with the same mesh as Jiménez and Hoyas [22, 23]. The second one (CHRC) uses a mesh considered enough for $Re_\tau = 3000$ [19, 24]. Two different codes were employed. OpenpipeFlow was used to run the pipe simulations

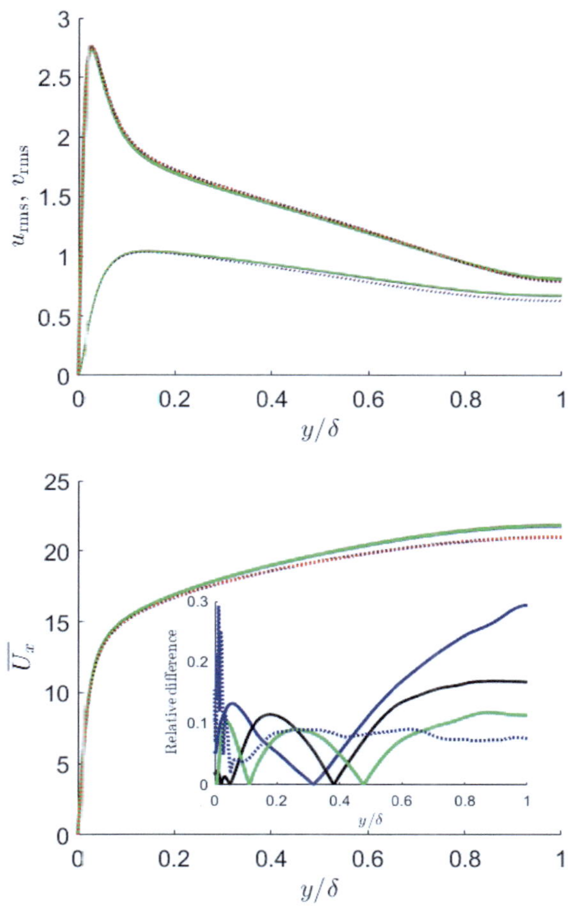

Fig. 2 (Top) Intensities in the wall-normal and streamwise directions for all cases. (Bottom) $\overline{U_x}$. The box shows the difference between the finer case and the others. Colors as in Table 1. Continuous lines, pipes. Dotted lines, channels

[21, 25]. The code LISO was used for running the channel flow [2, 26, 27]. Both codes employ Fourier-decomposition techniques in the streamwise and spanwise directions. Openpipeflow uses a seventh-order finite-difference scheme in the wall-normal direction, while LISO uses a tenth-order compact-finite-difference scheme [26, 28].

In the case of the pipe flow, the difference among the four simulations appears small when analyzing the streamwise and wall-normal fluctuations, as displayed in Fig. 2 (left), and the streamwise velocity, Fig. 2 (right). As shown in the latter, the difference between the coarser cases and the finer ones is below 0.3% everywhere. Such accuracy, however, is not sufficient for utilizing the indicator function Ξ to extract reliable coefficients of overlap region. As shown in Fig. 3 (left), the different curves of Ξ do not collapse exactly in the region between $y/\delta = 0.3$ and $y/\delta = 0.5$, where the slope value of Ξ is critical. This is more clearly appreciated in Fig. 3 (right),

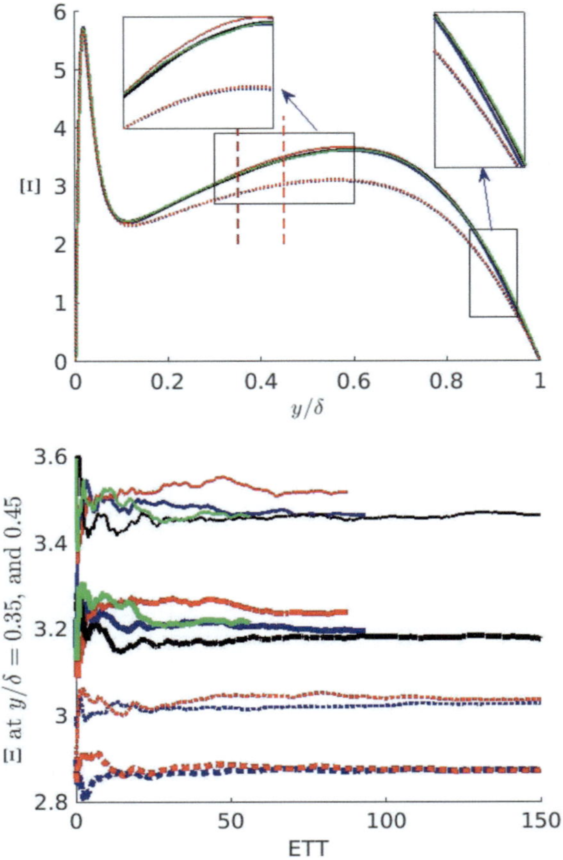

Fig. 3 (Top) Ξ for the six case under study. (Bottom) Values of Ξ at $y/\delta = 0.35$ and 0.45 as a function of the averaging time t_a expressed in eddy-turnover times (ETT). Colors as in Table 1. Continuous lines, pipes. Dotted lines, channels. Thick lines, $y/\delta = 0.35$. Thin lines, $y/\delta = 0.45$

where the differences across cases can be more clearly assessed. Notice that the finer mesh, PHRF case, corresponds to the solid red line. The other three cases have not converged to the PHRF's value even after 100 ETT. A better agreement is found in the case of the channels. Here, the convergence at $y/h = 0.35$ is obtained after approximately 15 ETT. However, at $y/h = 0.45$ the differences are of the order of 2% after 80 ETT, which is an extremely lengthy computation for high-Reynolds-number simulations.

The effect of this lack of adequate resolution can be better evaluated by the time-averaged values of the overlap coefficients κ, B, and S_0 as functions of ETT. After 50 ETTs, the differences among the cases are still above 3%, as shown in Fig. 4. Again, the convergence for the channel case is better, but there remains a small difference up to ETT of 150, which corresponds to very expensive computations. In any case, at least 20 ETTs are needed to obtain some level of convergence. Such differences in the indicator function Ξ should be smaller for any reliable simulation to predict the values of κ, B, and S_0.

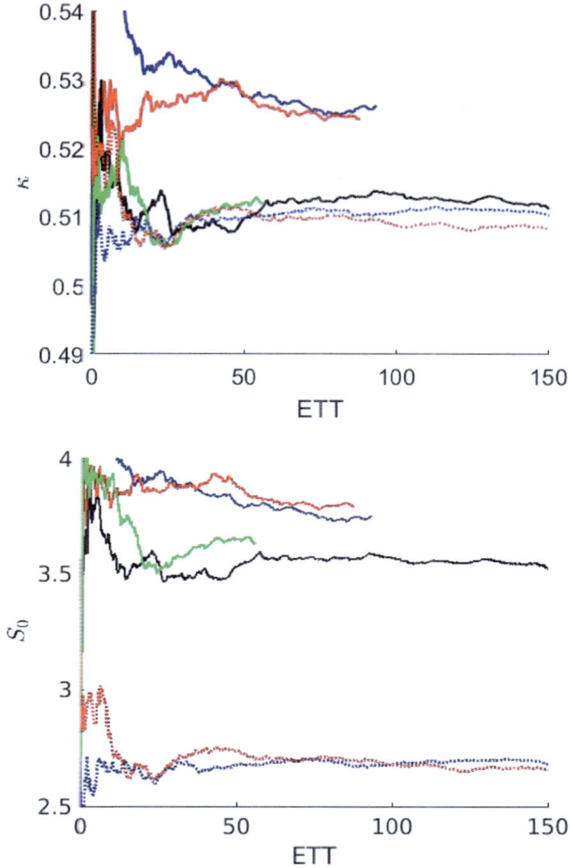

Fig. 4 κ (Top) and S_0 (Bottom) for the six cases under study as functions of the averaging time t_a expressed in ETT. Colors as in Table 1. Continuous lines, pipes. Dotted lines, channel

Finally, in the case of pipe flow, the values of these coefficients appear to depend on the azimuthal resolution, raising questions about the mesh aspect ratio of the grids. A careful study of such mesh variation effects in flows near their singularities, as in case around the axis of pipes, is highly recommended. Such a study should also examine the sesnetivisty of the indicator function Ξ to the order of the finite difference scheme used in the DNS. In general, the results demonstrate that classical and commonly used mesh size in the wall-normal direction for DNSs are not sufficiently fine to extract coefficients of the overlap region in wall-bounded turbulent flows, such as the the von Kármán coefficient, κ.

3 Summary

In conclusion, DNSs still stand as a powerful tool for the exploration of turbulent flows. However, the accuracy of DNS to compute certain important derivatives such as those required for the indicator function Ξ is challenged. To calculate Ξ, accurate values of the wall-normal derivative of the streamwise velocity must be established. This is a highly critical issue for experiments and should be addressed in all future measurements with more densely and equally spaced measurement locations in the wall-normal direction. Our investigation reveals that the indicator function exhibits sensitivity around the classic mesh size commonly used in DNSs. In particular, several coefficients of the overlap region required for the Monkewitz and Nagib's model of wall-bounded flows under streamwise pressure gradients [17] cannot be found with less than 3% difference among the various meshes we tested. For example, this creates a new uncertainty source in the estimation of the von Kármán coefficient that is very important for modeling and predictiong wall-bounded turbulent flows. This uncertainty can be also generalized to any quantity needing wall-normal derivatives.

4 Data Availabilty

The data used for this paper can be obtained by contacting S. Hoyas at serhocal@mot.upv.es.

Acknowledgements For computer time, this research used the resources of the Supercomputing Laboratory at King Abdullah University of Science & Technology (KAUST) in Thuwal, Saudi Arabia, Project K1652 RV acknowledges the financial support from ERC grant no. '2021-CoG-101043998, DEEPCONTROL'. Views and opinions expressed are however those of the author(s) only and do not necessarily reflect those of European Union or European Research Council. Neither the European Union nor granting authority can be held responsible for them. SH is funded by project PID2021-128676OB-I00 by Ministerio de Ciencia, innovación y Universidades/FEDER.

References

1. J. Kim, P. Moin, R. Moser, Turbulence statistics in fully developed channels flows at low Reynolds numbers. J. Fluid Mech. **177**, 133 (1987)
2. S. Hoyas, J. Jiménez, Scaling of the velocity fluctuations in turbulent channels up to $Re_\tau =$ 2003. Phys. Fluids **18**, 011702 (2006)
3. P.R. Spalart, J.H. Watmuff, Experimental and numerical study of a turbulent boundary layer with pressure gradients. J. Fluid Mech. **249**, 337 (1993)
4. J. Sillero, J. Jiménez, R.D. Moser, One-point statistics for turbulent wall-bounded flows at Reynolds numbers up to $\delta^+ = 2000$. Phys. Fluids **25**, 105102 (2013)
5. S. Pirozzoli, J. Romero, M. Fatica, R. Verzicco, P. Orlandi, One-point statistics for turbulent pipe flow up to $Re_\tau = 6000$. J. Fluid Mech. **926**, A28 (2021)

6. R. Vinuesa, S.M. Hosseini, A. Hanifi, D.S. Henningson, P. Schlatter, Pressure-gradient turbulent boundary layers developing around a wing section. Flow Turbul. Combust. **99**, 613 (2017)
7. S. Hoyas, M. Oberlack, F. Alcántara-Ávila, S.V. Kraheberger, J. Laux, Wall turbulence at high friction Reynolds numbers. Phys. Rev. Fluids **7**, 014602 (2022)
8. T. von Kármán, Turbulence and skin friction. J. Aero. Sci. **1**, 1 (1934)
9. C. M. Millikan, A critical discussion of turbulent flows in channels and circular tubes, in *Proc. 5th Int. Congr. Appl. Mech.* (Wiley, NY, 1938)
10. W.K. George, L. Castillo, Recent advancements toward the understanding of turbulent boundary layers. AIAA J. **44**, 2435 (2006)
11. G.I. Barenblatt, A.J. Chorin, V.M. Prostokishin, A note on the intermediate region in turbulent boundary layers. Phys. Fluids **12**, 2159 (2000)
12. P.A. Monkewitz, K.A. Chauhan, H.M. Nagib, Comparison of mean flow similarity laws in zero pressure gradient turbulent boundary layers. Phys. Fluids **20**, 105102 (2008)
13. R. Vinuesa, P. Schlatter, H.M. Nagib, Role of data uncertainties in identifying the logarithmic region of turbulent boundary layers. Exp. Fluids **55**, 1751 (2014)
14. H.M. Nagib, K.A. Chauhan, Variations of von Kármán coefficient in canonical flows. Phys. Fluids **20**, 101518 (2008)
15. P. Luchini, Universality of the turbulent velocity profile. Phys. Rev. Lett. **118**, 224501 (2017)
16. M. Oberlack, S. Hoyas, S. V. Kraheberger, F. Alcántara- Ávila, and J. Lawx, Turbulence statistics of arbitrary moments of wall-bounded shear flows: A symmetry approach. Phys. Rev. Lett. **128**, 024502 (2022)
17. P. Monkewitz, H. Nagib, The hunt for the Kármán 'constant' revisited. J. Fluid Mech. **967**, A15 (2023)
18. R. Vinuesa, C. Prus, P. Schlatter, H. Nagib, Convergence of numerical simulations of turbulent wall-bounded flows and mean cross-flow structure of rectangular ducts. Meccanica **51**, 3025 (2016)
19. A. Lozano-Durán, J. Jiménez, Effect of the computational domain on direct simulations of turbulent channels up to Re? = 4200. Phys. Fluids **26**, 011702 (2014)
20. F. Lluesma-Rodríguez, S. Hoyas, and M. Peréz-Quiles, Influence of the computational domain on DNS of turbulent heat transfer up to Re_τ = 2000 for Pr = 0.71, Int. J. Heat Mass Transf. **122**, 983 (2018)
21. J. Yao, S. Rezaeiravesh, P. Schlatter, F. Hussain, Direct numerical simulations of turbulent pipe flow up to Re_τ = 5200. J. Fluid Mech. **956**, A18 (2023)
22. S. Hoyas, J. Jiménez, Reynolds number effects on the Reynolds-stress budgets in turbulent channels. Phys. Fluids **20**, 101511 (2008)
23. J. Jiménez, S. Hoyas, Turbulent fluctuations above the buffer layer of wall-bounded flows. J. Fluid Mech. **611**, 215 (2008)
24. F. Alcántara-Ávila, S. Hoyas, and M. Pérez-Quiles, Direct Numerical Simulation of thermal channel flow for Re_τ = 5000 and Pr = 0.71, J. Fluid Mech. **916**, A29 (2021)
25. A.P. Willis, The openpipeflow navier-stokes solver. SoftwareX **6**, 124 (2017)
26. F. Lluesma-Rodríguez, F. Alcántara-Ávila, M. Pérez-Quiles, and S. Hoyas, A code for simulating heat transfer in turbulent channel flow. Mathematics 9, https://doi.org/10.3390/math9070756 (2021)
27. L. Yu, M. Z. Yousif, M. Zhang, S. Hoyas, R. Vinuesa, and H.-C. Lim, Three-dimensional esrgan for super-resolution reconstruction of turbulent flows with tricubic interpolation-based transfer learning. Phys. Fluids **34** (2022)
28. S.K. Lele, Compact finite difference schemes with spectral-like resolution. J. Comput. Phys. **103**, 16 (1992)

A Laboratory and Theoretical Framework for Systematic Non-equilibrium Turbulence Studies

C. M. Velte, A. Hodžić, H. Abitan, P. J. Olesen, M. Schiødt, S. L. Ribergaard, and Y. Zhang

Abstract The cornerstone assumption of equilibrium of the small and intermediate scales in the classical view of turbulence (K41–the combined efforts of Kolmogorov, Batchelor and Richardson) is under ever increased scrutiny. Although the K41 based models do appear to apply well to some flows, there exist many important flows that are problematic for these turbulence models. In particular, it is interesting to note that the most challenging applications appear to have one thing in common–rapid changes of the flow in the mean in time and/or space. It is thus interesting to systematically investigate what the bounds of validity of the classical K41-view of turbulence are, if any. And if the K41-picture of turbulence does indeed break down, what are the non-linear spectral energy transfer mechanisms that lead to non-equilibrium turbulence behavior (local vs. non-local)? Does the non-linear energy exchange between scales divert from the classically assumed Richardson cascade? And is the constancy of the spectral flux across the inertial range interrupted? In order to answer these questions, a new facility for the systematic study of non-equilibrium turbulence in a controlled setting has been established along with an accompanying theoretical framework that is tailored for addressing these specific issues.

1 Rationale

In the proceedings of the Marseille 2011 meeting, William K. George [1] identified the implicit but central assumption of equilibrium of the small and intermediate scales to be the 'Achilles heel' of the classical K41 picture of turbulence. Furthermore, previous workers have also touched upon this, including John L. Lumley [2], who expressed the issue as follows: "What part of modeling is in serious need of work? Foremost, I would say, is the mechanism that sets the level of dissipation in a turbulent flow, particularly in changing circumstances."

C. M. Velte (✉) · A. Hodžić · H. Abitan · P. J. Olesen · M. Schiødt · S. L. Ribergaard · Y. Zhang
Technical University of Denmark, Department of Civil and Mechanical Engineering, Nils Koppels Allé B. 403, Kgs. Lyngby, Denmark
e-mail: cmve@dtu.dk

R. Örlü et al. (eds.), *Progress in Turbulence X*, Springer Proceedings in Physics 404,
https://doi.org/10.1007/978-3-031-55924-2_2

The underlying idea of K41 is that energy flowing into the system, the (assumed constant) spectral energy flux, and the dissipation rate are, on average, balanced. And that the small and intermediate scales are always in a state of equilibrium (and even universal!), no matter what the dynamics are at the large scales. This implied balance is, in principle, quite delicate and can be readily invalidated using various thought experiments, e.g., under the following conditions:

- Strong non-stationarity, e.g., a suddenly starting jet.
- Strong non-local interactions that bypass the (assumed) Richardson cascade and thereby creates a non-constant spectral flux that does not equal to the dissipation rate, in general.
- The K41 theory rests on a fluid-flow analogy where the small and intermediate scales are effectively decoupled from the large scales, like molecules in a thermodynamic equilibrium. But even molecules can be pushed out of equilibrium (e.g., shocks, compressibility effects, etc.). Hence, there is no reason to believe that the (significantly larger) small and intermediate scales cannot be subject to non-equilibrium.

The validity of this local (i.e., small and intermediate scale) assumption remains to be systematically tested. K41 assumes that this local equilibrium is a result of the (Richardson) cascade resulting from local interactions. If this assumption fails, the theory is left with only one hope of remaining valid, i.e., that all scales are in equilibrium. Thus, one may argue that the K41 theory is in essence an equilibrium solution for turbulence.

A simple shear layer can be imagined in yet another thought experiment. Let two large co-rotating structures rotate sufficiently close to each other. As a consequence of continuity, they should be able to generate smaller scales through non-local interactions in between, see Fig. 1. Non-equilibrium in the large scales can thus readily be directly transferred to the small scales. Let these large structures become large, and one eventually obtains a mixing layer, which furthermore is also known from solutions of instability theory to be unstable to perturbations of *all* wavelengths. In contrast to homogeneous isotropic turbulence, shear layers have significant importance in both industrial applications and nature. If non-local interactions are by default present in shear flows it is important to understand under what conditions these are energetically amplified, resulting in higher degrees of non-equilibrium turbulence. From simulations and quantitative and qualitative flow visualisations, we know today that non-local interactions constitute the rule rather the exception, not least in shear flows and unsteady flows of various kinds.

2 Experimental Facility

So how do we advance our understanding? The round turbulent jet has been chosen as the test bed for several reasons, among the central ones being that the jet is near equilibrium when stationary and that it can be pushed out of equilibrium in a practical,

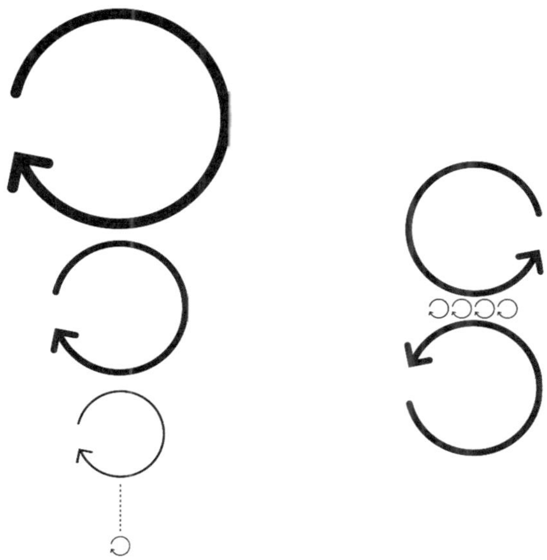

Fig. 1 Examples of mechanisms setting the level of dissipation. (left) Local interactions (Richardson cascade) and (right) Non-local interactions representing a thought experiment of a shear flow. In the latter case, the dynamics of the large scales can be directly transferred between the large and small scales, hence non-equilibrium can bypass the cascade and cause non-equilibrium in the small and intermediate scales

reproducible and arbitrary manner. The scales are also sufficiently large and slow such that the smallest relevant (dissipative) scales can be resolved. Further, the global development of the jet can be measured since the round turbulent jet evolves over practical distances.

Indeed, we are pushing the limits for what can be measured with currently available technology. Still, the experiments had to be divided up into two separate ones: Direct measurement of the full instantaneous dissipation rate are carried out to quantify the degree of non-equilibrium of the modulated jet flow. And global measurements of the jet in four dimensions to decompose the energetic and inertial range of the flow field into waves to analyze whether non-stationarity energizes different (non-local) non-linear interactions than the classical Richardson cascade, see e.g., Fig. 1. In this way, we can map out the mechanisms that set this level of dissipation. The measurements will be carried out for various degrees of non-equilibrium and Reynolds numbers, which are the main known parameters to investigate.

Both experiments are based on particle tracking velocimetry (PTV) and the Shake-the-box method [3]. The facility consequently consists of two separate test cells and a control room in the middle from where the experiments and acquisition are controlled. The experiments are challenging and have required in-house development of optical solutions, including improved high-power LEDs, in-house designed Scheimpflug camera mounts and dedicated optics to improve laser based light budgets.

2.1 Quantifying Degree of Non-Equilibrium from the Dissipation Rate

The left test cell is specifically designed for measuring the full instantaneous dissipation rate. The dissipation rate is highly interesting in its own right, since this is a key parameter both in turbulence theory and modeling. In the present setup, it is also used as a measure to quantify the degree of non-equilibrium in the measured turbulence. The resolution of the particle tracking measurements is sufficient to not only resolve the Kolmogorov scale; we can in fact measure both the dissipation rate and the dissipation of the dissipation. Hence, we are able to measure the balance of the full dissipation transport equation, which is of central importance for both theory and modeling of turbulence.

The inner jet used in this setup has an exit diameter of 100 mm (and an outer concentric one with 145.5 mm, both have the same jet exit area), which is an order of magnitude larger than that in the other (global dynamics) experiment. With this setup we are slightly beyond 1:1 imaging, which sets quite strict requirements on light source and cameras. We are illuminating $\sim 1\,\mu$m DEHS oil droplet tracers using an Innolas Blizz 40W laser with a beam diameter of a few millimeters (depending on measurement configuration) in the area of measurement. Four Phantom Highspeed v2640 cameras are used to image the scattering from the tracer particles. To be able to correct the focus for the perspective, we have developed two-axis Scheimpflug adaptors that can adjust the normal of the focal plane in two directions (Fig. 2).

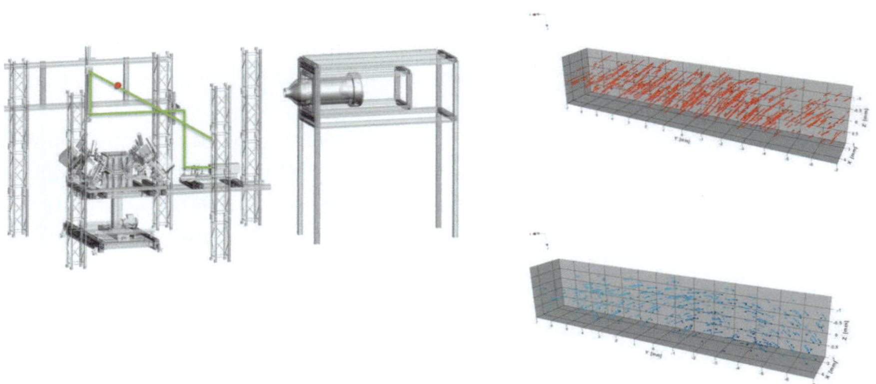

Fig. 2 (left) Drawing of dissipation rate measurements setup. (right) Snapshot of particle tracks measured at the centerline (top figure) and the maximum shear region (bottom figure) of the jet at 30 jet exit diameters downstream of the jet exit. The measurement volume is $2.0 \times 2.5 \times 14$ mm

2.2 *Unveiling the Mechanisms Setting the Level of Dissipation*

The right test cell is designed for capturing the global dynamics of the turbulence in the jet and to unveil the mechanisms that lead to the resulting value of the dissipation rate. Again, PTV is used in conjunction with shake-the-box. With this setup, we will be able to capture the largest turbulent scales and down to varying degrees along the spectrum, depending on the Reynolds number of the flow.

Again, we use four Phantom Highspeed v2640 cameras to record the signal. Soap bubbles are used as tracers, since 1 μm oil droplets would scatter far too little light. Since we need to measure a volume of $\sim 20 \times 20 \times 8$ cm, lasers are not a viable illumination source and instead LEDs are used. Unfortunately, available commercial LEDs are far too weak and expensive in terms of the required light, which is why we decided to develop our own LEDs tailored for the experiment. We have developed three generations of LEDs, each more than doubling the available pulse energy and reducing the spreading angle. The following numbers represent the LED pulse energy and spreading half angle, respectively, for each generation for a 25 μs pulse at 4 kHz:

$$\begin{aligned} \text{Commercial}: &\quad \sim 0.1\,\text{mJ}, \quad 10^\circ \\ \text{Generation1}: &\quad \sim 0.2\,\text{mJ}, \quad 4.9^\circ \\ \text{Generation2}: &\quad \sim 0.4\,\text{mJ}, \quad 2.2^\circ \\ \text{Generation3}: &\quad \sim 0.9\,\text{mJ}, \quad 2.2^\circ \end{aligned}$$

It is important to reduce the spreading angle of the LEDs, in order to decrease the forward scattering angle to the cameras while maintaining high optical density at the volume of interest, which substantially improves the light budget (Fig. 3).

Fig. 3 Drawing of setup for measuring the global dynamics of the jet that will unveil the mechanisms setting the level of dissipation

3 Theory Framework

Equally important to the physical laboratory facilities is the developed theoretical framework that enables robust analysis of the acquired measurements. We will primarily use the Proper Orthogonal Decomposition (POD) and similar techniques to analyze the measured flows. Some of the most recent and important methods developed are briefly described and referenced below:

- **Phase POD**: Provides modes as a function of phase (using phase averaging across ensembles rather than time averaging). This enables POD to be applied to non-stationary flows as well as stationary ones [4].
- **Dissipation POD**: In addition to the classical energy optimized POD, we have developed a dissipation optimized POD [5].
- **Combined POD**: Allows combining proper orthogonal decomposition (POD) bases optimized with respect to different norms, e.g., the energy and dissipation optimized POD, into a single complete basis [6].
- **Particle POD**: Applicable for the analysis of the modal dynamics of Lagrangian particle tracks in both stationary and non-stationary flows, e.g., tracer particles in PTV, inertial particles or droplets dispersed in a carrier fluid [7].
- **Discrepancies between POD and Fourier modes on aperiodic domains**: It is often desired to apply the Fourier transform in POD to significantly reduce the computational cost (by applying the Fast Fourier Transform) and to obtain analytical POD modes. However, neither in homogeneous nor truncated (typical measured) signals are Fourier modes solutions to the POD eigenvalue problem [8]. In contrast, in periodic homogeneous domains, Fourier modes are indeed POD modes. Since the round turbulent jet is periodic only in the axisymmetric direction, the general POD, rather than Fourier analysis, is implemented in the remaining spatial directions and time [8].

Acknowledgements This project has received funding from the European Research Council (ERC) under the European Union's Horizon 2020 research and innovation programme (grant agreement No 803419). The objective of the project is to test the bounds of validity of the classical theory of turbulence. Financial support from the Poul Due Jensen Foundation (Grundfos Foundation) for this research is gratefully acknowledged. The objective of the project is to develop new models and theory for non-equilibrium turbulence.

References

1. W.K. George, Reconsidering the local equilibrium hypothesis and implications for 'K41'-based theories of turbulence, in *Turbulence Colloquium Marseille 2011 EDP Sciences* (2014)
2. J.L. Lumley, Some comments on turbulence. Phys. Fluids **4**, 201–211 (1992)
3. Schanz et al., *Shake-the-box: Lagrangian particle tracking at high particle image densities* (Exp, Fluids, 2016)

4. Y. Zhang, A. Hodžić, F. Evrard, B. Van Wachem, C.M. Velte, Phase proper orthogonal decomposition of non-stationary turbulent flow. Phys. Fluids **35**(4), 045109 (2023). 1 April 2023
5. P.J. Olesen, A. Hodžić, S.J. Andersen, N.N. Sørensen, C.M. Velte, Dissipation-optimized proper orthogonal decomposition. Phys. Fluids **35**(1), 015131 (2023)
6. P.J. Olesen, A. Hodžić, C.M. Velte, Combined Proper Orthogonal Decompositions of Orthogonal Subspaces. arXiv:2303.16823 [physics.flu-dyn] (2023)
7. M. Schiødt, A. Hodžić, F. Evrard, M. Hausmann, B. van Wachem, C.M. Velte, Characterizing Lagrangian particle dynamics in decaying homogeneous isotropic turbulence using proper orthogonal decomposition. Phys. Fluids **34**(6), 063303 (2022)
8. A. Hodžić, P.J. Olesen, C.M. Velte, On the Discrepancies between POD and Fourier Modes on Aperiodic Domains. arXiv:2207.02550 [physics.flu-dyn] (2023)

Multiscale Dynamics in Turbulent Wakes

N. Biswas and O. R. H. Buxton

Abstract In this work we demonstrate the energy exchange processes to/from coherent structures present in a multiscale flow consisting of a main cylinder and a control rod. Apart from the vortex shedding frequencies of the two wake-generating bodies, new frequencies are observed that correspond to the sum and difference of the former. We show that the sheddings of the wake-generating bodies are energised by the mean flow, hence we refer to them as 'primary modes'. Contrastingly, the new frequencies are found to be entirely energised through non-linear interaction between the primary modes, establishing the notion of 'secondary modes' that are not directly energised by the mean flow. The same analysis is extended for a wind turbine wake which is inherently of multiscale nature. Preliminary results show that there not only exist modes that resemble the primary and secondary modes observed in the prior cylinder–control rod flow, but there also exist modes of mixed nature (energised by both mean flow and non-linear interactions) and modes that are solely driven by convection.

1 Introduction

Real life flows are often multiscale in nature, where multiple length/time scales can be present simultaneously. Typical examples could be flow through offshore platforms, a forest or cityscape or even flow through a wind turbine. In a wind turbine wake for instance, forcings are introduced at many different length/time scales from the blades (such as the tip and root vortices and trailing sheet vortices), from the nacelle and the tower as well as the large scale structures, such as wake meandering, that scale with the whole turbine *i.e.* rotor diameter D. All these structures may have a unique role in determining the spatio-temporal evolution of the wake [1–3]. Unravelling the physics

N. Biswas (✉) · O. R. H. Buxton
Imperial College London, London SW7 2BX, UK
e-mail: n.biswas20@imperial.ac.uk

O. R. H. Buxton
e-mail: o.buxton@imperial.ac.uk

R. Örlü et al. (eds.), *Progress in Turbulence X*, Springer Proceedings in Physics 404,
https://doi.org/10.1007/978-3-031-55924-2_3

of such a complicated multiscale system however, requires prior understanding from simpler multiscale flows. Such a simple example of a multiscale flow could be the flow through the combination of a large cylinder (so called main cylinder) and a small cylinder (so called control-rod). Recent PIV experiments by [4] have shown that such a flow can exhibit a number of interesting flow states depending on the precise location of the control rod. We first use the data from [4] to analyse this simplified flow and then take preliminary strides to develop a deeper understanding of the multiscale nature of a wind turbine wake.

Methodology: To understand the spatio-temporal nature of the different coherent structures present in the flow, a triple decomposition of the instantaneous velocity field can be sought in the following form.

$$u(x, y, t) = \bar{u}(x, y) + \sum_{l} \tilde{u}_l(x, y, t) + u^{'}(x, y, t) \tag{1}$$

Here, the flow field is decomposed as a combination of the mean component ($\bar{u}$), sum of contributions from different coherent motions ($\sum_l \tilde{u}_l$) and the incoherent or stochastic part (u'). The coherent part can be extracted through several modal decomposition methods such as Proper Orthogonal decomposition (POD), Dynamic mode decomposition (DMD), Optimal mode decomposition (OMD) etc. In the present study we employ OMD, which is a more generalised version of DMD [5, 6]. More details of the OMD-based triple decomposition technique can be found in [7]. The origin of the different frequencies can then be understood by studying the energy budgets for each of the various coherent modes, which were developed by [8]. The coherent energy budget ($\tilde{k}_l$) equation can be represented in a symbolic form as

$$\frac{\partial \tilde{k}_l}{\partial t} = -\tilde{C}_l + \tilde{P}_l - \hat{P}_l + \left(\tilde{T}_l^+ - \tilde{T}_l^-\right) - \tilde{\epsilon}_l + \tilde{D}_l. \tag{2}$$

In equation 2, the source terms on the right hand side consist of convection ($\tilde{C}_l$), production from the mean flow ($\tilde{P}_l$), production of stochastic turbulent kinetic energy ($\hat{P}_l$), triadic energy production ($\tilde{T}_l^+ - \tilde{T}_l^-$), dissipation ($\tilde{\epsilon}_l$) and diffusion ($\tilde{D}_l$).

2 Cylinder and Control Rod

A schematic of the flow contaning the main cylinder and a control rod is shown in Fig. 1a. The diameter of the control rod is 10 times smaller than that of the main cylinder. When we probe the region where the wakes of the two bodies interact, we not only see the shedding frequencies of the two cylinders (denoted by f_m and f_c), but also new frequencies are observed which correspond to the sum and difference of the two shedding frequencies ($f_c \pm f_m$). An example spectrum is shown in Fig. 1b. Such new frequencies were also observed by [8] for a different multiscale geometry

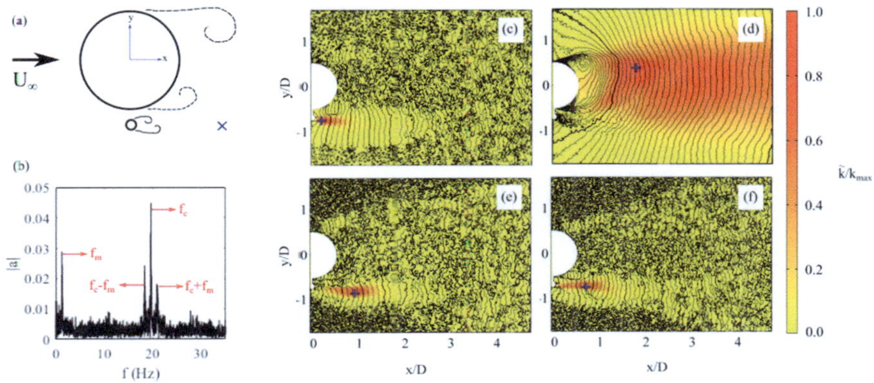

Fig. 1 **a** Schematic of a multiscale system consisting of a main cylinder and a control rod. **b** Frequency spectrum obtained at the point shown by the **x** sign. Energy contours and iso-phase lines associated with **c** the control rod's shedding mode (f_c), and **d** the main cylinder's shedding mode (f_m). The corresponding modes for the low and high frequency secondary modes ($f_c \mp f_m$) are shown in fig. **e** and **f** respectively The '+' sign shows the location where kinetic energy associated to each particular mode is maximum

and uncanny similarities in the properties of these frequencies were observed in two different studies [8, 9].

In Fig. 1c–f we show the modes associated with f_c, f_m and $f_c \mp f_m$ when the control rod is placed at a centre-centre separation of $0.7D$ such that the line joining the centres of the two bodies is perpendicular to the freestream. The filled contours show the time averaged kinetic energies of the modes, while the contour lines represent the iso-phase lines of transverse velocity component obtained from the modes (which is complex in nature, see [9] for more details). Figure 1c, d show the modes corresponding to the shedding of the control rod and main cylinder respectively, which is evident by the presence of a phase correlated region behind the respective wake-generating bodies in which the kinetic energies associated with the modes are concentrated. The location where the kinetic energy of the individual modes attains a maximum is shown by a '+' sign. The modes associated with the new frequencies, as shown in Fig. 1e–f, show a phase-correlated region in the vicinity of the control rod although not as coherent as f_m or f_c. Note that the kinetic energy of these two modes peak downstream between the corresponding location for f_c and f_m. It is pertinent to ask at this point where does this energy come from? In order to answer this question we estimate the different terms in equation 2 for the different modes we observe. Note that in equation 2, apart from classical kinetic energy budget terms, there is a new term denoted as $\tilde{T}_l^+ - \tilde{T}_l^-$ *i.e.* the triadic energy production term. The term is defined as

$$\tilde{T}_l^+ - \tilde{T}_l^- = -\frac{1}{2}\sum_{f_s, f_t} \overline{u_i^{f_l} u_j^{f_t} \frac{\partial u_i^{f_s}}{\partial x_j}} + \frac{1}{2}\sum_{f_s, f_t} \overline{u_i^{f_s} u_j^{f_t} \frac{\partial u_i^{f_l}}{\partial x_j}}. \tag{3}$$

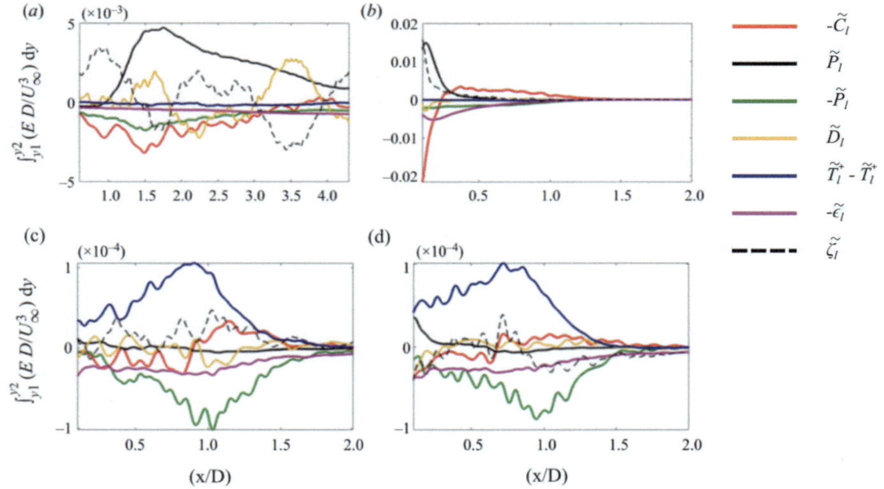

Fig. 2 Streamwise evolution of the spanwise averaged kinetic energy budget terms for **a** the main cylinder's and **b** the control rod's fundamental shedding modes. **c** and **d** show the same for the frequencies $f_c - f_m$ and $f_c + f_m$ respectively

which accounts for the net non-linear triadic energy transfer to/from the l_{th} coherent mode. In 3 the frequencies f_s and f_t correspond to the other frequencies which might influence the non-linear energy gain/loss of f_l. Baj et al. [8] argued that this term could be non-negligible only when the frequencies involved formed a triad, i.e. $f_l \pm f_s \pm f_t = 0$. To obtain the stochastic component of the flow, the coherent modes of physical significance (those associated with f_m and its first two harmonics, f_c and $f_c \pm f_m$) were subtracted from the mean-subtracted velocity field. The streamwise evolution of the spanwise averaged kinetic energy budget terms ($\int_y ED/U_\infty dy$, where E is any kinetic energy budget term) is shown in Fig. 2. For f_m and f_c (Fig. 2a, b), the main source term is the energy production from the mean flow ($\tilde{P}_l$) while for $f_c \mp f_m$ (Fig. 2c, d), the $\tilde{P}_l$ term is negligible. Instead, the major source of energy is the $\tilde{T}_l^+ - \tilde{T}_l^-$ term which shows that $f_c \mp f_m$ are fundamentally different from either f_c or f_m. As these modes are not directly energised by the mean flow in this multiscale flow, it is imperative to term them as 'secondary modes' as opposed to f_c and f_m which are the 'primary modes', being solely energised by the mean flow.

3 Wind Turbine Wake

A wind turbine wake is a more complex multiscale flow due to inherent three-dimensionality and the presence of many more coherent structures, simultaneously. We conducted extensive Particle Image Velocimetry (PIV) experiments on a small-scale wind turbine consisting of a rotor, a nacelle and a tower, mimicking a real

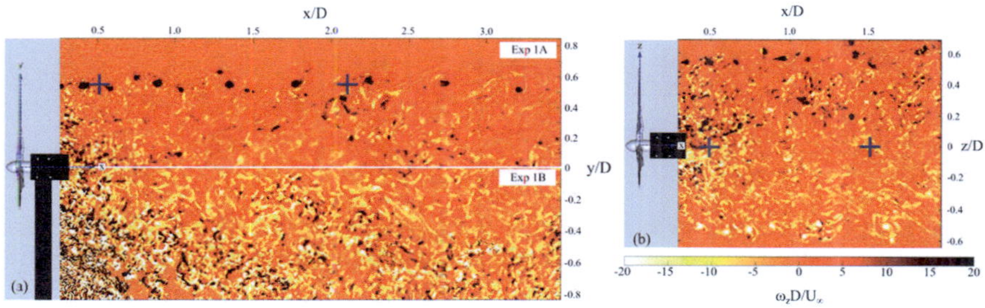

Fig. 3 Instantaneous vorticity field of a model wind turbine wake in **a** the $xy(z = 0)$ plane and **b** the $xz(y = 0)$ plane for $\lambda = 6$. For sub figure **a**, the top $(y > 0)$ and bottom $(y < 0)$ parts of the field of view correspond to two non-synchronous experiments

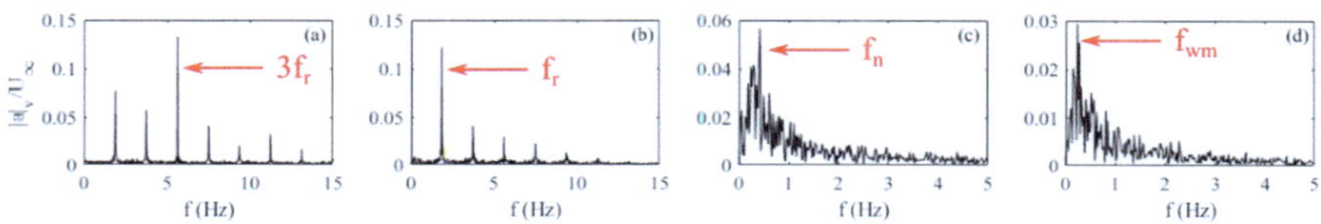

Fig. 4 **a**, **b** Frequency spectra obtained at the points shown by the + sign in Fig. 3a. Sub figures **c**, **d** show the same for Fig. 3b

scale wind turbine. The details of the design of the wind turbine model as well as the PIV experiments can be found in [3]. Figure 3 shows instantaneous vorticity fields at two different planes for tip speed ratio $\lambda = 6$ ($\lambda = \omega R/U_\infty$, where R is the turbine radius, ω is the rotational speed, and U_∞ is the free stream velocity). Figure 3a shows a 'side-view' of the wake at a plane aligned with the tower's axis, while Fig. 3b shows a representative 'top-view' at a plane aligned with the nacelle's axis and perpendicular to the tower's axis. An array of tip vortices can be seen along the outer edge of the wake ($y/D \approx +0.5$, $z/D \approx \pm 0.5$). Contrastingly, for $y < 0$ *i.e.* behind the tower, the flow field is more turbulent and chaotic due to the interaction between the tip vortices and the tower's vortex shedding. Frequency spectra are obtained at the points shown by the + signs and shown in Fig. 4a–d for the xy and xz planes respectively. A number of frequencies could be observed in different regions of the flow, including the blade passing frequency ($3f_r$), rotor frequency (f_r) and their harmonics, nacelle shedding frequency (f_n), wake meandering frequency (f_{wm}) etc, which essentially highlights the multiscale nature of the wake. Understanding the energy exchanges to and from these frequencies can be critical in developing a better understanding of the spatio-temporal development of wind-turbine wakes. Figure 5 shows the energy budget terms for f_r, $3f_r$ and their harmonics. The analysis is only performed for the upper $(y > 0)$ part of the integrated field of view shown in Fig. 3a. The different sectors correspond to the different terms in equation 2 integrated over the entire domain of investigation. The radial lengths of the sectors represent the relative contribution of the different terms in a logarithmic scale and the red and blue colours represent loss and gain of energy. Note that some of the modes are ener-

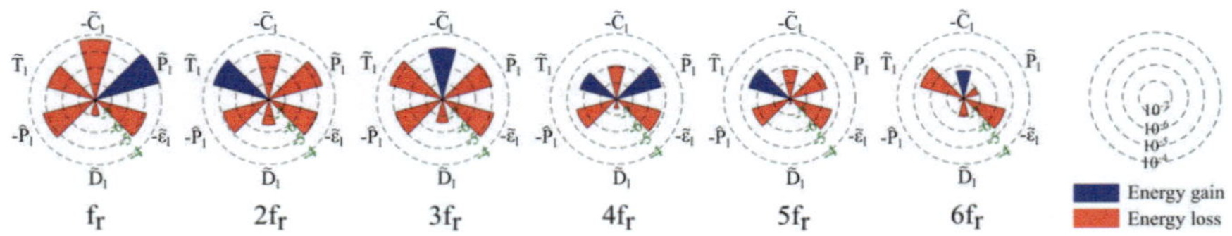

Fig. 5 Energy budget terms for the frequencies f_r–$6f_r$. A red sector denotes loss of energy while a blue sector denotes gain of energy. The dashed lines represent magnitudes of energy loss/gain in a logarithmic scale as shown in the right most sub figure. $\tilde{T}_l^+ - \tilde{T}_l^-$ is represented in short by $\tilde{T}_l$

gised by the mean flow while some are entirely energised by non-linearity, a scenario qualitatively similar to that observed in the simplified example containing a cylinder and a control rod. In addition, there are also some modes which are energised by both production from the mean flow and non-linearity and some solely due to convection. Future work will include a detailed analysis on the energy transfers to/from each of the modes for different operating conditions and turbine designs.

References

1. V.L. Okulov, I.V. Naumov, R.F. Mikkelsen, I.K. Kabardin, J.N. Sørensen, A regular Strouhal number for large-scale instability in the far wake of a rotor. J. Fluid Mech. **747**, 369–380 (2014)
2. L.E.M. Lignarolo, D. Ragni, F. Scarano, C.S. Ferreira, G.J.W. Van Bussel, Tip-vortex instability and turbulent mixing in wind-turbine wakes. J. Fluid Mech. **781**, 467–493 (2015)
3. N. Biswas, O.R.H. Buxton, Effect of tip speed ratio on multiscale coherent dynamics in the near wake of a model wind turbine. ArXiv preprint arXiv:2302.13154 (2023)
4. M.M. Cicolin, O.R.H. Buxton, G.R.S. Assi, P.W. Bearman, The role of separation on the forces acting on a circular cylinder with a control rod. J. Fluid Mech. **915**, A33 (2021)
5. P.J. Schmid, Dynamic mode decomposition of numerical and experimental data. J. Fluid Mech. **656**, 5–28 (2010)
6. A. Wynn, D.S. Pearson, B. Ganapathisubramani, P.J. Goulart, Optimal mode decomposition for unsteady flows. J. Fluid Mech. **733**, 473–503 (2013)
7. P. Baj, P.J. Bruce, O.R.H. Buxton, The triple decomposition of a fluctuating velocity field in a multiscale flow. Phys. Fluids **27**(7) (2015)
8. P. Baj, O.R.H. Buxton, Interscale energy transfer in the merger of wakes of a multiscale array of rectangular cylinders. Phys. Rev. Fluids **2**(11), 114607 (2017)
9. N. Biswas, M.M. Cicolin, O.R.H. Buxton, Energy exchanges in the flow past a cylinder with a leeward control rod. J. Fluid Mech. **941**, A36 (2022)

Linear Amplification of Large Scale Structures in Adverse Pressure Gradient Turbulent Boundary Layers Through Resolvent Analysis

S. R. Gomez and B. J. McKeon

Abstract Incompressible adverse pressure gradient (APG) turbulent boundary layers (TBL) are characterized by their non-parallel effects due to their streamwise deceleration. A biglobal resolvent operator that accounts for the streamwise and wall-normal inhomogeneity of the mean flow field is used to study the linear response in an APG TBL. The resolvent amplification reveals increased amplification in the outer region of the flow for large-scale spanwise wavenumbers with an increase in APG strength. A rank-1 resolvent model for the streamwise fluctuations that considers the near-wall scaling is also presented. The resolvent amplification is shown to grow linearly with the APG history, measured through the streamwise averaged APG strength. This linear growth relates to a recently used hybrid velocity scale that suppresses the APG effects in the streamwise fluctuations.

1 Introduction

Adverse pressure gradients (APG) are generated by a streamwise deceleration of the freestream velocity. In a turbulent boundary layer (TBL), the APG introduces non-negligible non-parallel effects. The APG TBL may be characterized by three parameters— the friction Reynolds number, Re_τ, the APG history, and the Clauser parameter, $\beta = \delta^*/\tau_w dP_\infty/dx$, where δ^*, τ_w, and dP_∞/dx denote the displacement thickness, shear stress at the wall, and free stream pressure gradient [6]. The APG history, $\overline{\beta}$, is defined here as the streamwise average of β, similar to [6], though with a different streamwise average. An increase in β results in an increased wake velocity deficit, stronger streamwise fluctuations, $\overline{uu}$, in the outer region, and energization of

S. R. Gomez (✉)
California Institute of Technology, Pasadena, California, USA
e-mail: gomezsr@stanford.edu

B. J. McKeon
Stanford University, Stanford, California, USA
e-mail: bjmckeon@stanford.edu

R. Örlü et al. (eds.), *Progress in Turbulence X*, Springer Proceedings in Physics 404,
https://doi.org/10.1007/978-3-031-55924-2_4

structures with large spanwise lengthscales [1, 2, 4, 5, 8]. History effects also change the turbulent statistics—flows with larger upstream β generally have increased $\overline{uu}$ [6].

The resolvent analysis framework of [7] has been shown to find key features of turbulent motions. Here, resolvent analysis is used to investigate linear amplification mechanisms associated with the APG TBL mean flow fields. The mean flow fields used are the APG TBL large eddy simulations (LES) of [1, 2] and the zero pressure gradient (ZPG) TBL LES of [3]. These flows are characterized by similar Re_τ with different β and $\overline{\beta}$. In Sect. 2, the biglobal resolvent analysis framework is described along with the numerical implementation. The effects of β and $\overline{\beta}$ on the resolvent amplification are studied in Sects. 3 and 4.

2 Methodology

The state, $\breve{\mathbf{q}} = [\breve{\mathbf{u}}, \breve{p}]$, satisfies the incompressible Navier Stokes Equations with the no-slip boundary conditions over a flat plate with an imposed APG. Using a Reynolds decomposition, the state can be split into $\breve{\mathbf{q}} = \overline{\mathbf{q}} + \mathbf{q}$, where $\overline{\mathbf{q}} = [\overline{U}(x, y), \overline{V}(x, y), 0, \overline{P}(x, y)]$ is a known mean flow field and $\overline{\mathbf{q}}$ is the fluctuation. Assuming homogeneity in the spanwise direction and stationarity in time, $\mathbf{q}$ may be expressed as a sum of Fourier modes as

$$\mathbf{q}(x, y, z, t) = \sum_{\omega, k_z} \widehat{\mathbf{q}}(x, y; k_z, \omega) \exp^{-i\omega t + k_z z} \tag{1}$$

where ω and k_z denote the temporal frequencies and spanwise wavenumbers. Following [7], the terms linear in $\widehat{\mathbf{q}}$ are separated from the nonlinear terms. The linear operator acting on $\widehat{\mathbf{q}}$ is inverted to create the input-output relationship

$$\widehat{\mathbf{q}} = (-i\omega + \mathbf{L})^{-1}\widehat{\mathbf{f}} = \mathbf{R}\widehat{\mathbf{f}}, \tag{2}$$

where $\mathbf{R}$ is the resolvent operator and $\widehat{\mathbf{f}}$ denotes the nonlinear terms. Here, $\widehat{\mathbf{f}}$ is treated as an uncorrelated forcing. The inner product is chosen as the kinetic energy,

$$\langle \mathbf{a}, \mathbf{b} \rangle = \int_{x_0}^{x_0+L} \int_0^{y_{max}} m(x, y)(a_u^* b_u + a_v^* b_v + a_w^* b_w) dy dx, \tag{3}$$

where L is the domain length, y_{max} is the maximum wall-normal height, and m is a spatial mask that is equal to 0 or 1 depending on which regions of the domain are included in the inner product. With this inner product, $\mathbf{R}$ is decomposed with a singular value decomposition (SVD) for the discretized system, as

$$\mathbf{R}(x, y; k_z, \omega) = \sum_{i=1} \sigma_i(k_z, \omega) \boldsymbol{\psi}_i(x, y; k_z, \omega) \boldsymbol{\phi}_i^*(x, y; k_z, \omega) \tag{4}$$

where $\boldsymbol{\psi}_i$ and $\boldsymbol{\phi}_i$ are the response and forcing modes and σ_i is their linear gain.

These modes are discretized using fourth order finite difference differentiation schemes with N_x equispaced points in x and N_y points in y that are stretched to increase the resolution near the wall. The inlet and outlet domains are treated with sponge zones. An Arnoldi algorithm is used to compute the SVD. The inverse is never explicitly computed; instead, an LU decomposition is used at each step of the Arnoldi iteration. The large eddy simulation (LES) mean flow fields of [1, 3, 6] are interpolated onto the grid.

3 Effects of β on the Resolvent Amplification

To query the effect of β on the resolvent amplification, we consider domains centered at x_r where $Re_\tau(x_r) = 537$. Variables without superscripts are nondimensionalized with the edge velocity, $U_e(x_r)$, and local boundary layer thickness, $\delta_{99}(x_r)$. Variables with + superscripts are nondimensionalized with the viscous length and velocity scales at x_r. The mean flow fields used are a ZPG ($\overline{\beta} = 0$) [3], and APG mean flow fields corresponding to cases b1n ($\overline{\beta} = .95$), b2n ($\overline{\beta} = 1.64$), and m18n ($\overline{\beta} = 2.81$) of [1]. These flow fields are characterized by large regions where β is approximately constant to isolate the effects of β on the amplification. The domain sizes are $L = 43.4$ long for all the mean flow fields, except m18n which is 36.9 long. To compare the amplification at x_r, masking is applied to the inner product involving $\boldsymbol{\psi}_i$. The mask, m, is chosen so that only the streamwise region $x \in [x_r - 2, x_r + 2]$ is considered in the response. The inner product for $\boldsymbol{\phi}_i$ is not constrained. The spatial masking allows both small and large scales to be accounted for at x_r.

The effect of the masking is shown in Fig. 1 for representative small and large scales using the $\overline{\beta} = 1.64$ mean flow field. When masking is not applied, the small scales are localized upstream where the shear stress is largest while the large scales are supported across the entire domain. When masking is applied, the small scales

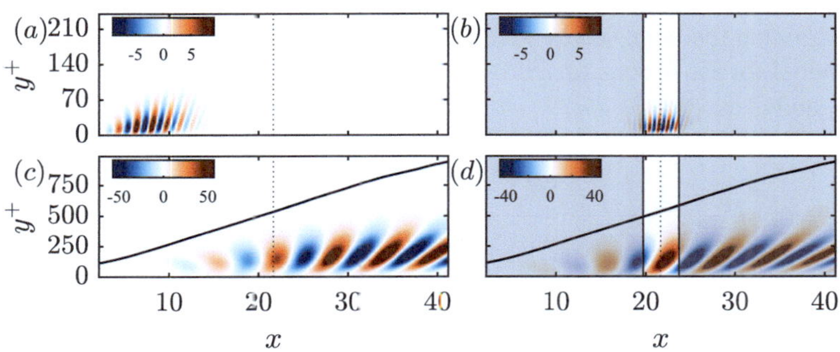

Fig. 1 Real part of $\sigma_1 \psi_{u,1}$ for $\lambda_z^+ = 50, \lambda_t^+ = 50$ **a**, **b** and $\lambda_z^+ = Re_\tau, \lambda_t^+ = 200$ **c**, **d** where masking in the response is applied in **b**, **d** with $\overline{\beta} = 1.64$. The dotted black lines denote x_r. The solid black line denotes $\delta_{99}(x)$

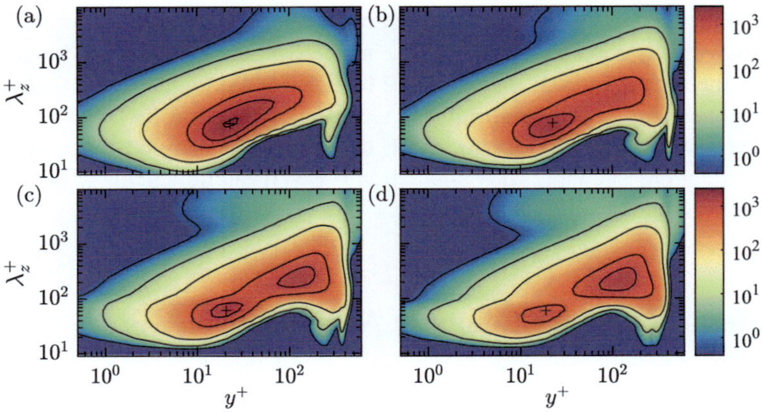

Fig. 2 E^+_{uu} for $\overline{\beta} = 0$ **a**, $\overline{\beta} = .95$ **b**, $\overline{\beta} = 1.62$ **c**, and $\overline{\beta} = 2.81$ **d** at $x = x_r$. The black contour lines denote $E^+_{uu} = .62, 6.17, 61.66, 616.62, 1950$. The black crosses denote the near wall peaks

are centered at x_r. The mode shape of the large scales is only slightly affected by the masking since these modes are amplified by the convective non-normality which amplifies downstream responses via upstream forcing.

The premultiplied kinetic energy spectra showed that the energization of large scales in the outer region of the flow increased with β [8]. To model this using the resolvent amplification, the premultiplied linear amplification is defined as

$$E^+_{uu}(k_z, y^+) = k_z^+ \int_\omega \left|\sigma_1^+(k_z^+, \omega^+)\psi^+_{u,1}(x_r, y^+, k_z^+, \omega^+)\right|^2 d\omega^+, \tag{5}$$

computed using a sweep over k_z and ω, and is plotted in Fig. 2. E^+_{uu} has a near wall peak with small λ_z^+ for all the mean flow fields. The APG cases reveal secondary peaks for large scale structures whose amplification increases with $\overline{\beta}$ that are abesent in the ZPG case. This is consistent with the observations made in experiment or simulation that the large scales are energized with increased β [8].

A model for $\overline{uu}$ is constructed by incorporating a scale factor and integrating in both k_z and ω as

$$K^+_{uu} = \int\int \left|\xi\sigma_1^+(k_z^+, \omega^+)\psi^+_{u,1}(x, y; k_z^+, \omega^+)\right|^2 d\omega^+ dk_z^+, \tag{6}$$

where ξ sets the units for K^+_{uu}. Here, $\xi = U_e(x_r)/(\sqrt{15}u_\tau(x_r))$ is chosen to normalize K^+_{uu} and ensure that K_{uu} scales with $U_e u_\tau$. This mixed scaling has been used in the past to collapse the inner region of $\overline{uu}$ [5]. K^+_{uu} is compared to $\overline{uu}^+$ and is shown to agree qualitatively with $\overline{uu}^+$ from the LES in Fig. 3. There is a distinct secondary peak in the APG flow fields, absent in the ZPG. The secondary peak is over predicted

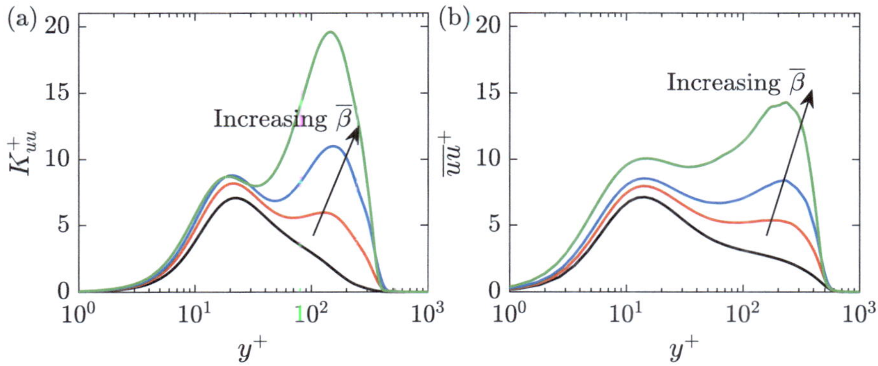

Fig. 3 K^+_{uu} **a** and $\overline{uu}^+$ from the LES data **b** at $x = x_r$. The black, red, blue, and green lines correspond to $\overline{\beta} = 0$, .95, 1.62, and 2.81

with this model since the scaling of ξ only accounts for the near wall scaling. Further work will be required to define a scale dependent ξ.

The similarities between the predictions made by the resolvent amplification and the increased streamwise kinetic energy in simulation and experiment suggest that the energization of the large scales is a linear mechanism occurring because of the change in the mean flow field.

4 History Effects on Resolvent Amplification

The ZPG [3] and APG simulations of [1, 2] are used to study the history, the streamwise variation of $\beta(x)$, on the linear amplification. These flows all have different β variations in x. The domains are chosen with respect to the reference location x_r such that $Re_\tau(x_r) = 777$ and $x \in [x_r - .81L, x_r + .21L]$ where $L = 15.89$. A streamwise average of β was shown to account for history effects in [6]. Following this line of work, $\overline{\beta}$ is defined as the streamwise average of β over the domain. The focus is on the large scale structures which are most susceptible to the APG effects.

In Fig. 4a, $\sigma_1(k_z, \omega; \overline{\beta})/\sigma_1(k_z, \omega; \overline{\beta}_0)$ is plotted for each of the available datasets, with the λ_z and λ_t specified in the legend and $\overline{\beta}_0 = .88$. The lines of best fit

$$\frac{\sigma_m(k_z, \omega, \overline{\beta})}{\sigma_1(k_z, \omega; \overline{\beta}_0)} = \frac{1 + m(k_z, \omega)\overline{\beta}}{1 + m(k_z, \omega)\overline{\beta}_0}, \tag{7}$$

determine $m(k_z, \omega)$ by fitting $\sigma_1(k_z, \omega; \overline{\beta})$ for $\overline{\beta} > 0$ and are plotted in Fig. 4a. The linear growth with $\overline{\beta}$ is similar to the linear growth with β seen in the hybrid velocity scale of [4]. The hybrid velocity scale has been shown to mitigate APG-related effects in $\overline{uu}$ by noticeably reducing the secondary peaks in the outer region. Here,

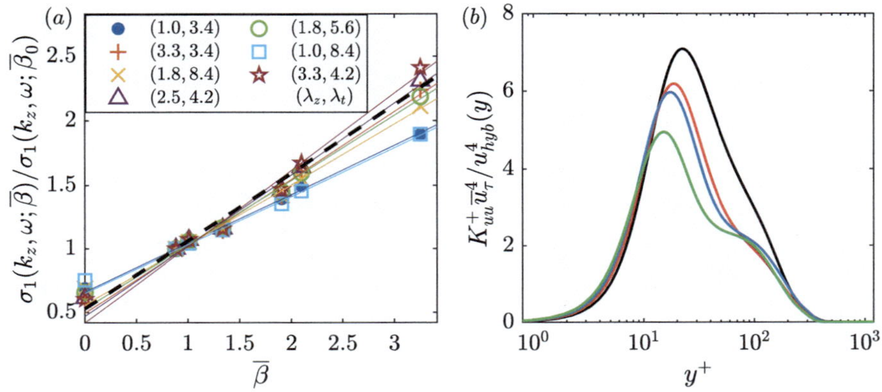

Fig. 4 **a** $\sigma_1(k_z, \omega; \overline{\beta})$, normalized by $\sigma_1(k_z, \omega; \overline{\beta}_0)$, with the symbols and colors labeled in the legend. Solid lines are lines of best fit for each spectral pair. The black dashed line is $z(1+\overline{\beta})/(1+\overline{\beta}_0)$. **b** K_{uu}^+ from Fig. 3, rescaled by $\overline{u}_\tau^4/u_{hyb}^4$

an augmented hybrid velocity scale is defined as

$$u_{hyb}^2 = \overline{u}_\tau^2 \left(1 + \frac{y}{\overline{\delta}^*}\overline{\beta}\right) \tag{8}$$

to account for the integrated β. The original hybrid velocity scale, u_h, is recovered by removing the overbars and using the local quantities. The form of Equation 8 can be recovered in Equation 7 if the wall-normal location, $y/\overline{\delta}^*$, is fixed as m. The linear fits are compared to Equation 7 with $m = 1$. The latter requires no fitting and agrees well with all but $\lambda_z = 1$.

In Fig. 4b, K_{uu}^+ is rescaled with $(1 + y/\overline{\delta}^*\overline{\beta})^2$ to account for the σ_1^2 dependence in Equation 6. Despite the differences in the domain lengths and x_r between the sweeps in Sects. 3 and 4, the u_{hyb} scaling tracks the changes in amplification. The u_{hyb}^4 scaling differs from the expected u_h^2 scaling from [4] because ξ is scale-independent, accounting only for an expected near-wall scaling.

5 Conclusion

The biglobal resolvent analysis described herein captures non-parallel effects associated with APG TBL flows. In particular, resolvent amplification increases in the outer region for the large λ_z modes, consistent with the energization of large scales in experiment and simulation. The accumulated effect of $\beta(x)$ over a domain leads to increased resolvent amplification for the large scale modes, similar to how increased upstream β leads to increased $\overline{uu}^+$, even if β and Re_τ are matched. The growth in σ_1 grows linearly with $\overline{\beta}$ and relates to u_h.

The modeled streamwise fluctuations shown in Sect. 3 relied on a scale-independent nonlinear weight to reconstruct statistics. Future work will focus on using physics-based scaling for the large scale modes to improve comparisons and predictions with experiment and simulation.

Acknowledgements Support from the U.S. Office of Naval Research through grants N00014-17-1-2307 and N00014-17-1-3022 is gratefully acknowledged.

References

1. A. Bobke, R. Vinuesa, R. Örlü, P. Schlatter, History effects and near equilibrium in adverse-pressure-gradient turbulent boundary layers. J. Fluid Mech. **820**, 667–692 (2017)
2. R. Pozuelo, Q. Li, P. Schlatter, R. Vinuesa, An adverse-pressure-gradient turbulent boundary layer with nearly constant $\beta \simeq 1.4$ up to $Re_\theta \simeq 8700$. J. Fluid Mech. **939**, A34 (2022)
3. G. Eitel-Amor, R. Örlü, P. Schlatter, Simulation and validation of a spatially evolving turbulent boundary layer up to $Re_\theta = 8300$. Int. J. Heat Fluid Flow **47**, 57–69 (2014)
4. S. Romero, S. Zimmerman, J. Philip, J. Klewicki, Stress equation based scaling framework for adverse pressure gradient turbulent boundary layers. Int. J. Heat Fluid Flow **93**, 108885 (2022)
5. C. Aubertine, J. Eaton, Turbulence development in a non-equilibrium turbulent boundary layer with mild adverse pressure gradient. J. Fluid Mech. **532**, 345–364 (2005)
6. R. Vinuesa, R. Örlü, C. Sanmiguel Vila, A. Ianiro, S. Discetti, P. Schlatter, Revisiting history effects in adverse-pressure-gradient turbulent boundary layers. Flow Turb. Comb. **99**, 565–587 (2017)
7. B. McKeon, A. Sharma, A critical-layer framework for turbulent pipe flow. J. Fluid Mech. **658**, 336–382 (2010)
8. J. Lee, Large-scale motions in turbulent boundary layers subjected to adverse pressure gradients. J. Fluid Mech. **810**, 323–361 (2017)

Statistical Symmetries in Second Order Turbulence Modeling

F. C. Putz and M. Oberlack

Abstract Early turbulence models, such as zero- and one-equation models, exhibit a lack of symmetries, which are axiomatic properties of flow physics. Conversely, more comprehensive second moment closure models contain all of the symmetries of the Navier-Stokes equations. Recently, new symmetries in statistical descriptions of turbulence have been discovered, yet current turbulence models fail to account for these symmetries. We show how commonly used turbulence models fail to preserve classical symmetries and propose a second moment closure turbulence model and demonstrate its invariance under statistical symmetries. To achieve this, an auxiliary velocity and pressure field must be defined.

1 Introduction

Reynolds averaged Navier-Stokes (RANS) turbulence models have historically progressed from simple zero-equation models like Prandtl's mixing length model to more advanced two-equation models, and further to increasingly complex second moment closure models.

However, their limitations are often due to the lack of symmetries such as the Galilean invariance, which are axiomatic properties of flow physics, as exemplified by Prandtl's mixing length model. Introducing further symmetries into the model incorporates more physics and thus generalizes the model, being applicable to a wider range of flows.

In this context, symmetries refer to variable transformations that maintain a given equation form invariant. The incompressible Navier-Stokes equation's symmetries include the generalized Galilean transformation and scaling symmetries.

F. C. Putz (✉) · M. Oberlack
Chair of Fluid Dynamics, Otto-Berndt-Str. 2, 64287 Darmstadt, Germany
e-mail: putz@fdy.tu-darmstadt.de

M. Oberlack
e-mail: oberlack@fdy.tu-darmstadt.de

R. Örlü et al. (eds.), *Progress in Turbulence X*, Springer Proceedings in Physics 404,
https://doi.org/10.1007/978-3-031-55924-2_5

Recently, new symmetries of statistical descriptions of turbulence have been found [3, 5] and due to their importance for turbulent scaling laws in boundary layer flows, turbulence models should also fulfill these symmetries.

With the aim of such a turbulence model, the text is categorized as follows. First, a brief mathematical explanation of Lie symmetries is provided. Next, starting with the earliest turbulence models, their symmetries are analyzed and compared to the complete set of symmetries in the Navier-Stokes equations. Finally, this report presents the proposed turbulence model and demonstrates that it is statistically symmetric invariant.

2 Symmetries of the Navier-Stokes Equations

A one parameter Lie group of point transformations for the set of independent $\mathbf{x} = (x_1, \ldots, x_n)$ and dependent variables $\mathbf{y} = (y_1, \ldots, y_m)$ is given by

$$\mathcal{G}: \quad \tilde{\mathbf{x}} = \Phi(\mathbf{x}, \mathbf{y}; \varepsilon), \qquad \tilde{\mathbf{y}} = \Psi(\mathbf{x}, \mathbf{y}; \varepsilon) \tag{1}$$

with group parameter $\varepsilon \in \mathbb{R}$ and the law of composition ϕ, which is a binary operation that combines two elements of the group to produce another element of the group. For $\mathcal{G}$ to be a group, and even more restrictively, a Lie group, ϕ must satisfy several axioms, as described in [1].

Then, a symmetry of a differential equation is a variable transformation that leaves the form of the equation unchanged when written in transformed variables. Recalling (1), the transformation is a Lie symmetry of the p-th order partial differential equation $F(\mathbf{x}, \mathbf{y}, \mathbf{y}^{(1)}, \ldots, \mathbf{y}^{(p)}) = 0$, if its form is preserved, i.e.

$$F(\tilde{\mathbf{x}}, \tilde{\mathbf{y}}, \tilde{\mathbf{y}}^{(1)}, \ldots, \tilde{\mathbf{y}}^{(p)}) = 0 \tag{2}$$

holds after the transformation.

2.1 Classical Symmetries

Given the incompressible Navier-Stokes equations

$$\frac{\partial U_i}{\partial x_i} = 0, \tag{3}$$

$$\frac{\partial U_i}{\partial t} + U_j \frac{\partial U_i}{\partial x_j} + \frac{\partial P}{\partial x_i} - \nu \frac{\partial^2 U_i}{\partial x_j^2} = 0, \tag{4}$$

where the pressure P is divided by the density, the Lie symmetries can be calculated algorithmically. In the limit of $\nu \to 0$, i.e. $Re \to \infty$, the equations admit 6 Lie symmetry groups, dubbed classical symmetries, that each carry the following physical meaning: The symmetry of time translation indicates the absence of any absolute origin on time. Similarly, in the case of incompressible flows, there is no fixed pressure, which results in the existence of pressure translation symmetry. When using the proper tensor formulation, rotational symmetry is always maintained. Furthermore, the Galilean group postulates that physical laws remain unchanged under constant velocity translation. Finally, in the inviscid case, two scaling symmetries (one related to space and the other to time) are linked to dimensional correctness. When there is no vanishing viscosity, it breaks the symmetry and reduces the two scaling symmetries to one. For details see [5].

Taking the ensemble average of the Navier-Stokes equations leads to the RANS equations

$$\frac{\partial \overline{U}_i}{\partial x_i} = 0, \tag{5}$$

$$\frac{\partial \overline{U}_i}{\partial t} + \overline{U}_j \frac{\partial \overline{U}_i}{\partial x_j} + \frac{\partial \overline{P}}{\partial x_i} - \nu \frac{\partial^2 \overline{U}_i}{\partial x_j^2} + \frac{\partial R_{ij}}{\partial x_j} = 0, \tag{6}$$

with the unknown Reynolds stress tensor R_{ij}, the variance of the mean velocity. All the classical symmetries are preserved by the averaged equations [5, 6].

The RANS equations, however, allow for additional symmetries, which we will discuss in the next section.

2.2 *Statistical Symmetries*

Consider (6) written not with the variance R_{ij}, but instead in the second moment of the instantaneous velocity, i.e. $H_{ij} = R_{ij} + \overline{U}_i \overline{U}_j$

$$\frac{\partial \overline{U}_i}{\partial t} + \frac{\partial H_{ij}}{\partial x_j} + \frac{\partial \overline{P}}{\partial x_i} - \nu \frac{\partial^2 \overline{U}_i}{\partial x_j^2} = 0. \tag{7}$$

There are two key differences of this equation compared to (6): it is linear and each term is a derivative of a dependent variable. Similar to (6), (7) is not closed. Instead, it represents the first equation of an infinite hierarchy of moment equations, where the next equation is a transport equation for H_{ij}. Of utmost relevance is that the linearity, combined with its divergence form, remain for the infinite hierarchy. This allows for two additional types of symmetries, which are coined statistical symmetries, as they only appear in the statistical descriptions of turbulence. In addition to the RANS (or more generally in the multi-point correlation equations, MPCE) equations, they are also present in other exact descriptions of turbulence, e.g. the Lundgren probability density function (PDF) hierarchy [3, 5].

The first additional symmetry group is a scaling of each variable by the group parameter a_{S_c}

$$\begin{aligned} T_{\mathrm{Sc},stat}: t^* = t,\ x_i^* = x_i,\ \overline{U}_i^* = \overline{U}_i e^{a_{Sc}}, \\ R_{ij}^* = R_{ij}e^{a_{Sc}} + \overline{U}_i\overline{U}_j\left(e^{a_{Sc}} - e^{2a_{Sc}}\right),\ H_{ij}^* = H_{ij}e^{a_{Sc}},\ \ldots. \end{aligned} \tag{8}$$

This transformation also applies to any higher moment in the infinite hierarchy of equations for the instantaneous averaged Navier-Stokes equations.

The second type of statistical symmetry is a translation of the averaged variables by a constant group parameter

$$\begin{aligned} T_{\mathrm{Tr,stat},1}: t^* = t,\ x_i^* = x_i,\ \overline{U}_i^* = \overline{U}_i + a_{\mathrm{Tr,stat},1,i},\ H_{ij}^* = H_{ij} \\ R_{ij}^* = R_{ij} - \overline{U}_i a_{\mathrm{Tr,stat},1,j} - \overline{U}_j a_{\mathrm{Tr,stat},1,i} - a_{\mathrm{Tr,stat},1,i} a_{\mathrm{Tr,stat},1,j},\ \ldots \end{aligned} \tag{9}$$

$$T_{\mathrm{Tr,stat},2}: t^* = t,\ x_i^* = x_i,\ \overline{U}_i^* = \overline{U}_i,\ H_{ij}^* = H_{ij} + a_{\mathrm{Tr,stat},2,ij},\ \ldots \tag{10}$$

Wacławczyk et al. investigated the physical meaning of these symmetries using the PDF approach [3]. They discovered a connection between statistical translation symmetry and non-Gaussianity, as well as statistical scaling and intermittency. More recently, turbulent scaling laws based on these symmetries showed good agreements to DNS data for a channel flow [6]. An example of a turbulent scaling law that depends on the statistical translation symmetry is the well-known log-law for the mean velocity [4]. Best-fits of the second-moment turbulent scaling laws in a channel flow's logarithmic region at $Re_\tau = 10^4$ are shown in Fig. 1.

Fig. 1 Components of the Reynolds stress tensor and turbulent kinetic energy for a channel flow at $Re_\tau = 10^4$, normalized by the square of friction velocity. Squares mark the DNS data by [2]

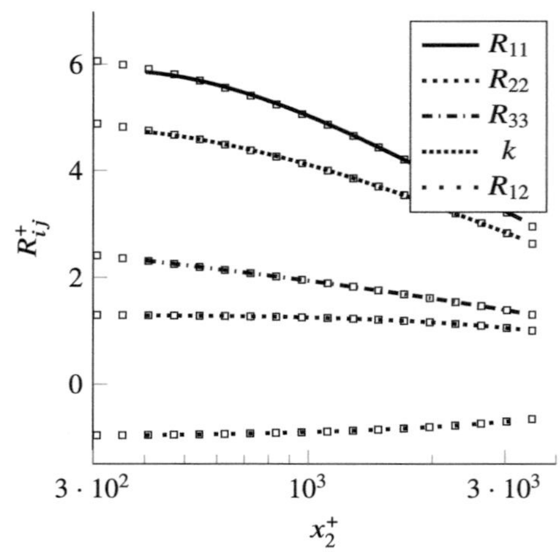

3 Classical Turbulence Models

The goal of a turbulence model is to find a model for the Reynolds stress tensor R_{ij} in (6), with the earliest being the Prandtl's mixing length model based on the Boussinesq approximation. It introduces ν_t, the turbulent viscosity, modeled via an analogy to the mean free path of molecules in a gas and the turbulent kinetic energy k. In that case, a flow specific external length scale, the mixing length, is needed, breaking the scaling symmetry. The fact that even this simple turbulence model produce good results for a well defined mixing length (which has to be newly specified for each case) shows that symmetry breaking does not necessarily render the model useless, but only limits its applicability.

Besides the k-equation, two-equation models resolve this by introducing a second scale providing variable, most commonly the dissipation ε or ω, being an inverse integral timescale of turbulence. However, without additional correction terms, these models are invariant under a rotation of the frame of reference. In that case, these models have too many symmetries, as the external rate of rotation acts symmetry breaking in the exact equations [4].

Second order turbulence models model the unknown terms on the transport equation of the Reynolds stress and thus avoid the Boussinesq approximation. They have the same classical symmetries as the exact equations, if the unclosed terms are modeled correctly, and are thus more general than two-equation models. However, partly due to the model introduced for the triple-correlation $R_{ijk} = \overline{u_i u_j u_k}$, which has a very complicated transformation under the statistical translation symmetries (see (9)), they are not invariant under statistical symmetries.

To build a turbulence model, which is not only invariant under the classical, but also under statistical symmetries, Klingenberg et al. [7] showed, that the variable space of the dependent variables has to be expanded to include an additional velocity field with a corresponding pressure field, dubbed model velocity $\hat{U}_i$ and model pressure $\hat{P}$, which are invariant under all statistical symmetries.

4 Statistical Invariant Reynolds Stress Model

Written in fluctuating variables, the proposed turbulence model

$$\frac{\partial R_{ij}}{\partial t} + \overline{U}_k \frac{\partial R_{ij}}{\partial x_k} = \mathcal{I}_{ij}^{(R)} + \mathcal{P}_{ij} - \Pi_{ij} - \varepsilon_{ij} + \mathcal{V}_{ij}, \tag{11}$$

consists of the material derivative on the left hand side and on the right hand side from left to right: an invariance tensor, production, velocity pressure-gradient correlation, dissipation and diffusion. It will now be shown, that the model is invariant under both the classical and statistical symmetries.

As even common second moment models fulfill all the classical symmetries, only the material derivative under the Galilean transformation is discussed here. A translation of the velocity results in a transformation of the time derivative $(\partial/\partial t)^* = \partial/\partial t - f'_{\text{Gal}_i}(t)\partial/\partial x_i$, which exactly cancels out the additional term that is created due to the velocity translation.

To show that the model is invariant under the statistical symmetries, it is written in its instantaneous formulation

$$\begin{aligned}\frac{\partial H_{ij}}{\partial t} + \hat{U}_k \frac{\partial H_{ij}}{\partial x_k} &= -\hat{U}_i \frac{\partial H_{jk}}{\partial x_k} - \hat{U}_j \frac{\partial H_{ik}}{\partial x_k} + \hat{U}_j \hat{U}_k \frac{\partial \overline{U}_i}{\partial x_k} + \hat{U}_i \hat{U}_k \frac{\partial \overline{U}_j}{\partial x_k} + \mathcal{P}_{ij} \\ &- \overline{U_i \frac{\partial P}{\partial x_j}} - \overline{U_j \frac{\partial P}{\partial x_i}} - \varepsilon_{ij} + \nu \frac{\partial^2 H_{ij}}{\partial x_k^2} + 2\nu c_1 \left(\frac{\partial \hat{U}_i}{\partial x_k} \frac{\partial \overline{U}_j}{\partial x_k} + \frac{\partial \hat{U}_j}{\partial x_k} \frac{\partial \overline{U}_i}{\partial x_k} \right).\end{aligned} \tag{12}$$

It is apparent that, like the instantaneous formulation of the RANS equation (7), it is both linear in mean variables and every first or second moment occurs in a partial derivative. The first property allows for the statistical scaling symmetry to be fulfilled while the second allows for the moments to be translated individually, e.g. $\overline{U}_i^* = \overline{U}_i + a_{\text{Tr,stat},1,i}$ or $H_{ij}^* = H_{ij} + a_{\text{Tr,stat},2,ij}$, verifying the statistical translation symmetry.

The first terms on the right hand side of (12) are necessary due to the complex transformation of H_{ij} under the Galilean symmetry and the production term is similar to the exact production in the transport equation of the Reynolds stress tensor, except that the Reynolds stress is replaced by a model Reynolds stress $\hat{R}_{ij}$. This is necessary as the production would not be linear otherwise. A classical turbulence model is chosen for $\hat{R}_{ij}$, as it does not transform under statistical symmetries. Furthermore, the dissipation is modeled via an ε equation, modified to account for the additional symmetries. Finally, the velocity pressure-gradient correlation is given as a Poisson equation, with additional model terms that resemble classical return-to-isotropy and rapid models.

References

1. G.W. Bluman, S.C. Anco, *Symmetry and Integration Methods for Differential Equations* (Springer, New York, 2002)
2. S. Hoyas, M. Oberlack, F. Alcántara-Ávila, S.V. Kraheberger, J. Laux, Wall turbulence at high friction Reynolds numbers. Phys. Ref. Fluids **7**, 014602 (2022)
3. M. Wacławczyk, N. Staffolani, M. Oberlack, A. Rosteck, M. Wilczek, R. Friedrich, Statistical symmetries of the Lundgren-Monin-Novikov hierarchy. Phys. Rev. E **90**, 13–22 (2014)
4. M. Oberlack, A unified approach for symmetries in plane parallel turbulent shear flows. J. Fluid Mech. **427**, 299–328 (2001)
5. M. Oberlack, A. Rosteck, New statistical symmetries of the multi-point equations and its importance for turbulent scaling laws. Discr. Cont. Dyn-S **3**(3), 451–471 (2010)

6. M. Oberlack, S. Hoyas, S.V. Kraheberger, F. Alcántara-Ávila, J. Laux, Turbulence statistics of arbitrary moments of wall-bounded shear flows: A symmetry approach. Phys. Rev. Lett. **128**(2), 024502–024508 (2022)
7. D. Klingenberg, M. Oberlack, D. Pluemacher, Symmetries and turbulence modeling. Phys. Fluids **32**, 025108 (2020)

Infinite Lundgren Hierarchy of Turbulence: Isotropy, Super-isotropy and a Finite Dimensional Eigenvalue Problem

S. Görtz, D. Plümacher, J. H. Conrad, and M. Oberlack

Abstract We analyze Lundgren's infinite multipoint hierarchy of probability density functions (PDFs) in the context of homogeneous isotropic turbulence (HIT). To account for the physical properties of HIT, we express the PDF in terms of the scalar invariants under the SO(3) group of rotations, yielding new coordinates for a homogeneous isotropic PDF and leading to a dimensional reduction of the problem. Furthermore, we transform Lundgren's hierarchy to a description in spatial increment variables in order to obtain a description consistent with the approach chosen by von Kármán and Howarth for the description of two-point velocity moments. For further dimensional reduction compared to the homogeneous isotropic case, a small scale approximation is introduced, which greatly simplifies the hierarchy. For the simplified hierarchy, a solution approach of superposed eigenfunctions is introduced. Together with the side conditions of the hierarchy, this leads to a formally closed eigenvalue problem.

1 Introduction

The key task in fundamental turbulence research lies in the prediction of the flow behavior. It is common that there is a lack of understanding of main properties of turbulent flows based on first principles, to date, empirical models and closure approaches in engineering are state of the art. Our modern understanding of turbulence is focused on its statistical behavior, which is based on a statistical treatment of the underlying equations of motion, i.e. the Navier-Stokes equations. Their nonlinearity and nonlocality lead to the statistical equations being unclosed [2]. Solving the statistical equations has only been successful in a few cases, for example for Kolmogorov's famous four-fifth law [7].

In order to obtain a complete description of all turbulent statistics, it is common to work with kinetic equations, investigating transport equations for the probability density functions of the instantaneous velocities. Pioneering work on this has

S. Görtz (✉) · D. Plümacher · J. H. Conrad · M. Oberlack
Chair of Fluid Dynamics (FDY), Otto-Berndt-Str. 2, 64287 Darmstadt, Germany
e-mail: goertz@fdy.tu-darmstadt.de

R. Örlü et al. (eds.), *Progress in Turbulence X*, Springer Proceedings in Physics 404,
https://doi.org/10.1007/978-3-031-55924-2_6

been done by Lundgren [4], Monin [5] and Novikov [6] by formulating the infinite multi-point hierarchy of PDF equations (LMN hierarchy). Based on that, extensive research has been conducted on this hierarchy. With a focus on combustion, Pope [7] introduced working with conditional averages, which also allows to formally close the hierarchy.

The linearity of the LMN hierarchy as outstanding property provides the reasoning for a solution approach presented in this work. An approach for the PDF as a superposition of products is introduced ('product approach') and the question for a solution of the LMN hierarchy is reformulated in terms of an eigenvalue problem.

The following notation will be used: Points are declared by a prefix, i.e. ${}_{(k)}\mathbf{x}$ refers to the absolute position of the k-th point. As we work with velocity PDFs and the physical velocity ${}_{(k)}\mathbf{u}({}_{(k)}\mathbf{x}, t)$, shortened by ${}_{(k)}\mathbf{u}$, is treated as a random variable, the corresponding independent sample space velocity variable is denoted by ${}_{(k)}\mathbf{v}$. The PDF in absolute coordinates can thus be defined weakly by the ensemble average $\langle(\cdot)\rangle$ of any (test-)function Q depending on the velocities, i.e.

$$\langle Q(\mathbf{U})\rangle = \int_{\mathbb{R}^{3N}} Q(\mathbf{V})\,{}_{(N)}f\left({}_{(1)}\mathbf{x}, \ldots, {}_{(N)}\mathbf{x}, \mathbf{V}, t\right) \mathrm{d}\mathbf{V}, \tag{1}$$

where we use the short-hand notations $\mathbf{U} = {}_{(1)}\mathbf{u}, \ldots, {}_{(N)}\mathbf{u}$ and $\mathbf{V} = {}_{(1)}\mathbf{v}, \ldots, {}_{(N)}\mathbf{v}$ for the set of (sample space) velocity vectors and $\mathrm{d}\mathbf{V} = \mathrm{d}_{(1)}\mathbf{v} \ldots \mathrm{d}_{(N)}\mathbf{v}$. With that the LMN hierarchy is derived from the Navier-Stokes equations after replacing the pressure using Green's function as a solution of the Poisson equation [8]. Lundgren [4] uses the definition of the fine grained pdf for the derivation.

The LMN hierarchy comes with a number of side conditions, ensuring a physically and mathematically proper solution for the PDF [8].

2 LMN Hierarchy in Spatial Increments

The description of HIT by von Kármán and Howarth [3] yields absolute two-point velocity moments depending on spatial increments ${}_{(k)}\mathbf{r} = {}_{(k)}\mathbf{x} - \mathbf{x}$, with $\mathbf{x} = {}_{(1)}\mathbf{x}$ as base point. For consistency, an absolute velocity PDF is defined in spatial increments, similar to the approach chosen by Ulinich and Lyubimov [8] for spatial and velocity increments. We introduce the weak definition of the spatial increment PDF similar to (1) by

$$\langle Q(\mathbf{U})\rangle = \int_{\mathbb{R}^{3N}} Q(\mathbf{V})\,{}_{(N)}F\left(\mathbf{x}, {}_{(2)}\mathbf{r}, \ldots, {}_{(N)}\mathbf{r}, \mathbf{V}, t\right) \mathrm{d}\mathbf{V}. \tag{2}$$

The latter definition allows the introduction of homogeneity, i.e. a dimensional reduction by 3 independent variables, by simply dropping dependence from the base point $\mathbf{x}$. The spatial increment and absolute coordinate PDFs are related by

$$ {}_{(N)}F = \int_{\mathbb{R}^{3N}} {}_{(N)}f\, \delta\left(\mathbf{x} - {}_{1}\mathbf{x}\right) \prod_{i=2}^{N} \delta\left({}_{(i)}\mathbf{r} - {}_{(i)}\mathbf{x} + {}_{(1)}\mathbf{x}\right) \mathrm{d}_{(1)}\mathbf{x} \ldots \mathrm{d}_{(N)}\mathbf{x}. \tag{3} $$

Using this definition, the LMN hierarchy in absolute coordinates can be transformed to a hierarchy in increments by manipulating it according to

$$ \int_{\mathbb{R}^{3N}} (\mathrm{LMN}) \cdot \delta\left(\mathbf{x} - {}_{1}\mathbf{x}\right) \prod_{i=2}^{N} \delta\left({}_{(i)}\mathbf{r} - {}_{(i)}\mathbf{x} + {}_{(1)}\mathbf{x}\right) \mathrm{d}_{(1)}\mathbf{x} \ldots \mathrm{d}_{(N)}\mathbf{x}, \tag{4} $$

where the sifting property of the delta distribution is used. The resulting hierarchy reads in a cartesian framework

$$ \begin{aligned} 0 = {} & \frac{\partial_{(N)}F}{\partial t} + \left({}_{(1)}\mathbf{v} \cdot \nabla + \sum_{k=2}^{N} \left({}_{(k)}\mathbf{v} - {}_{(1)}\mathbf{v}\right) \cdot {}_{(k)}\nabla\right) {}_{(N)}F \\ & + {}_{{}_{(1)}\mathbf{v}}\nabla \cdot \int \mathbf{G}'\left({}_{(N+1)}\mathbf{r}\right) \left({}_{(N+1)}\mathbf{v} \cdot {}_{(N+1)}\nabla\right)^2 {}_{(N+1)}F \, \mathrm{d}_{(N+1)}\mathbf{v} \, \mathrm{d}_{(N+1)}\mathbf{r} \\ & - \sum_{k=2}^{N} {}_{{}_{(k)}\mathbf{v}}\nabla \cdot \int \left(\mathbf{G}'\left({}_{(k)}\mathbf{r} - {}_{(N+1)}\mathbf{r}\right) - \mathbf{G}'\left({}_{(N+1)}\mathbf{r}\right)\right) \\ & \left({}_{(N+1)}\mathbf{v} \cdot {}_{(N+1)}\nabla\right)^2 {}_{(N+1)}F \, \mathrm{d}_{(N+1)}\mathbf{v} \, \mathrm{d}_{(N+1)}\mathbf{r} \\ & + \nu \sum_{k=1}^{N} {}_{{}_{(k)}\mathbf{v}}\nabla \int \delta\left({}_{(k)}\mathbf{r} - {}_{(N+1)}\mathbf{r}\right) {}_{(N+1)}\Delta\left({}_{(N+1)}\mathbf{v} \, {}_{(N+1)}F\right) \mathrm{d}_{(N+1)}\mathbf{v} \, \mathrm{d}_{(N+1)}\mathbf{r} \end{aligned} \tag{5} $$

where $\mathbf{G}'(\mathbf{x}) = \nabla\left(4\pi\left|\mathbf{r}\right|\right)^{-1}$ denotes the gradient of Green's function as a solution of the Poisson equation for an unbounded domain. Note that ${}_{(1)}\mathbf{r} = 0$ and ${}_{(k)}\nabla$ denotes the nabla operator w.r.t spatial increments ${}_{(k)}\mathbf{r}$ for $k = 2 \ldots N$ and ∇ w.r.t the base point $\mathbf{x}$.

3 Increment PDFs for HIT

In order to describe HIT, we seek for a functional form/dependence of the PDF. For homogeneity we demand independence from the base point $\mathbf{x}$. Expressed in symmetries of the PDF, homogeneity means a shift-symmetry in the variable $\mathbf{x}$ and isotropy is equivalent to symmetry under the group of rotations SO(3) acting on the vectors ${}_{(k)}\mathbf{r}$ and ${}_{(k)}\mathbf{v}$, where $k = 1, \ldots, N$. Since both homogeneity and isotropy are three dimensional groups, they lead for the PDF to a dimensional reduction by six, compared to the general case. Concluding, a homogeneous isotropic PDF as a scalar density is independent from the base point $\mathbf{x}$ and can only depend on scalars which

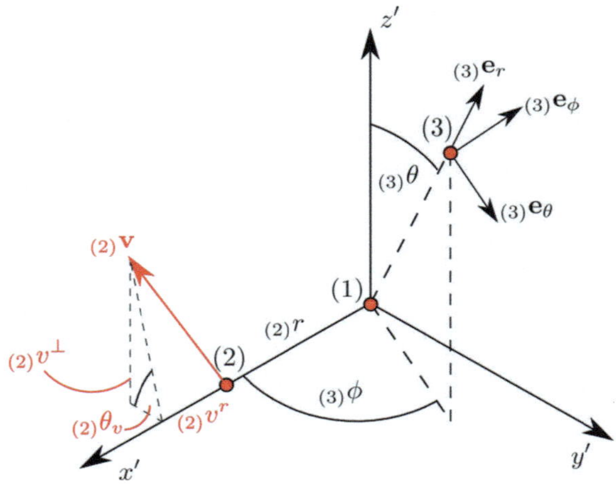

Fig. 1 Adapted coordinate system to construct the scalar invariants as new coordinates of a homogeneous isotropic PDF

are invariant under the group of rotations. The aim of this chapter is therefore to give scalar invariants from the independent variables ${}_{(1)}\mathbf{v}$, ${}_{(2)}\mathbf{v}$, ${}_{(2)}\mathbf{r}, \ldots, {}_{(n)}\mathbf{v}$, ${}_{(n)}\mathbf{r}$, yielding a new set of variables a homogeneous isotropic PDF depends on. The choice of invariants in not unique, a suitable one depends on the problem. We try to find a set of invariants which can be easily identified as new coordinates in a homogeneous isotropic hierarchy emerging from (5). In order to construct the scalar invariants, we introduce a set of new coordinates, adapted to the direction of the vectors ${}_{(2)}\mathbf{r}$, ${}_{(1)}\mathbf{v}$, constructed as follows: The x'-direction is chosen to be parallel to the vector ${}_{(2)}\mathbf{r}$. Further, the system is rotated that the sample space velocity at the base point, ${}_{(1)}\mathbf{v}$, lies in the $y' - z'$-plane. The $x' - y' - z'$-system and corresponding minimal set of coordinates can be taken from Fig. 1. For clarity, the sample space velocities at points (1) and (3) are not displayed. For the description of the sample space velocity we locally introduce a cylindrical system, describing ${}_{(2)}\mathbf{v}$ by the two components ${}_{(2)}v^r$ (corresponding to the axial component in the cylindrical system or to the radial component in a global spherical system) and ${}_{(2)}v^{\perp}$ (corresponding to the radial component in the cylindrical system) and by the cylindrical angle ${}_{(2)}\theta_v$. A homogeneous isotropic one- and two-point PDF then admit the following forms,

$$ {}_{(1)}F = {}_{(1)}F\left(|{}_{(1)}\mathbf{v}|, t\right). \tag{6a} $$

$$ {}_{(2)}F = {}_{(2)}F\left({}_{(2)}r, {}_{(1)}v^r, {}_{(1)}v^{\perp}, {}_{(2)}v^r, {}_{(2)}v^{\perp}, {}_{(2)}\theta_v, t\right), \tag{6b} $$

where ${}_{(2)}r = |{}_{(2)}\mathbf{r}|$. Here we have for the two-point PDF already obtained a dimensional reduction by six variables, bringing us to the conclusion that for PDFs of higher order, we always have to add the full set of six coordinates per additional point, i.e. ${}_{(3)}F$ depends additionally on ${}_{(3)}r$, ${}_{(3)}\theta$, ${}_{(3)}\phi$ and three sample space velocity components describing ${}_{(3)}\mathbf{v}$. Using a coordinate-free covariant form of (5), it is possible to derive a hierarchy for the homogeneous isotropic PDFs in the new coordinates, which are at the same time coordinates in a spherical system. This is part of undergoing research and due to its complexity not shown here.

4 Small Scale Approximation and Super-isotropy

In order to simplify the equations and obtain further dimensional reduction, we introduce a small scale approximation for isotropic turbulence, where we drop the dependence of all angles. This consideration is carried out for the hierarchy in absolute coordinates, since it simplifies the treatment of the hierarchy on all levels. The hierarchy can for example be taken from [4] The isotropic small-scale PDF in absolute coordinates, defined by (1), is assumed to have the following dependence:

$$ {}_{(N)}f = {}_{(N)}f\left({}_{(1)}r, {}_{(1)}v^r, {}_{(1)}v^\perp, \ldots, {}_{(N)}r, {}_{(N)}v^r, {}_{(N)}v^\perp, t\right). \tag{7} $$

Note that, unlike the other chapters, ${}_{(k)}r$ denotes the spherical radius of the absolute coordinates, i.e. ${}_{(k)}\mathbf{x} = {}_{(k)}r\,{}_{(k)}\mathbf{e}^r$. The set of variables included in (7) corresponds to the scalar invariants of N different groups of rotations SO(3), meaning that we assume invariance of the PDF under the groups given by

$$ \left({}_{(k)}\mathbf{x}^*, {}_{(k)}\mathbf{v}^*\right) = \left({}_{(k)}\mathbf{R}\,{}_{(k)}\mathbf{x}, {}_{(k)}\mathbf{R}\,{}_{(k)}\mathbf{v}\right), \qquad k = 1, \ldots, N. \tag{8} $$

Since we add an additional set of invariance conditions compared to the classical homogeneous isotropic PDF, we call the PDF (7), depending only on the scalar invariants of (8), “super-isotropic”. We argue that this concept of super-isotropy can be seen as an extension of isotropy on small scales. The set of invariants of (8) was calculated in [1].

In the following, the LMN hierarchy in absolute coordinates is written in the scalar invariants of the super-isotropic group and a solution approach is discussed. As it was shown by Conrad [1], the classical form of the hierarchy is not coordinate independent. Since the scalar invariants (7) of (8) are given in spherical coordinates, we use the covariant and therefore coordinate-free form of LMN, derived in [1], to express the hierarchy in spherical coordinates and identify the invariants (7) for the super-isotropic case. For sake of brevity, the result is not shown here. The arising integro-differential equation is linear, therefore we propose a functional form for (7) consisting of superposed products of eigenfunctions h_l, depending on the m-th eigenvalue λ_m. The superposition of eigenfunctions with different eigenvalues is necessary in order to ensure statistical correlation of the processes and to fulfill all side conditions of the hierarchy. The functional form is given by

$$ {}_{(N)}f = \sum_{m=1}^{\infty} \prod_{l=1}^{N} h_l\left(t, {}_{(l)}r, {}_{(l)}v^r, {}_{(l)}v^\perp, \lambda_m\right). \tag{9} $$

The permutation side condition, ensuring invariance of the PDF under permutations of ${}_{(k)}\mathbf{x}$, ${}_{(k)}\mathbf{v}$ with ${}_{(l)}\mathbf{x}$, ${}_{(l)}\mathbf{v}$, leads to $h_k = h_l = h$ with $k, l = 1 \ldots N$. Using this, the hierarchy simplifies to the closed nonlinear first-level integro differential equation

$$0 = \frac{\partial h}{\partial t} + v^r \frac{\partial h}{\partial r} - 2\pi \frac{\partial h}{\partial v^r} \int_0^r \left(\frac{\tilde{r}}{r}\right)^2 \int \left(\left(\tilde{v}^r\right)^2 \partial_{\tilde{r}}{}^2 + \frac{1}{\tilde{r}} \left(\tilde{v}^\perp\right)^2 \partial_{\tilde{r}} \right) h \, \mathrm{d}\tilde{v}^r \, \mathrm{d}\left(\tilde{v}^\perp\right)^2 \mathrm{d}\tilde{r}$$
$$+ \frac{2\pi}{\mathrm{Re}} \lim_{\tilde{r} \to r} \frac{\partial h}{\partial v^r} \int \tilde{v}^r \left(\partial_{\tilde{r}}{}^2 + \frac{2}{\tilde{r}} \partial_{\tilde{r}} \right) h \, \mathrm{d}\tilde{v}^r \, \mathrm{d}\left(\tilde{v}^\perp\right)^2. \tag{10}$$

λ_m does not occur explicitly in this equation. We state that the eigenvalue problem including λ_m is formed by the solution of (10) together with all side conditions. Regarding the coincidence side condition [4], it becomes clear why the super-isotropic group (8) is only an approximation. This condition ensures equality of the velocities if two points coincide, i.e. it is a condition for ${}_{(k)}\mathbf{x} \to {}_{(l)}\mathbf{x}$. From dropping all angles in the functional form of the super-isotropic PDF (7), we can only formulate a condition for ${}_{(k)}r \to {}_{(l)}r$, which does not necessarily correspond to both points coinciding. Nevertheless, this error becomes small on small scales, meaning small r. We therefore argue that super-isotropy is a good approximation for isotropic turbulence on small scales and greatly simplifies the equations.

References

1. J.H. Conrad, Gradient statistics in homogeneous isotropic turbulence. MA Thesis, TU Darmstadt (2023)
2. R. Friedrich et al., The Lundgren–Monin–Novikov hierarchy: kinetic equations for turbulence. Comptes Rendus Physique **13**(9), 929–953 (2012)
3. T. de Karman, L. Howarth, On the statistical theory of isotropic turbulence. Proc. R. Soc. Lond. Ser. A - Math. Phys. Sci. **164**(917), 192–215 (1938)
4. T.S. Lundgren, Distribution functions in the statistical theory of turbulence. Phys. Fluids **10**, 969–975 (1967)
5. A.S. Monin, Equations of turbulent motion. Prikl. Mat. Mekh. **31**(6), 1057 (1967)
6. E.A. Novikov, Kinetic equations for a vortex field. Sov. Phys. Dokl. **12**, 1006–1008 (1968)
7. S.B. Pope, *Turbulent Flows* (Cambridge University Press, 2000)
8. F.R. Ulinich, B.Y. Lyubimov, The statistical theory of turbulence of an incompressible fluid at large Reynolds numbers. Soviet Phys. Jetp **28**, 494–500 (1969)

Symmetry Theory and Turbulent Jet Scaling Laws of a Spatially Evolving Turbulent Round Jet

C. T. Nguyen and M. Oberlack

Abstract Using symmetry methods, we calculate similarity-type scaling laws for arbitrarily high moments of velocity from the infinite set of moment equations. Most centrally, symmetry theory provides moment-based scaling on instantaneous rather than fluctuation velocities. To prove its validity, a large-scale direct numerical simulation (DNS) of a turbulent jet flow was conducted at a Reynolds number $Re = 3500$ based on the jet diameter D and jet-inlet bulk velocity U_b in a box with the length of $z/D = 75$. Almost 200 washouts are calculated for a very good statistical convergence. Virtually perfect similarity is observed in the $z/D = 25 - 65$ range and this is especially true for U_z-moments up to order $n = 10$. Integration constants in the prefactor of the scaling laws are independent of the moment order n and therefore universal. Further, in matching theory and DNS data, we found that the U_z-moments show Gaussian-like curves that get increasingly narrower with n, and this n-dependence is non-linear. The two statistical symmetries describing non-Gaussianity and intermittency, which were central for high-order moments in near-wall turbulence, are broken for turbulent jets.

1 Background

In Oberlack [1], symmetries have been applied for the first time to turbulent shear flows to generate invariant solutions, which in this context is equivalent to turbulent scaling laws. However, this approach was limited to the mean velocity at that time. In turbulence, usually the Reynolds decomposition $U_i = \overline{U}_i + u_i$ is used where $\overline{(\cdot)}$ denotes an ensemble average and u_i the fluctuation. In the Reynolds-averaged Navier-Stokes (RANS) equation, the moment $H_{ij} = \overline{U_i U_j}$ is decomposed into $H_{ij} = \overline{U}_i \overline{U}_j + \overline{u_i u_j}$ where $R_{ij} = \overline{u_i u_j}$ is the Reynolds stress tensor. With H_{ij}

C. T. Nguyen (✉) · M. Oberlack
Chair of Fluid Dynamics, TU Darmstadt, Otto-Berndt-Strasse 2, 64287 Darmstadt, Germany
e-mail: nguyen@fdy.tu-darmstadt.de

M. Oberlack
e-mail: oberlack@fdy.tu-darmstadt.de

R. Örlü et al. (eds.), *Progress in Turbulence X*, Springer Proceedings in Physics 404,
https://doi.org/10.1007/978-3-031-55924-2_7

and R_{ij}, the moments based on the instantaneous velocities shall be called the H-approach and the classical approach is called the R-approach. Instantaneous multi-point moments of arbitrary order n are denoted by $H_{i_{\{n\}}}$ and one-point moments by $\overline{U_i^n}$. The corresponding multi-point moment equations (MPMEs) define an infinite set of *linear* equations with the only coupling between "neighboring" equations. Just recently [2, 3], turbulent scaling laws of arbitrary H-moments for the log and the core region of a turbulent channel flow were derived and validated using a new DNS at a friction Reynolds number of $Re_\tau = 10^4$. Key ingredients are statistical symmetries [4, 5] as they are only admitted by the MPMEs and not by the Navier-Stokes equations (NSE).

2 Symmetry Theory for a Spatially Evolving Turbulent Round Jet

The H-MPMEs are formed by the NSEs and symmetries of the NSEs directly translate to the MPMEs. The MPMEs, which can be viewed in [3], are derived by multiplying the momentum equations with $n-1$ velocities at $n-1$ different locations $\boldsymbol{x}_{(l)}$ with a subsequent statistical averaging. The H-MPMEs are considerably more simple for a symmetry analysis compared to the R-equations, since the system of equations is linear and the equations are only coupled between the order n and $n+1$. The following symmetries are relevant for the turbulent round jet flow. Translation in space

$$\overline{T}_x: \quad t^* = t, \quad x_i^* = x_i + a_{x_i}, \quad \overline{U_i}^* = \overline{U_i}, \quad H^*_{i_{\{n\}}} = H_{i_{\{n\}}}, \tag{1}$$

and the scaling in space (a_{Sx}) and time (a_{St})

$$\begin{aligned} \overline{T}_S \;:\quad t^* = \mathrm{e}^{a_{St}} t, \quad x_i^* = \mathrm{e}^{a_{Sx}} x_i, \quad \overline{U_i}^* = \mathrm{e}^{a_{Sx}-a_{St}} \overline{U_i}, \\ H^*_{i_{\{n\}}} = \mathrm{e}^{n(a_{Sx}-a_{St})} H_{i_{\{n\}}}, \quad \overline{P}^* = \mathrm{e}^{2(a_{Sx}-a_{St})} \overline{P}, \end{aligned} \tag{2}$$

have their foundation in the NSEs in the limit of $Re \to \infty$ and are transferred to the H-approach. A third scaling symmetry of the MPMEs

$$\overline{T}_{Ss}: \quad t^* = t, \quad x_i^* = x_i, \quad \overline{U_i}^* = \mathrm{e}^{a_{Ss}} \overline{U_i}, \quad H^*_{i_{\{n\}}} = \mathrm{e}^{a_{Ss}} H_{i_{\{n\}}}, \tag{3}$$

defines a measure of intermittency [5] and is of statistical nature, i.e. it has no counterpart in the NSEs as pointed out above.

Subsequently, we focus on one-point statistics, i.e. $\overline{U_{[i]}^n}$ where [ ] means that no summation is implied. With the symmetries (1)–(3) the characteristic condition for the invariant solution of a spatially developing turbulent round jet can be generated

$$\frac{\mathrm{d}r}{a_{Sx}r} = \frac{\mathrm{d}z}{a_{Sx}z + a_z} = \frac{\mathrm{d}\overline{U_{[i]}}}{[(a_{Sx} - a_{St}) + a_{Ss}]\overline{U_{[i]}}} = \ldots = \frac{\mathrm{d}\overline{U^n_{[i]}}}{[n(a_{Sx} - a_{St}) + a_{Ss}]\overline{U^n_{[i]}}}, \tag{4}$$

see e.g. [3, 4]. Further, a turbulent jet admits the momentum integral invariant here with non-zero streamwise pressure gradient [6]

$$I = 2\pi \int_0^\infty \left[\overline{U_z^2} - \frac{1}{2}\left(\overline{u_r^2} + \overline{u_\varphi^2}\right)\right] r\,\mathrm{d}r. \tag{5}$$

Considering $\overline{U^2_{[i]}} = \overline{U_{[i]}}^2 + \overline{u^2_{[i]}}$ and implementing (2)–(3) into (5) infers that $a_{St} = 2a_{Sx}$ and $a_{Ss} = 0$ have to hold for the latter to be invariant. These conditions lead to the classical scaling to be shown below in (6)–(7). Most important the symmetry (3), which denotes intermittency and is a key ingredient for near-wall scaling laws, is apparently broken here. This seems strange at first, since turbulent jets are known to be strongly intermittent [7]. This could mean that the intermittency is "hidden" and not all intermittency effect are measured by (3). The above scaling symmetry breaking leads in (2) to a scaling symmetry identical to the scaling of the NSE with $\nu \neq 0$. However, this is merely a coincidence because the actual cause is the integral invariant (5) and the viscous diffusion is up to 2 orders of magnitude smaller than all the other terms. The term is usually not shown with this large Re number present. Integrating (4) gives the invariants, which form the turbulent scaling laws

$$\eta = \frac{r}{z - z_0} \tag{6}$$

$$\widetilde{\overline{U^n_{[i]}}}(\eta)\mathrm{e}^{c_{i,n}n} = \frac{\overline{U^n_{[i]}}(r, z)}{(z - z_0)^{-n}} \tag{7}$$

where $z_0 = -a_z/a_{Sx}$. In doing so one must note that the constant of integration, i.e. invariant η in (6) arises from the integration of the first two terms and where (7) arises from the integration of the remaining terms. The resulting integration constants $c_{i,n}$ represent the invariants related to the moments and get into the exponent after exponentiation. Here we have to be careful, because the invariants $c_{i,n}$ are in general functions of other invariants and here this is η. At this point, however, we want to take advantage of an important result which becomes visible only further on in Fig. 1, namely that $c_{i,n}$ do not depend on n, and thus represent invariants with respect to the moment order.

3 Validation of the Turbulent Jet Scaling Laws

In order to validate the scaling laws (6)–(7), we conducted a large-scale DNS of a turbulent round jet at $Re = 3500$. The grid has about 2.4×10^8 degrees of freedom and extends $75D$ axially and up to $64D$ radially. For very good statistical convergence we calculated almost 200 washouts. For validation purposes, we first set $r = 0$ into (7), which is then rearranged to

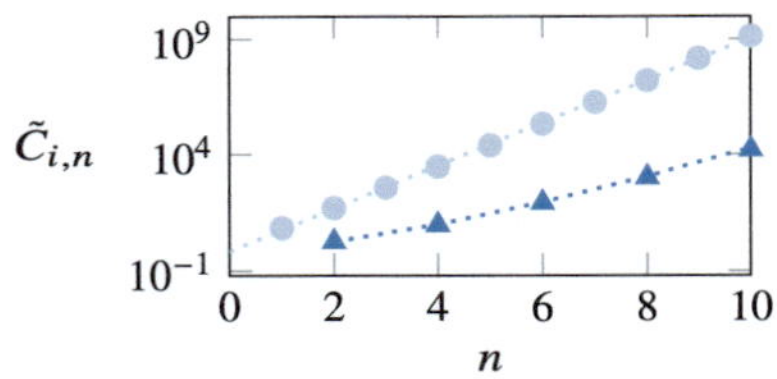

Fig. 1 $\tilde{C}_{i,n} = \alpha_{i,n} e^{c_i n}$ according to (8) for each moment up to $n = 10$ for each direction compared to DNS data: $\tilde{C}_{r,n}$ (▲), $\tilde{C}_{z,n}$ (●), $\tilde{C}_{r,n}$ scaling (......), $\tilde{C}_{z,n}$ scaling (......)

$$\overline{U^n_{i,c}} = \overline{U^n_{[i]}}(r = 0, z) = \frac{\widetilde{\overline{U^n_{[i]}}}(\eta = 0)\alpha_{i,n} e^{c_i n}}{(z - z_0)^n} \tag{8}$$

where $\widetilde{\overline{U^n_{[i]}}}(\eta = 0) = 1$. We extract from the present DNS data that the virtual orifice is at $z_0 = 0$. For the axial moments, we obtain $c_z = 2.123$ and $\alpha_{z,n} = \alpha_z = 0.673$. The DNS data follow the exponential prefactors $\alpha_z e^{c_z n}$ in (8) quite accurately and are displayed in Fig. 1. This is very similar to what is observed for the turbulent scaling laws of a turbulent channel flow [3]. The r and φ moments can be modelled as a centered Gaussian process in a very good approximation as the DNS data shows in Fig. 1. This is due to the fact that all the odd r and φ moments on the centerline are zero and therefore the instantaneous moments are equal to the corresponding fluctuations. This means all the higher moments n are determined by the second moment [8] with

$$\alpha_{r,n} e^{c_r n} = \alpha_{\varphi,n} e^{c_\varphi n} = \left(\alpha_{r,2} e^{2c_r}\right)^{\frac{n}{2}} (n-1)!! \tag{9}$$

where $\alpha_{r,2} e^{2c_r} = 1.81$, $(\cdot)!!$ refers to the double factorial and $\alpha_{r,n} = 0$ for odd n. With Stirling's formula for factorials, we obtain $c_{r,n} = c_r = -0.5$ and $\alpha_{r,n} = \left(\alpha_{r,2} e^{2c_r}\right)^{\frac{n}{2}} \sqrt{2e}(n-1)^{n/2}$. As noted above, c_z and c_r are independent of n which implies that in (7) the integration constants are $c_{i,n} = c_i$ and therefore universal.

One of the key results of our contribution is that we presently identify a Gaussian-type behavior in η not only in the mean but also for all higher instantaneous axial velocity moments. Based on (7), we find

$$\frac{\overline{U^n_z}}{\overline{U^n_{z,c}}} = \widetilde{\overline{U^n_z}}(\eta) = e^{-\gamma_n \eta^2}, \tag{10}$$

being an excellent representation of the moments. At first, this was an empirical observation but could be later cast into the group theoretical framework which we will show in an imminent publication. The self-preservation of a total of six axial moments up to order ten at five stages between $z/D = 25 - 65$ can be taken from

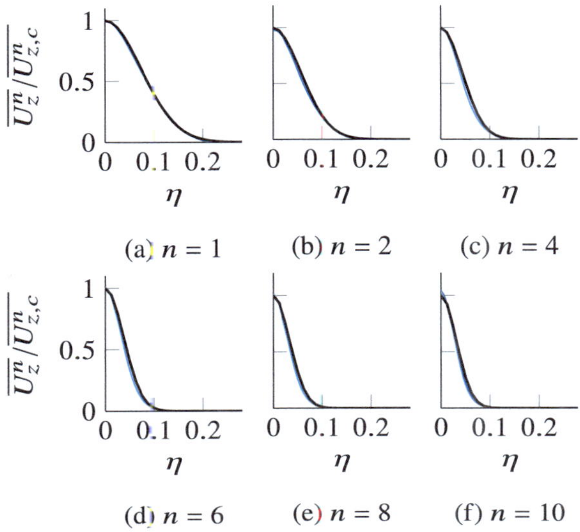

Fig. 2 nth axial velocity moment profiles normalized with (7) at different distances from the orifice: $z/D = 25$ ( ), 35 ( ), 45 ( ), 55 ( ), 65 ( ). The black solid lines indicate the Gaussian behavior from (10) using γ_n from (11)

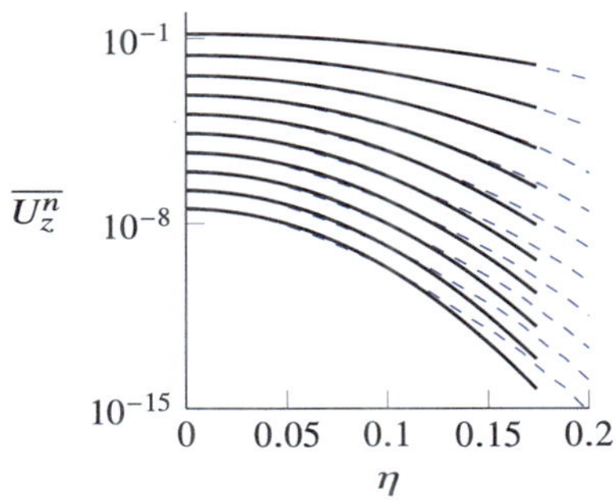

Fig. 3 1st (top) up to the 10th (bottom) axial velocity moment profiles (blue, dashed) and the corresponding curves (black) from (10) shown in a semi-logarithmic plot at $z/D = 45$

Fig. 2 and exhibits an almost perfect collapse of the data by normalizing with (8). As expected, the curves become increasingly narrower as the moment order increases.

A particularly sensitive representation focuses on the exponent of (10) alone and Fig. 3 compares the corresponding DNS data with a parabolic fit, while the deviation of the DNS data at the edge of the jet is most likely due to insufficient statistics of extremely small values. A quadratic fitting of γ_n in n for the axial moments can be taken from Fig. 4 and we find

$$\gamma_n = -1.54n^2 + 61.95n + 27.10, \tag{11}$$

i.e. higher moments n deviate stronger from a linear behavior. An interesting analogy arises with the moments of the structure function for small-scale turbulence [9]

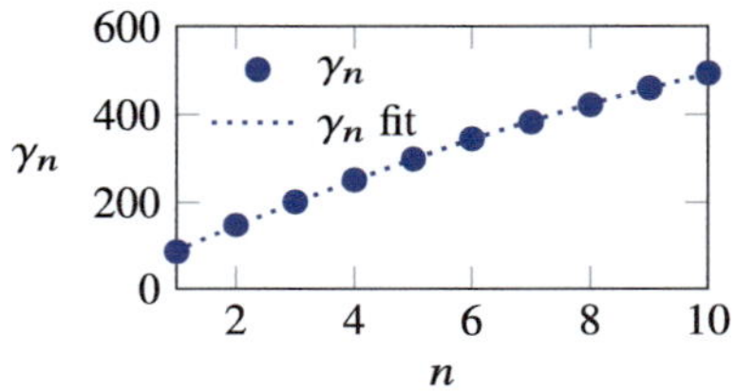

Fig. 4 Constants γ_n from (10) are shown for each moment up to order $n = 10$ determined by fitting to the DNS

where due to internal intermittency [10] the experimental data deviate from the pure dimensional scaling given in (2) leading to a linear Kolmogorov scaling of the exponent in n.

4 Conclusion

Turbulent jet scaling laws have been constructed with Lie symmetry analysis that agree with classical scaling. Further, these scaling laws have been validated with a large-scale jet DNS. The integration constants in the prefactor of the scaling laws are independent of the moment order n and therefore universal. Additionally, the radial profiles of the axial velocity moments show that the higher moments cannot be approximated by a trivial exponent which in turn seem to indicate intermittency.

Acknowledgements The authors gratefully acknowledge the Gauss Centre for Supercomputing e.V. (www.gauss-centre.eu) for funding this project under the project number pn73fu by providing computing time on the GCS Supercomputer SUPERMUC-NG at Leibniz Supercomputing Centre (www.lrz.de). The authors also thank Hamed Sadeghi for the fruitful discussions. CTN is supported by the Graduate School CE at TU Darmstadt and wants to acknowledge the funding from the German Academic Scholarship Foundation (German: Studienstiftung des deutschen Volkes).

References

1. M. Oberlack, A unified approach for symmetries in plane parallel turbulent shear flows. J. Fluid Mech. **427**, 299–328 (2001)
2. S. Hoyas, M. Oberlack, F. Alcántara-Ávila, S.V. Kraheberger, J. Laux, Wall turbulence at high friction Reynolds numbers. Phys. Rev. Fluids **7**(1), 014602 (2022)
3. M. Oberlack, S. Hoyas, S.V. Kraheberger, F. Alcántara-Ávila, J. Laux, Turbulence statistics of arbitrary moments of wall-bounded shear flows: a symmetry approach. Phys. Rev. Lett. **128**(2), 024502 (2022)
4. M. Oberlack, A. Rosteck, New statistical symmetries of the multi-point equations and its importance for turbulent scaling laws. Discrete Contin. Dyn. Syst. - S. **3**(3), 451–471 (2010)
5. M. Waclawczyk, N. Staffolani, M. Oberlack, A. Rosteck, M. Wilczek, R. Friedrich, Statistical symmetries of the Lundgren-Monin-Novikov hierarchy. Phys. Rev. E **90**(1), 013022 (2014)
6. H.J. Hussein, S.P. Capp, W.K. George, Velocity measurements in a high-Reynolds-number, momentum-conserving, axisymmetric, turbulent jet. J. Fluid Mech. **258**, 31–75 (1994)

7. I. Wygnanski, H. Fiedler, Some measurements in the self-preserving jet. J. Fluid Mech. **38**(3), 577–612 (1969)
8. A. Papoulis, S.U. Pillai, *Probability, Random Variables, and Stochastic Processes*, 4th edn. (McGraw Hill, Boston, 2002)
9. K.R. Sreenivasan, R.A. Antonia, The phenomenology of small-scale turbulence. Annu. Rev. Fluid Mech. **29**(1), 435–472 (1997)
10. M. Nelkin, Universality and scaling in fully developed turbulence. Adv. Phys. **43**(2), 143–181 (1994)

Reynolds Number Induced Growth of Large-scale Rolls in Plane Couette Flow and Invariant Scaling Laws for Added Wall-transpiration Using Resolvent Analysis

T. Dokoza and M. Oberlack

Abstract In DNS of turbulent plane Couette flow (PCF) long streamwise rolls further grow as the Reynolds number increases [1]. To analyse this, we utilise resolvent analysis, focusing on high Reynolds numbers ($Re \to \infty$) and small streamwise wavenumbers ($\alpha \to 0$) with the limit $Re_\alpha = Re \cdot \alpha = O(1)$. We observe that Re_α acts as a local invariant in the amplification mechanism of the PCF characterised through the first singular value σ_1 of the resolvent operator within this limit. To maintain constant streamwise structures for increasing Reynolds numbers, the streamwise wavenumber has to decrease, agreeing with DNS studies about increasing length of the streamwise structures with the Reynolds number. We further investigate the plane Couette flow with constant wall-tranpiration (PCFT) with emphasis on the effect of the wall-transpiration Reynolds number Re_{V0} on the coherent structures which move closer to the upper wall and become narrower for growing Re_{V0}.Invariant scaling laws within the amplification mechanims can be found for various flow parameters.

1 Reynolds Number Induced Growth of Large-scale Rolls in Plane Couette Flow

1.1 Governing Equations

For the analysis to follow subsequently, the Navier-Stokes equations for the fluctuations will be the basis, i.e.,

$$\frac{\partial u_i'}{\partial t} + U_1(x_2)\frac{\partial u_i'}{\partial x_1} + \delta_{i1}u_2'\frac{dU_1(x_2)}{dx_2} + \frac{\partial p'}{\partial x_i} - \frac{1}{Re}\frac{\partial^2 u_1'}{\partial x_k^2} = f_i', \qquad \frac{\partial u_i'}{\partial x_i} = 0. \quad (1)$$

T. Dokoza (✉) · M. Oberlack
Technical University of Darmstadt, Chair of Fluid Dynamics, Department of Mechanical Engineering, Otto-Berndt-Str. 2, 64287 Darmstadt, Germany
e-mail: dokoza@fdy.tu-darmstadt.de

R. Örlü et al. (eds.), *Progress in Turbulence X*, Springer Proceedings in Physics 404,
https://doi.org/10.1007/978-3-031-55924-2_8

where a flow decomposition was applied with $u_1 = U_1(x_2) + u_1', \quad u_2 = u_2', u_3 = u_3', p = p'$, wherein u_i refers to the velocity in each spatial direction x_i, with $i = 1, 2, 3$, p denotes the pressure and Re represents the Reynolds number. $U_1(x_2) = x_2$ is the non-dimensionalised base velocity for PCF in streamwise direction, while u_i' and p' refer to the fluctuations of velocity and pressure. The non-linear term of the fluctuations in (1) is being summarised as an intrinsic forcing on the right-hand side as $f_i' = -u_j' \frac{\partial u_i'}{\partial x_j}$. We enforce no-slip boundary conditions, i.e., $\left(u_1', u_2', u_3'\right)^{\mathsf{T}} (x_1, x_2 = \pm 1, x_3) = 0$. Further we use a Normal-mode ansatz for the velocity fluctuations with $u_i' = \tilde{u}_i'(x_2) e^{\mathrm{i}(\alpha x_1 + \beta x_3 - \omega t)}$, where $\alpha \in \mathbb{R}$ an $\beta \in \mathbb{R}$ are the streamwise and spanwise wavenumber, while $\omega \in \mathbb{C}$ represents the frequency of the fluctuations. By applying the resolvent analysis, the Navier-Stokes equations for the fluctuations (1) can be formulated in an input-output formulation with $\tilde{\mathbf{q}}' = \mathbf{C}(-\mathrm{i}\omega \mathbf{I}_{2\times 2} - \mathbf{L})^{-1} \mathbf{B} \tilde{\mathbf{f}}' = \mathbf{H} \tilde{\mathbf{f}}'$, where $\mathbf{I}$ represents the unity matrix. In this formulation the Fourier transformed forcing $\tilde{\mathbf{f}}' = (\tilde{f}_1', \tilde{f}_2', \tilde{f}_3')$ is amplified by the linear resolvent operator $\mathbf{H}$ resulting in the response mode $\tilde{\mathbf{q}}' = \left(\tilde{u}_1', \tilde{u}_2', \tilde{u}_3'\right)^{\mathsf{T}}$. $\mathbf{L}$ represents the Orr-Sommerfeld and Squire operator, i.e.,

$$\mathbf{L} = \mathbf{M}^{-1} \begin{pmatrix} -\mathrm{i} Re \alpha x_2 \left(\frac{d^2}{dx_2^2} - k^2\right) + \frac{d^4}{dx_2^4} - 2k^2 \frac{d^2}{dx_2^2} + k^4 & 0 \\ -\mathrm{i} Re \beta & -\mathrm{i} Re \alpha x_2 + \frac{d^2}{dx_2^2} - k^2 \end{pmatrix}, \tag{2}$$

where the mass-matrix $\mathbf{M}$ and the forcing matrix $\mathbf{B}$ hereby read

$$\mathbf{M} = \begin{pmatrix} Re\left(\frac{d^2}{dx_2^2} - k^2\right) & 0 \\ 0 & Re \end{pmatrix}, \qquad \mathbf{B} = \mathbf{M}^{-1} \begin{pmatrix} -\mathrm{i} Re \alpha \frac{d}{dx_2} & -Re k^2 & -\mathrm{i}\beta Re \frac{d}{dx_2} \\ \mathrm{i} Re \beta & 0 & -\mathrm{i} Re \alpha \end{pmatrix} \tag{3}$$

and the continuity operator is represented by

$$\mathbf{C} = \frac{1}{k^2} \begin{pmatrix} \mathrm{i}\alpha \frac{d}{dx_2} & -\mathrm{i}\beta \\ k^2 & 0 \\ \mathrm{i}\beta \frac{d}{dx_2} & \mathrm{i}\alpha \end{pmatrix}, \qquad k^2 = \alpha^2 + \beta^2. \tag{4}$$

Resolvent analysis identifies key directions for amplifying $\tilde{\mathbf{f}}'$ via $\mathbf{H}$, yielding responses in $\tilde{\mathbf{q}}'$. This is done through a singular value decomposition (SVD) of $\mathbf{H}$, with σ_i indicating amplification gain linking forcing and response modes.

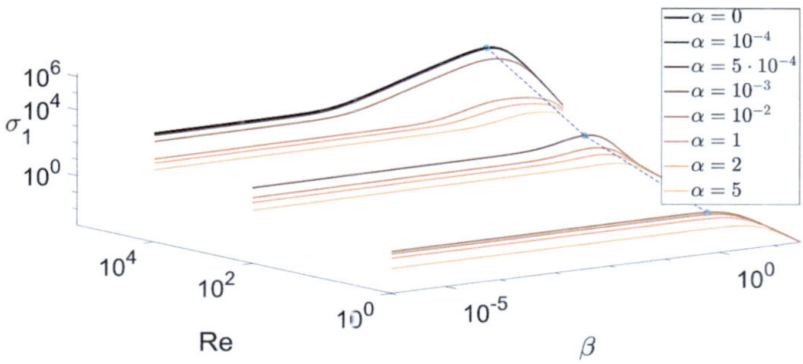

Fig. 1 σ_1 is shown for $\omega = 0$ and $\alpha = [0, 10^{-4}, 5 \cdot 10^{-4}, 10^{-3}, 10^{-2}, 1, 2, 5]$ for $Re = [10^0, 10^2, 10^4]$ over β

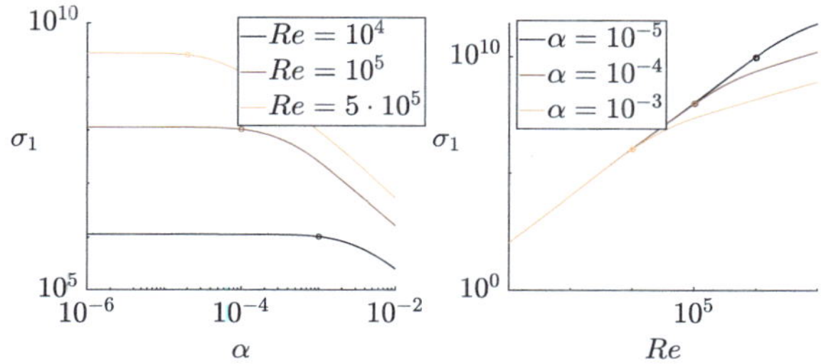

Fig. 2 σ_1 value is shown for $\omega = 0, \beta = 2$ and $Re = [10^4, 10^5, 5 \cdot 10^5]$ over α on the left and $\alpha = [10^{-5}, 10^{-4}, 10^{-3}]$ over Re on the right

1.2 Results

Only harmonic forcing, i.e., $\omega = 0$ is considered in the subsequent analysis. We are particularly interested in the coherent structures within the PCF, which can be observed as being spanwise-periodic and their dependence on α and Re. Due to the spanwise periodicity the most amplified PCF structures must possess a nonzero optimal spanwise wavenumber β_{max}. In Fig. 1 the first singular value, representing the largest amplification gain is analysed for $Re = [1, 10^2, 10^4]$ over β for various α. Only at high Reynolds numbers does the leading singular value occur at $\beta_{max} \neq 0$, enabling spanwise-periodic structures, while they are only significantly amplified for small α. Thus, we will focus our analysis on the distinguished limit (DL) of $Re \to \infty$ and $\alpha \to 0$ with $Re_\alpha = Re \cdot \alpha = O(1)$. In [3] the author found that the amplification mechanism for the coherent structures in PCF with streamwise invariance ($\alpha = 0$) is to a great extent encoded in the Orr-Sommerfeld operator alone, independent of the Squire operator. Further analyses in our work, which, however shall not be presented here, show that this statement is also true in the limiting case of $\alpha \to 0$. The Orr-Sommerfeld operator, represented by the first row of (2), is only dependent on Re_α. Thus Re_α plays a significant role in the amplification mechanism of the PCF withtin the DL.

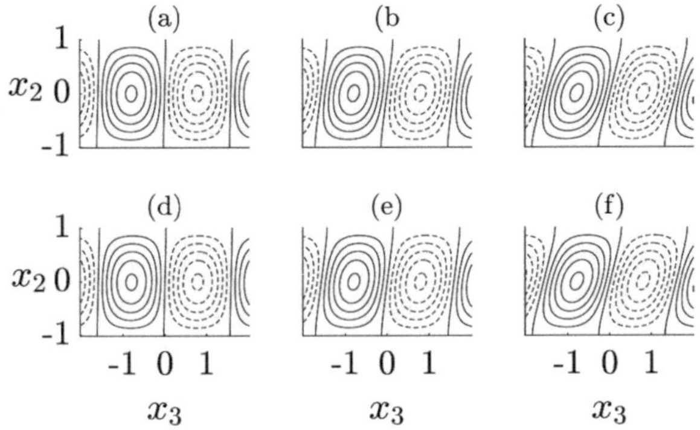

Fig. 3 Streamwise structures are shown for $[\omega, \beta] = [0, 2]$ and **a** $Re_\alpha = 1, \alpha = 10^{-4}$, **b** $Re_\alpha = 5$, $\alpha = 5 \cdot 10^{-4}$, **c** $Re_\alpha = 10, \alpha = 10^{-3}$, **d** $Re_\alpha = 1$, $Re = 10^4$, **e** $Re_\alpha = 5$, $Re = 5 \cdot 10^4$, **f** $Re_\alpha = 10$, $Re = 10^5$. Solid lines represent fluctuations with a positive sign, while dashed lines represent fluctuations with a negative sign

1.2.1 Distinguished Limit

We will now look at the influence of α and Re separately within the DL. Re_α is controlled by adjusting Re for fixed α and vice versa, maintaining $Re_\alpha = O(1)$. Figure 2 displays σ_1 for $\omega = 0$, $\beta = 2$, and $Re = [10^4, 10^5, 5 \cdot 10^5]$ over α on the left, and $\alpha = [10^{-5}, 10^{-4}, 10^{-3}]$ over Re on the right. σ_1 decreases monotonously over α, remaining almost constant for small α-values before declining significantly. Interestingly, Re_α remains roughly constant before this decrease at $Re_\alpha \approx 10$ for each constant Re line. On the right, σ_1 increases with Re^2 over Re until reaching $Re_\alpha \approx 10$ once again, then transitions to a saturation curve. In the DL, Re_α acts as a local invariant, determining σ_1's behavior over both α and Re independent of both parameters individually within this DL. Figure 3 shows the streamwise structures over the spanwise and wall-normal direction (x_3, x_2) for a variation of α and a fixed Re in the upper row and for various Re and a fixed α in the lower row, where Re_α on each column is equal. The structures are increasingly inclined for growing Re_α, while the effect of an increasing Re and α on their inclination angles is the same as long as $Re_\alpha = const.$, being evident from the identical inclination angles of the structures in (a) and (d), (b) and (e), and (c) and (f), where each Re_α is the same. Thus, Re_α affects the inclination and the appearance of these structures rather than α or Re alone, further substantiating its role as a local invariant within the DL. Thus to maintain constant streamwise structures for growing Re, α must decrease, aligning with the DNS findings of growing structures for increased Reynolds numbers in [1].

2 Invariant Scaling Laws for Added Wall-transpiration

In plane Couette flow with constant wall-transpiration (PCFT), suction at the upper wall and blowing at the lower wall with a uniform transpiration velocity V_0 is applied onto the PCF. The velocity profile for the laminar PCFT in streamwise and wall-normal direction are given as $U_1(x_2) = \frac{2e^{Re_{V0}x_2} - e^{Re_{V0}} - e^{-Re_{V0}}}{e^{Re_{V0}} - e^{-Re_{V0}}}$, $U_2(x_2) = \frac{Re_{V0}}{Re} = \gamma$, where due to V_0 two relevant Reynolds number are defined as $Re = \frac{U_w h}{\nu}$ and $Re_{V0} = \frac{V_0 h}{\nu}$. The inclusion of wall-transpiration changes the Orr-Sommerfeld-Squire operator (2) to

$$\mathbf{L} = \mathbf{M}^{-1} \begin{pmatrix} L_{11} & 0 \\ -\mathrm{i}\beta U_1'(x_2) & L_{22} \end{pmatrix}, \tag{5}$$

with $L_{11} = \left(-\mathrm{i}\alpha U_1(x_2) - U_2 \frac{\mathrm{d}}{\mathrm{d}x_2}\right) \Delta + \mathrm{i}\alpha U_1''(x_2) + \frac{1}{Re}\Delta^2$, $L_{22} = -\mathrm{i}\alpha U_1(x_2) - U_2 \frac{\mathrm{d}}{\mathrm{d}x_2} + \frac{1}{Re}\Delta$, $\Delta = \frac{d^2}{dx_2^2} - k^2$ and $U_1'(x_2) = dU_1(x_2)/\,dx_2$, $U_1''(x_2) = d^2U_1(x_2)/dx_2^2$.

2.1 Results

PCFT's amplification mechanisms are studied via a SVD of $\mathbf{H}$. σ_1 is shown for Re_{V0}=[0, 0.1, 2, 5, 20, 50, 75, 100, 150, 200], $[\omega, Re] = [0, 10^5]$ over β for $\alpha = 0$ and over α for $\beta = 2$ in Fig. 4. Initially, σ_1 increases linearly with β, peaking at β_{max}, and then decreasing with further growing β, the same as for PCF in Fig. 1. A growing Re_{V0} broadens the positive gradient of σ_1 over β, leading to larger β_{max} and narrower coherent structures. An invariant relationship between Re_{V0} and β_{max} is found as $\beta_{max} = a \cdot Re_{V0}^2 + b \cdot Re_{V0} + c$. For small Re_{V0}, the most amplified structures are found at $\alpha_{max} = 0$, with a monotonous decrease of σ_1 over α. Yet, a σ_1 peak emerges at $\alpha_{max} \neq 0$ starting from $Re_{V0} = Re_{V0}^* = 50$. For $Re_{V0} > Re_{V0}^*$ an invariant relationship between α_{max} and Re_{V0} is given as $\alpha_{max} = a \cdot Re_{V0} + b$. This indicates a change in the behavior of the most amplified structures for large enough Re_{V0}, contrasting with PCF's streamwise independence. In Fig. 5, σ_1 is shown over Re for various ratios $\gamma = \frac{Re_{V0}}{Re}$ for $[\omega, \beta, \alpha] = [0, 2, 0]$. Introducing wall-transpiration alters σ_1's behavior, from linearly growing over Re to forming peaks at specific Re_{max}

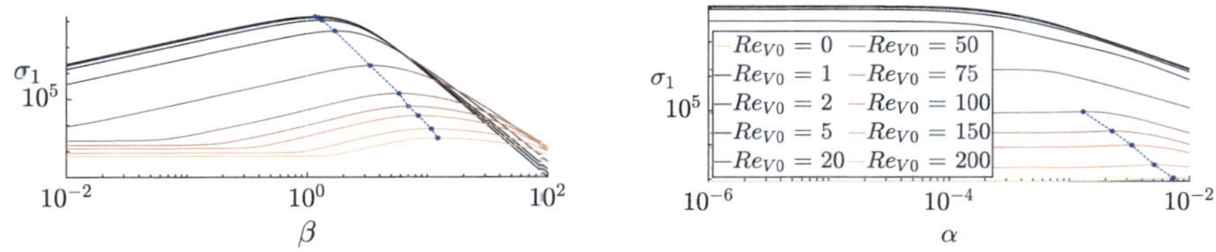

Fig. 4 σ_1 is shown for Re_{V0}=[0, 0.1, 2, 5, 20, 50, 75, 100, 150, 200], $[\omega, Re] = [0, 10^5]$ over β for $\alpha = 0$ on the left and over α for $\beta = 2$ on the right

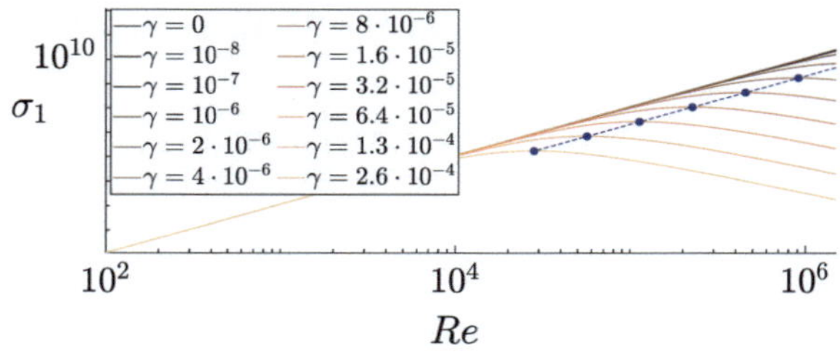

Fig. 5 σ_1 is shown over Re for $[\omega, \beta, \alpha] = [0, 2, 0]$ and $\gamma = [0, 10^{-8}, 10^{-7}, 10^{-6}, 2 \cdot 10^{-6}, 4 \cdot 10^{-6}, 8 \cdot 10^{-6}, 1.6 \cdot 10^{-5}, 3.2 \cdot 10^{-5}, 6.4 \cdot 10^{-5}, 1.28 \cdot 10^{-4}, 2.56 \cdot 10^{-4}]$

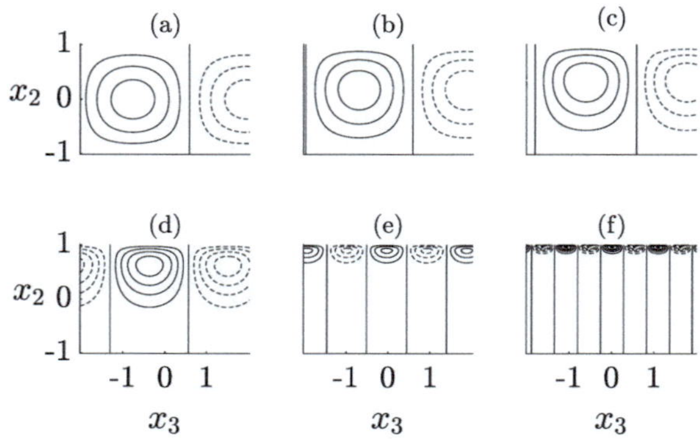

Fig. 6 Streamwise strcutures are shown for $[\alpha, \omega] = [0, 0]$ and $[Re_{max}, \beta_{max}]$ for each $\gamma = [2.56 \cdot 10^{-4}, 1.28 \cdot 10^{-4}, 3.2 \cdot 10^{-5}, 1.6 \cdot 10^{-5}, 4 \cdot 10^{-6}, 2 \cdot 10^{-6}]$ from top-left to bottom-right

with $Re_{max} = C \cdot \gamma^a$. Figure 6 shows the most amplified streamwise structures for $[\alpha, \beta, \omega, Re] = [0, \beta_{max}, 0, 10^5]$ and Re_{V0}=[0, 1, 2, 5, 20, 50]. As Re_{V0} increases, the structures localise nearer the upper wall, agreeing with DNS outcomes in [4] and become narrower in the wall-normal and spanwise direction.

References

1. M. Lee, R. Moser, Extreme-scale motions in turbulent plane Couette flows. J. Fluid Mech. **842**, 128–145 (2018)
2. T. Dokoza, M. Oberlack, Reynolds number induced growth of the large-scale rolls in plane Couette flow using resolvent analysis. J. Fluid Mech. **968**, A23 (2023)
3. S.J. Illingworth, Streamwise-constant large-scale structures in Couette and Poiseuille flows. J. Fluid Mech. **889**, A13 (2020)
4. S. Kraheberger, S. Hoyas, M. Oberlack, DNS of a turbulent Couette flow at constant wall transpiration up to $Re_\tau = 1000$. J. Fluid Mech. **835**, 421–443 (2018)

Wall-Bounded Flows

Augmented Restricted Nonlinear (ARNL) Model for High Reynolds Number Wall-Turbulence

B. A. Minnick, B. Viggiano, and D. F. Gayme

Abstract A restricted nonlinear (RNL) representation of wall-bounded turbulence accurately predicts key characteristics of low to moderate Reynolds number flows. This model limits nonlinear interactions to those producing streamwise constant dynamics (defined as large-scales), and restricts streamwise varying small-scales to lie within the region associated with peak outer-layer dissipation. Here, we propose an augmented RNL (ARNL) model that includes additional scales, but maintains order reduction by combining the large-scale dynamical restriction imposed in generalized quasilinear models with RNL band-limiting of the small-scales. This ARNL model with a single nonlinearly interacting streamwise varying mode is shown to yield more accurate predictions of low-order statistics versus the RNL model at $\mathrm{Re}_\tau = 550$ and 1000. At $\mathrm{Re}_\tau = 2000$, a second non-zero large-scale mode improves the fidelity of ARNL predictions.

1 Introduction

The ubiquity of coherent structures at a wide-range of scales in wall-bounded turbulence, see e.g., [1] and references therein, has inspired structure based reduced-order modeling techniques to investigate the underlying mechanisms and dynamical processes in a simplified setting. For example, early observations of the interactions of near-wall streaks [2], motivated mechanistic models of a self-sustaining near-wall cycle [3, 4]. This concept was later generalized to a scale-based hierarchy of similar self-sustaining processes [5]. Two-dimensional three-velocity component (2D/3C)

B. A. Minnick · D. F. Gayme (✉)
Johns Hopkins University, Baltimore, MD 21212, USA
e-mail: dennice@jhu.edu

B. A. Minnick
e-mail: bminnic2@jhu.edu

B. Viggiano
Polytechnique Montréal, Montréal, QC H3C 3A7, Canada
e-mail: bianca.viggiano@polymtl.ca

R. Örlü et al. (eds.), *Progress in Turbulence X*, Springer Proceedings in Physics 404,
https://doi.org/10.1007/978-3-031-55924-2_9

models designed to capture streamwise coherent motions have been used to describe skin-friction drag generation near the wall [6], as well as large scale momentum transfer in a range of wall-bounded shear flows, see e.g., [7, 8].

The restricted nonlinear (RNL) model builds upon the 2D/3C model wherein this nonlinear streamwise constant mean flow dynamics is coupled to a dynamically restricted streamwise varying perturbation field, see e.g., [9, 10] and references therein. This restriction of the Navier-Stokes (NS) equations to interactions between the streamwise constant nonlinear mean flow dynamics and coupled linear streamwise varying perturbation dynamics produces self-sustaining turbulence at vastly reduced computational cost [11]. This decomposition corresponds to a quasi-linear (QL) model [12, 13] with a streamwise rather than horizontally averaged mean flow. The parameterized RNL model improves upon the performance of this baseline RNL model by band-limiting the perturbation dynamics to include a limited number of smaller more dissipative scales [10, 14, 15]. The generalized QL (GQL) model is obtained through a spectral filter in the streamwise and spanwise directions with respective non-zero cut-off wavenumbers (modes in a Fourier representation), see [16] for details. The GQL model thus similarly provides improved predictions over the QL model by permitting nonlinear interactions between streamwise and spanwise varying modes. Permitting non-zero streamwise and spanwise wavenumbers to interact nonlinearly scatters energy, leading to a higher order model wherein all simulated small-scales remain active [17].

Parametrized RNL models have proven successful in reproducing key features of wall-bounded turbulence in low to moderate Reynolds number regimes [14, 15]. However, the limitation to streamwise constant nonlinear interactions and restriction of streamwise varying small-scales to those associated with peak dissipation in the outer-layer leads to a representation that does not sufficiently account for the range of scales and scale interactions characteristic of higher Reynolds number flows. This lack of sufficient scale interactions leads to a breakdown of model performance with increasing Reynolds number. This breakdown was also noted in RNL-Large Eddy Simulations (RNL-LES) at arbitrarily high Reynolds numbers, where cross-plane grid resolution acts as an analog to Reynolds number [18]. Based on an analysis of LES data, Bretheim et al. [18] suggested that as grid resolution becomes finer additional scales become important. The authors tested this notion through the addition of an intermediate scale mode within the linear dynamics of the perturbation field. Their results showed that this approach can improve performance if this mode is only active farther from the wall. They concluded that the lack of nonlinear interactions associated with this mode led to nonphysical dynamics close to the wall, which motivates a similar approach wherein this intermediate mode is allowed to interact nonlinearly with the mean flow. GQL models provide a mechanism for including such nonlinear interactions, however computational challenges can quickly arise due to the significant increase in the number active small scales as the number of large-scales that interact nonlinearly is increased.

In this work, we present a hybrid approach to overcome the shortcomings of both RNL and GQL models. In particular, we propose an augmented RNL (ARNL) framework that adopts a GQL like treatment for the large-scales that are expanded to include specific combinations of non-zero streamwise varying modes. An RNL based parametrization for the small-scales associated with the perturbation field is employed to maintain order reduction, thereby resulting in only a small increase in computational expense over the RNL model. The success of this newly proposed ARNL approach is demonstrated in channel flow simulations at $Re_\tau = 550$, 1000 and 2000. The next section describes the ARNL model while highlighting connections between QL, RNL and GQL approaches.

2 Model Equations

The governing equations for the QL, GQL, RNL and ARNL models are all obtained by decomposing the total velocity field $\mathbf{u}_\mathrm{T}$ into a large-scale mean component $\mathbf{U}$ and 'smaller-scale' perturbations about that mean $\mathbf{u}$, such that, $\mathbf{u}_\mathrm{T} = \mathbf{U} + \mathbf{u}$. The 'large scales' are defined based on a 'large-scale' spectral filter $\mathbf{U}(x, y, z, t) := \langle \mathbf{u}_\mathrm{T} \rangle_l$ whose form is specified by the particular model. Here x, y, z and t respectively denote the streamwise, wall-normal, spanwise, and temporal coordinates. The definition of the small-scales is similarly model dependent. An analogous decomposition and filtering is applied to the pressure field. Order reduction is achieved by neglecting nonlinear interactions that do not contribute to the modeled 'large scales', leading to dynamics of the form

$$\partial_t \mathbf{U} + \langle \mathbf{U} \cdot \nabla \mathbf{U} \rangle_l + \nabla P/\rho - \nu \nabla^2 \mathbf{U} = -\langle \mathbf{u} \cdot \nabla \mathbf{u} \rangle_l, \tag{1a}$$

$$\partial_t \mathbf{u} + \langle \mathbf{U} \cdot \nabla \mathbf{u} \rangle_s + \nabla p/\rho - \nu \nabla^2 \mathbf{u} = -\langle \mathbf{u} \cdot \nabla \mathbf{U} \rangle_s, \tag{1b}$$

where ρ is the fluid density, ν is the kinematic viscosity, and $\langle \cdot \rangle_s$ describes the spectral cut-off for small-scale modes. These equations differ from the NS equations in that they neglect the terms $\langle \mathbf{u} \cdot \nabla \mathbf{U} \rangle_l$ and $\langle \mathbf{U} \cdot \nabla \mathbf{u} \rangle_l$ in (1a), as well as $\langle \mathbf{U} \cdot \nabla \mathbf{U} \rangle_s$ and $\langle \mathbf{u} \cdot \nabla \mathbf{u} \rangle_s$ in (1b), i.e., those that do not contribute to the modeled large scales. The first three of these discarded terms are identically zero when the large-scale averaging satisfies the properties of Reynolds averaging as in the QL or RNL models [9]. In GQL models, neglecting these terms is part of the imposed dynamical restriction. Neglecting $\langle \mathbf{u} \cdot \nabla \mathbf{u} \rangle_s$ in equation (1b) restricts the perturbations, or small-scale dynamics, to be linear in $\mathbf{u}$. In all of these models, the nonlinear interactions are restricted to those of the large-scales, i.e., $\langle \mathbf{U} \cdot \nabla \mathbf{U} \rangle_l$, and perturbations resulting in a large-scale component, i.e., $\langle \mathbf{u} \cdot \nabla \mathbf{u} \rangle_l$. The retained term, $\langle \mathbf{U} \cdot \nabla \mathbf{u} \rangle_s + \langle \mathbf{u} \cdot \nabla \mathbf{U} \rangle_s$, in (1b) represents regulation of the small-scales by the large-scale mean.

Specialization of equations (1) to a particular model is based on specification of the spectral filter for the large-scales in (1a) and the small-scales supporting the dynamics in (1b). The spectral filter for the large-scales in the QL model typically represents a horizontal average [12, 13]; all other scales are considered small. The GQL model is obtained through a spectral filter in the streamwise and spanwise directions with respective non-zero cut-off wavenumbers κ_x and κ_z, see [16] for details. In turbulent channel flow the GQL model has been implemented with truncation only in the streamwise direction, i.e., take $\kappa_x < \infty$ and $\kappa_z \rightarrow \infty$ [19], which enables variation of the streamwise wavenumber cut-off as a parameter that can be used to adjust the nonlinear interactions considered in the dynamics.

A GQL model with a streamwise wavenumber cut-off that constrains the large scale dynamics to those associated with the streamwise wavenumber $K_x = 0$ is equivalent to applying a streamwise average, which leads to a type of QL model that has been referred to as the baseline RNL model [9]. This type of model only permits nonlinear interactions contributing to the streamwise constant dynamics. The small scales that support the dynamics are only those large enough to sustain energy, see e.g., [11] for further discussion. Bretheim et al. [14] introduced a parametrized RNL model that employs the same restriction of the large-scales to $K_x = 0$ but imposes a band-pass filter that restricts the small-scale modes (streamwise varying wavenumbers $k_x \neq 0$) to more dissipative scales. In what follows, we refer to this parameterized RNL as the RNL model and the baseline RNL as a QL model.

The ARNL model builds upon the RNL dynamics by introducing a streamwise-varying mode to be treated as a large-scale. The new large-scale $K_x \neq 0$ mode interacts with the $K_x = 0$ mode non-linearly through the term $\langle \mathbf{U} \cdot \nabla \mathbf{U} \rangle_l$ in (1a). Certain small-scale modes, $k_x^{(1)}$ and $k_x^{(2)}$ nonlinearly interact to contribute to the new large-scale $K_x \neq 0$ through the term $\langle \mathbf{u} \cdot \nabla \mathbf{u} \rangle_l$. These small-scales also interact with the new large-scale through the term $\langle \mathbf{U} \cdot \nabla \mathbf{u} \rangle_s + \langle \mathbf{u} \cdot \nabla \mathbf{U} \rangle_s$ in (1b). In the spirit of the GQL model, these non-zero wavenumbers interact in a consistent manner, such that $k_x^{(2)} = K_x^{(1)} + k_x^{(1)}$. However, as in the RNL paradigm the active wavenumbers are limited to those selected, i.e., the nonlinear interactions that produce new small-scales in the GQL are not permitted. The large-scales retained in the ARNL model also differ from those of the GQL in that they are not determined by a spectral cutoff but rather pre-selected.

The ARNL formulation proposed here is general enough to include an arbitrary number of non-zero large-scale modes interacting with small-scale modes in a self-consistent manner. For example, a permissible wavenumber support for two non-zero large-scales comprise large-scale modes $K_x^{(1)}$ and $K_x^{(2)} = K_x^{(1)} + K_x^{(1)}$ that can interact with at least three non-zero small-scale modes $k_x^{(1)}$, $k_x^{(2)}$, and $k_x^{(3)}$. Where consistency requires $k_x^{(2)} = K_x^{(1)} + k_x^{(1)}$ and $k_x^{(3)} = K_x^{(1)} + k_x^{(2)} = K_x^{(2)} + k_x^{(1)}$. Additional large-scales can be added in this manner to enable a hierarchy of large-scale streamwise modes interacting with small-scale modes to be included in analogy with Townsend's attached eddy hypothesis [20].

3 Simulation Set Up

The open source pseudo-spectral code LESGO (https://lesgo.me.jhu.edu) is used to solve equations (1) for a full-channel configuration with no-slip boundary conditions applied to the bottom ($y = 0$) and top ($y = 2\delta$) surfaces. Periodic boundary conditions are applied in the horizontal directions. Wall-normal derivatives are discretized using a second-order central finite differencing scheme and spectral derivatives are used in the homogeneous directions. The 3/2 rule is used for dealiasing. Time advancement employs the second-order Crank-Nicolson method for wall-normal diffusive terms and the second-order Adams-Bashforth approach for the remaining terms.

Our numerical implementation takes advantage of the simplified dynamics in the RNL and ARNL models by computing the nonlinear term as a convolution in (k_x, y, z, t) space instead of as a product in physical space. This approach directly encodes the nonlinear interactions and eliminates the need for transforms back and forth between physical and Fourier space in the streamwise direction, saving computational time and resources, see [18] for more details. This approach also obviates the need to prescribe a physical streamwise grid in the simulation.

The streamwise wavenumbers used in the simulations are listed in Table 1. The large-scales in both RNL and ARNL simulations consist of the streamwise mean, $K_x\delta = 0$. The ARNL simulations labeled A-1, A-2 and A-3.1 also include a large-scale mode $K_x\delta = 7$, which has a wavelength corresponding to the half-channel height. The RNL small-scale modes, $k_x\delta$, have wavelengths corresponding to the peak region of the surrogate dissipation spectra. ARNL small-scales have wavelengths in a similar range but are prescribed such that $k_x^{(1)} = K_x^{(1)} + k_x^{(2)}$. We use two non-zero large-scales in the ARNL simulations at $\mathrm{Re}_\tau = 2000$ with the additional large-scale mode $K_x\delta = 14$ corresponding to a wavelength that is half of the half-channel height. As described in the previous section, additional scales can be added to the ARNL model, leading to a limited scale hierarchy in the spirit of Townsend [20].

As indicated in Table 1, the cross-plane grid resolution is comparable to that used in DNS. The wall-normal grid is stretched using the hyperbolic tangent function described in [21]. Five grid points exist below $y^+ = 5$ and consecutive grid points are stretched with the ratio $\Delta y_{n+1}/\Delta y_n < 1.03$.

Table 1 Numerical details of RNL simulations at the various Re_τ considered

Label	Model	Re_τ	$K_x\delta \neq 0$	$\approx \lambda_x/\delta$	$k_x\delta$	$\approx \lambda_x^+$	L_z/δ	Δz^+
R-1	RNL	550	–	–	23	150.3	2π	6.00
A-1	ARNL	550	7	0.898	19.5, 26.5	177.2, 130.4	2π	6.00
A-2	ARNL	1000	7	0.898	38.5, 45.5	163.2, 138.1	2π	6.14
A-3.1	ARNL	2000	7	0.898	80.5, 87.5	156.1, 143.6	2π	6.14
A-3.2	ARNL	2000	7, 14	0.898, 0.449	77, 84, 91	163.2, 149.6, 138.1	2π	6.14

4 Results and Discussion

Figure 1 shows first and second-order statistics for all cases in Table 1. At $\mathrm{Re}_\tau = 550$ both the RNL and ARNL models are shown to predict accurate log-law behavior up to $y^+ \approx 100$. The ARNL mean velocity profile continues to closely approximate the log-law above $y^+ \approx 100$ and the $\overline{u'_\mathrm{T} v'_\mathrm{T}}$ Reynolds stress compares well with DNS. However, for $y^+ > 100$ the RNL model predicts a higher mean velocity and a corresponding decrease in the $\overline{u'_\mathrm{T} v'_\mathrm{T}}$ Reynolds stress. The accuracy of this prediction is comparable to that seen in the RNL study of [23], where a more energetic wavenumber support was used at $\mathrm{Re}_\tau = 940$.

The most notable differences between the RNL and ARNL model predictions at $\mathrm{Re}_\tau = 550$ are in the streamwise and wall-normal Reynolds stress components. These differences are most notable in the outer-layer, where the RNL model is shown to over-predict the streamwise component and under-predict the wall-normal com-

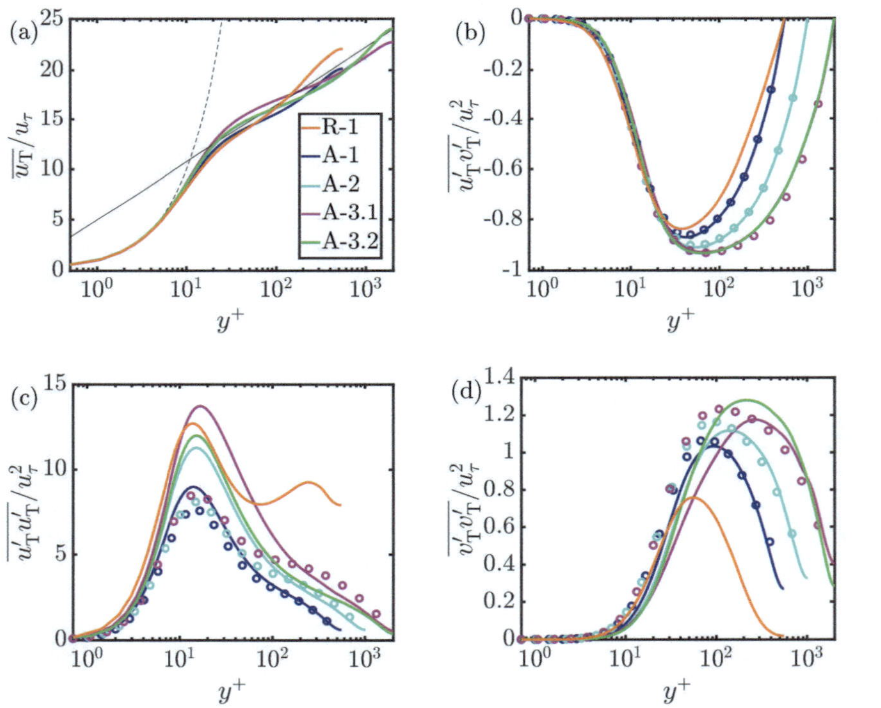

Fig. 1 **a** Total streamwise mean velocity. **b** Cross, **c** streamwise, and **d** wall-normal Reynolds stresses from RNL at $\mathrm{Re}_\tau = 550$ (orange) and ARNL at $\mathrm{Re}_\tau = 550$ (blue), 1000 (cyan), and 2000 (magenta). Profiles are compared to **a** $\overline{u_\mathrm{T}}/u_\tau = y^+$ (black dashed) and $\overline{u_\mathrm{T}}/u_\tau = (1/0.4)\log(y^+) + 5$ (solid black) and **b–c** DNS from [22] (circle markers). ARNL with two streamwise varying large-scales at $\mathrm{Re}_\tau = 2000$ shows improved accuracy (green)

ponent. The ARNL model shows significant improvements in the Reynolds stress prediction accuracy at this Reynolds number. For example, the location of peak of the streamwise component is closer to that of the DNS. The behavior in the outer-layer is also more consistent with DNS data. Similar improvements are seen in the other Reynolds stress components, although the peak of the wall-normal component is slightly lower and farther from the wall than that of the DNS data.

The fact that the RNL model predicts a clear secondary outer-layer peak in the streamwise Reynolds stress suggests that the streamwise constant RNL large-scale is inadequately dissipated by the small-scale perturbation component in that region. Results from the ARNL simulations indicate that including nonlinear interactions between a streamwise varying large-scale and the small scales, e.g., $k_x^{(2)} = K_x^{(1)} + k_x^{(1)}$ effectively diminishes the energy in the streamwise mean component in the outer layer, thereby reducing the streamwise Reynolds stress and increasing the wall-normal Reynolds stress far from the wall. This new nonlinear interaction thus leads to more accurate predictions of these statistics.

Figure 2a–c shows pre-multiplied spanwise spectra predicted by the RNL and ARNL models at $\mathrm{Re}_\tau = 550$. The peak of the RNL spectra for all components shown is located at smaller spanwise scales compared to DNS. This trend has been observed in previous work with the RNL system, and attributed to spanwise scales adjusting to compensate for the lack of streamwise scales [15]. Spanwise spectra generated by the ARNL model does not show this trend, and peaks closer to spanwise scales where the DNS peaks, except for the wall-normal component, which appears to peak at larger spanwise scales. The spread across wall-normal and spanwise scales where these spectra peak also differ. For both models the peak spans a larger range of scales for the streamwise component, and the RNL model peak ranges across fewer scales for the wall-normal component compared to the DNS. This behavior agrees with observations made previously for the Reynolds stresses in Fig. 1.

Consistent with corresponding Reynolds stress predictions, the RNL model predicts a outer-layer peak in the streamwise component and diminished wall-normal component that spreads only just to the outer-layer. The cross-component of the RNL spanwise spectra also show a outer-layer peak, albeit less pronounced. These observations are the result of the streamwise mean acting alone in the outer-layer. In the absence of energetic streamwise varying motions, outer-layer spanwise structures are strongly influenced. These trends are not observed in the ARNL model, instead predictions of these spanwise spectra are more comparable to the DNS at this Reynolds number, reconciled by the streamwise varying large-scale mode.

We now consider ARNL at $\mathrm{Re}_\tau = 1000$ and 2000. Figure 1a indicates that with increasing Reynolds number, the intercept of the predicted mean streamwise velocity increases. Despite this deviation from expected log-law behavior, the slope remains relatively constant with increasing Reynolds number and therefore the predicted cross-Reynolds stress required to satisfy the time-averaged momentum balance continues to agree well with DNS. More noticeable differences from the expected behavior are observed for the streamwise and wall-normal Reynolds stresses. As expected, the inner-layer peak of the streamwise Reynolds stress increases with Reynolds number, however the increases are more significant than is typical in DNS data.

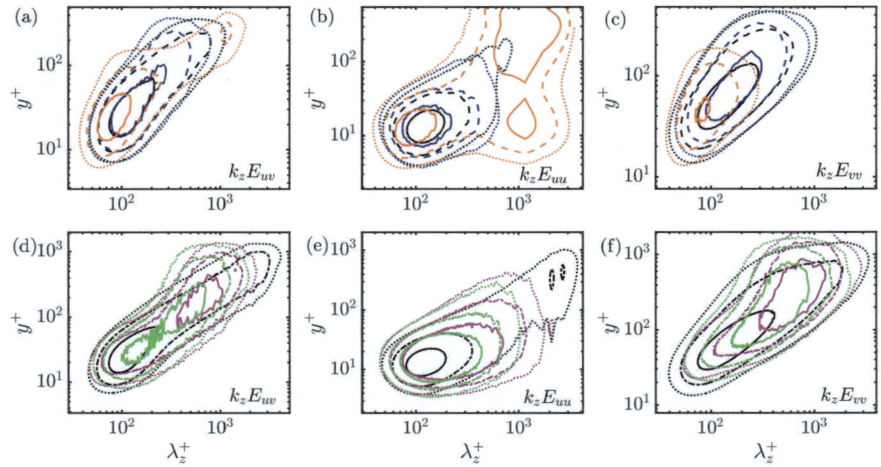

Fig. 2 Pre-multiplied spanwise spectra, **a**, **d** $k_z E_{uv}$, **b**, **e** $k_z E_{uu}$, and **c**, **f** $k_z E_{vv}$, from DNS [22] (black), R-1 (orange), A-1 (blue), A-3.1 (magenta), and A-3.2 (green). Spectra from $\text{Re}_\tau = 550$ shown in the top row **a–c** and from $\text{Re}_\tau = 2000$ is shown in the bottom row **d–f**. Contour lines correspond to 0.8 (solid), 0.5 (dashed), and 0.3 (dotted) times the maximum value of the respective DNS spectra

The manner in which this peak value decays in the log-region is also different in that a larger slope is predicted by the ARNL model. Predictions of the wall-normal Reynolds stress are found to be more comparable to DNS, although the location of the peak is again farther from the wall compared to that of the DNS and this difference increases with increasing Reynolds number.

In order to examine the effect of additional scale interactions, we performed an ARNL simulation that includes an additional streamwise varying large-scale mode (case A-3.2 in Table 1). Figure 1 indicates that the additional scales lead to a clear increase in model fidelity. The mean velocity profile is closer to the expected log-law, the over-predicted peak in the streamwise Reynolds stress lowers, and the peak of the wall-normal Reynolds stress is closer to the wall and more comparable to that of DNS. Although the outer-layer peak in the streamwise Reynolds stress that seems to be developing in the DNS data at $\text{Re}_\tau = 2000$ is not evident in either ARNL simulation, the additional scale changes the shape of the profile in the outer-layer to more closely follow the DNS trend. The wall-normal Reynolds stress predictions are also improved, however they do not approach the accuracy of the ARNL predictions at $\text{Re}_\tau = 550$ and 1000. These results suggest that additional nonlinearity may be required to predict the outer-layer peak and maintain accurate cross-plane Reynolds stresses.

Figure 2d–f shows the same pre-multiplied spanwise spectra in panels (a-c) for $\text{Re}_\tau = 2000$. Despite fairly good agreement with the second-order statistics shown in Fig. 1, the ARNL model with a single streamwise-varying large-scale mode shows significant differences from the DNS data. All components are shown to peak farther from the wall and at larger spanwise scales, most notably in the cross and wall-normal components. The outer-layer peak that is beginning to emerge in the streamwise-component of the DNS spectra is not evident in the ARNL spectra, instead the inner-layer peak is shown to spread across a larger range of scales.

A significant improvement in pre-multiplied spanwise spectra predictions is noted for the A-3.2 simulation, which includes an additional streamwise-varying large-scale. The cross and wall-normal components peak closer to scales where the DNS spectrum peaks and the streamwise component peaks over a more concentrated range. These results suggest that cross-plane fluctuations become more active near the wall and at smaller spanwise scales. However, a secondary peak is still not evident in the streamwise component of the spectra. Future work is needed to determine if this peak can be recovered by adding a third or fourth scale.

5 Conclusions and Future Work

The ARNL model blends the RNL and GQL modeling paradigms in order to provide a high Reynolds number wall-bounded turbulence model that accurately predicts low-order statistics and structural flow features at low computational cost. Results at $\mathrm{Re}_\tau = 550$ and 1000 demonstrate that supporting the ARNL dynamics with a set of streamwise wavenumbers involving a large-scale non-zero mode and two small-scale modes which scale with the outer-layer peak of the surrogate dissipation spectra provides improved accuracy with little increase in computational complexity. Analysis at $\mathrm{Re}_\tau = 2000$ indicates that the statistics can be further improved by introducing an additional large-scale mode in a hierarchy of nonlinearly interacting non-zero modes that adhere to a Townsend inspired hierarchy of streamwise scales.

The wide range of wall-normal and spanwise scales simulated by the reduced dynamics makes the ARNL modeling framework an ideal candidate to study scale separation and high Reynolds number wall-turbulence phenomena. Differences in the behavior of the RNL and ARNL models shown here have already provided insight into the physics of wall-turbulence. For example a nonphysical outer-layer secondary peak in the streamwise Reynolds stress is predicted by the RNL model at $\mathrm{Re}_\tau = 550$. The correction of this behavior in the ARNL model suggests additional scale interactions are necessary to induce a cascade of energy down to the dissipative scales in the outer layer. At $\mathrm{Re}_\tau = 2000$ two additional scales were insufficient to predict a outer-layer peak in the spectra. Further analysis is needed to determine whether or not including additional nonlinear interactions in the ARNL model would enable the prediction of this outer-layer peak while maintaining accurate cross-plane Reynolds stresses.

Future directions involve further study of different combinations of scale interactions. For example an analysis of amplitude modulation [24] in this reduced framework can enable isolation of the underlying mechanisms. An additional direction for future work would be to construct large-scale streamwise mode scale hierarchies to systematically probe the dynamics and examine the connections to Townsend's attached eddy hypothesis [20, 25].

Acknowledgements This work was partially supported by the Office of Naval Research (N00014-20-1-2535) and European Research Council (Caust grant ERC-AdG-101018287). Simulations employed the Advanced Research Computing at Hopkins (ARCH) core facility (rockfish.jhu.edu), which is supported by the National Science Foundation (OAC 1920103).

References

1. J. Jiménez, Coherent structures in wall-bounded turbulence. J. Fluid Mech. **842**, P1 (2018)
2. S.J. Kline, W.C. Reynolds, F.A. Schraub, P.W. Runstadler, The structure of turbulent boundary layers. J. Fluid Mech. **30**(4), 741–773 (1967)
3. J. Jiménez, A. Pinelli, The autonomous cycle of near-wall turbulence. J. Fluid Mech. **389**, 335–359 (1999)
4. K. Hamilton, J. Kim, F. Waleffe, Regeneration mechanisms of near-wall turbulence structures. J. Fluid Mech. **287**, 317–348 (1995)
5. C. Cossu, Y. Hwang, Self-sustaining processes at all scales in wall-bounded turbulent shear flows. Phil. Trans. R. Soc. A **375**, 20160088 (2016)
6. P. Orlandi, J. Jiménez, On the generation of turbulent wall friction. Phys. Fluids **6**(2), 634–641 (1994)
7. J.-L. Bourguignon, B.J. McKeon, A streamwise-constant model of turbulent pipe flow. Phys. Fluids **23**(9), 095111 (2011)
8. D.F. Gayme, B.J. McKeon, B. Bamieh, A. Papachristodoulou, J.C. Doyle, Amplification and nonlinear mechanisms in plane Couette flow. Phys. Fluids **23**(6), 065108 (2011)
9. B.F. Farrell, D.F. Gayme, P.J. Ioannou, A statistical state dynamics approach to wall turbulence. Phil. Trans. R. Soc. A **375**, 20160081 (2017)
10. D.F. Gayme, B.A. Minnick, Coherent structure-based approach to modeling wall turbulence. Phys. Rev. Fluids **4**, 110505 (2019)
11. V. Thomas, B.F. Farrell, P.J. Ioannou, D.F. Gayme, A minimal model of self-sustaining turbulence. Phys. Fluids **27**, 105104 (2015)
12. B.F. Farrell, P.J. Ioannou, Structure and spacing of jets in barotropic turbulence. J. Atmos. Sci. **64**, 3652–3665 (2007)
13. F. Ait-Chaalal, T. Schneider, B. Meyer, J.B. Marston, Cumulant expansions for atmospheric flows. New J. Phys. **18**, 025019 (2016)
14. J.U. Bretheim, C. Meneveau, D.F. Gayme, Standard logarithmic mean velocity distribution in a band-limited restricted nonlinear model of turbulent flow in a half-channel. Phys. Fluids **27**, 011702 (2015)
15. B.A. Minnick, D.F. Gayme, Characterizing energy transfer in restricted nonlinear wall-bounded turbulence, in *Int'l Symp on Turbulence and Shear Flow Phenomena* (2019)
16. J.B. Marston, G.P. Chini, S.M. Tobias, Generalized quasilinear approximation: Application to zonal jets. Phys. Rev. Lett. **116**, 214501 (2016)
17. C. Hernández, Y. Hwang, Spectral energetics of a quasilinear approximation in uniform shear turbulence. J. Fluid Mech. **904**, A11 (2020)
18. J.U. Bretheim, C. Meneveau, D.F. Gayme, A restricted nonlinear large eddy simulation model for high Reynolds number flows. J. Turbulence **19**, 141–166 (2018)
19. C.B. Kellam, *Generalized Quasilinear Simulation of Turbulent Channel Flow*, PhD Thesis (University of New Hampshire, Manchester, NH, USA, 2019)
20. A.A. Townsend, *The Structure of Turbulent Shear Flow*, 2nd edn. (Cambridge University Press, Cambridge, 1976)

21. M. Vinokur, On one-dimensional stretching function for finite-difference calculations. J Comp. Phys. **50**, 215–234 (1983)
22. M. Lee, R.D. Moser, Direct numerical simulation of turbulent channel flow up to $Re_\tau = 5200$. J. Fluid Mech. **774**, 7395–415 (2015)
23. B.F. Farrell, P.J. Ioannou, J. JimÃl'nez, N.C. Constantinou, A. Lozano-DurÃan, M.-A. Nikolaidis, A statistical state dynamics-based study of the structure and mechanism of large-scale motions in plane Poiseuille flow. J. Fluid Mech. **809**, 290–315 (2016)
24. N. Hutchins, I. Marusic, Evidence of very long meandering features in the logarithmic region of turbulent boundary layers. J. Fluid Mech. **579**, 1–28 (2007)
25. Y. Hwang, Statistical structure of self-sustaining attached eddies in turbulent channel flow. J. Fluid Mech. **767**, 254–289 (2015)

Law of the Wake and Scaling of the Mean Velocity Profile in Turbulent Pipe Flow

G. Bellani, L. Lazzarini, and A. Talamelli

Abstract In this paper we apply a recursive methodology to identify the wake function from mean velocity profiles of turbulent pipe flow ranging between 22000 < Re_τ < 40000. Unlike standard methodologies, the present method requires no a priori assumptions on the log-law. The wake function is obtained from pairs of velocity profiles obtained at different Reynolds numbers by applying a recursive procedure originally proposed by Luchini (2018). The velocity profiles have been acquired in the Long Pipe facility in CICLoPE by hot-wire anemometry. We report the law of the wake obtained from 3 different pairs of Reynolds numbers, using the higher Reynolds number case as a reference. We compare the results with those obtained from the standard procedure of subtracting a pre-determined log-law and we discuss the implications of the observed differences on the scaling of the log region.

1 Introduction

The last two decades have seen the advent of modern high-Reynolds number facilities [1–3] and of numerical simulations at increasingly high-Reynolds number [4, 5], focused on the study of turbulent pipe flow. Despite the relatively good agreement between different sets of experimental and numerical data, there is still quite a lot of debate around the models that best describe their asymptotic behaviour. In particular, around the so called log-region [6, 7] the debate is mainly fed by the contrasting views on the proper methodology used to extract the constants and other fitting parameters of the said models (minimum Reynolds number required, boundaries of log-region, need of higher order terms, etc.). In this work we focus the attention

G. Bellani (✉) · L. Lazzarini · A. Talamelli
Dipartment of Industrial Engineering/CIRI Aerospace, University of Bologna, Forlì, Italy
e-mail: gabriele.bellani2@unibo.it

L. Lazzarini
e-mail: lorenzo.lazzarini7@unibo.it

A. Talamelli
e-mail: alessandro.talamelli@unibo.it

R. Örlü et al. (eds.), *Progress in Turbulence X*, Springer Proceedings in Physics 404,
https://doi.org/10.1007/978-3-031-55924-2_10

on the possible relevance of the law of the wake on the estimation of the empirical parameters describing the log region. The law of the wake is typically estimated from empirical data by subtracting a pre-determined log-law. However, if the log-law is to be estimated, the law of the wake must be determined without a priori assumptions on the inner part. This can be done using an iterative procedure based on the subtraction of two data-sets obtained at different Reynolds numbers. This procedure has been proposed in [8]. However, this analysis was applied mainly to low-Reynolds number data and only to a limited extent to experimental data at higher Reynolds number. No results for the wake function were explicitly reported for experimental data.

In this paper we apply this methodology to mean velocity profiles of turbulent pipe flow ranging between $22000 < Re_\tau < 39000$. The velocity profiles have been acquired in the Long Pipe CICLoPE by hot-wire anemometry (see Fig. 1). We report the law of the wake obtained by applying the procedure reported in [8] from three different pairs of mean velocity profiles, using the higher Reynolds number case as a reference (see Fig. 1). We compare the results with those obtained from DNS and we discuss the sensitivity of this procedure to experimental uncertainty. Finally, we investigate the influence of the law of the wake on the extraction of the fitting parameters in the log region for various Reynolds numbers.

2 Experimental Setup and Methodology

The velocity profiles presented in this study have been acquired in the Long Pipe facility in CICLoPE. The Long Pipe facility is a wind tunnel-type system circulating the fluid in a closed-loop design with a pipe length of $L \simeq 111.15\,\text{m}$ and a diameter of $D \simeq 0.9\,\text{m}$, resulting in a test pipe radius of $R \simeq 0.45\text{m}$. Velocity time-series are acquired using conventional hot-wire probes of 0.5 mm of length at a streamwise location located at 120 diameters from the pipe inlet. Detailed description about the facility and the measurement setup can be found in [9, 11–14].

The data-set used in this paper includes velocity profiles measured at four different friction Reynolds numbers ranging between $22000 \leq Re_\tau \leq 39000$. Table 1 shows the list of the experimental cases and related parameters. The inner-scaled mean velocity profiles obtained from this data-set are shown in Fig. 1.

Table 1 List of experimental data-sets. The last column indicates the acquisition time T_{acq} normalized by the integral time scale T_Λ of the corresponding case

Re_τ	u_τ [ms^{-1}]	T_{acq}/T_Λ [–]
22000	0.75	4889
32000	1.08	4977
35000	1.19	4666
39000	1.32	5200

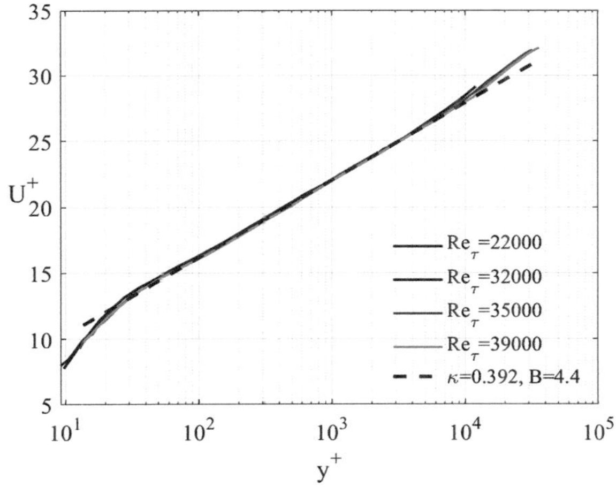

Fig. 1 Inner-scaled mean velocity profiles for $Re_\tau = 22000, 32000, 35000, 39000$, with line color going from black to light gray for increasing Reynolds number. The dashed line shows the log law assuming $\kappa = 0.392$ and B = 4.4

2.1 Recursive Method

The wake function $W(Y)$, where $Y = y/R$ is the wall-normal coordinate normalized by the pipe radius, is defined as: $W(Y) = u^+ - f(y^+)$. In the present work, W(Y) is estimated according to a recursive procedure described in details in Luchini [8] as:

$$W(Y) = \sum_{n=0}^{\infty} \Delta W_{12}(m^n Y), \tag{1}$$

where ΔW_{12} is the difference between two inner-scaled velocity profiles obtained at different Reynolds numbers and $m = \frac{Re_{\tau,1}}{Re_{\tau,2}}$, with $\mathrm{Re}_{\tau,1} < \mathrm{Re}_{\tau,2}$. The summation converges quickly to its asymptotic value. Here it is truncated to n=8 because of limited data points available in the very near-wall region ($Y \ll 1$). In order to limit the contamination due to measurement noise, a fifth-order fit is used to smooth ΔW_{12}. It is important to remark that this procedure does not imply any assumption on the law of the wall other than its Reynolds invariance.

In this work we used the data obtained at $\mathrm{Re}_{\tau,2} = 39000$ as a reference and we obtained an independent estimate of the wake function from each of the remaining cases listed in Table 1.

3 Results

Figure 1 shows the inner-scaled mean velocity profiles used in this analysis. The figure illustrates a good collapse of the inner part and a well-developed log region which appears to be best fit by the value of $\kappa = 0.392$, in agreement with other results in literature.

Figure 2 shows the wake function obtained from log-law subtraction assuming either $\kappa = 0.392$ (left Fig. 2) or $\kappa = 0.42$ (right Fig. 2), a value which is compatible with those obtained from previous analysis of centerline data [9]. The wake function obtained using assuming $\kappa = 0.39$ is well matched by Lewkowickz wake model with $\Pi = 0.2$. On the other hand, the same model seems to deviate from the wake function obtained assuming $\kappa = 0.42$ ($\Pi = 0.3$). The deviation is mostly evident in the near-wall part, where Lewkowickz model approaches zero much earlier in the near-wall region ($Y < 0.2$).

Figure 3 shows the wake profiles computed through recursive methods without assumptions on the log-law. The profiles show a good collapse from the near-wall to the outer region, indicating a Reynolds-number invariance of the wake function in this Re-number range. In the left figure, the wake function obtained here is compared to the results obtained in [8] by DNS at $360 < Re_\tau < 1000$. Although the wake profiles look fairly similar to those obtained here, the same invariance is not observed at lower Reynolds numbers. In the right figure, the wake functions are compared to the wake functions obtained assuming $\kappa = 0.42$. The two methods are in fairly good agreement, in particular in the region $Y < 0.2$. In this region, unlike standard wake models, the wake function has a non negligible value and in the region $Y < 0.3$ it is well approximated by the linear function $y^+ = \frac{3}{Re_\tau} y^+ - 0.15$, as shown by the solid blue line in Fig. 3. This could be an indication that the wake function may contaminate the estimate of the Von Kármán constant from individual velocity profiles and this may explain some Reynolds number trends and inconsistency observed in previous studies.

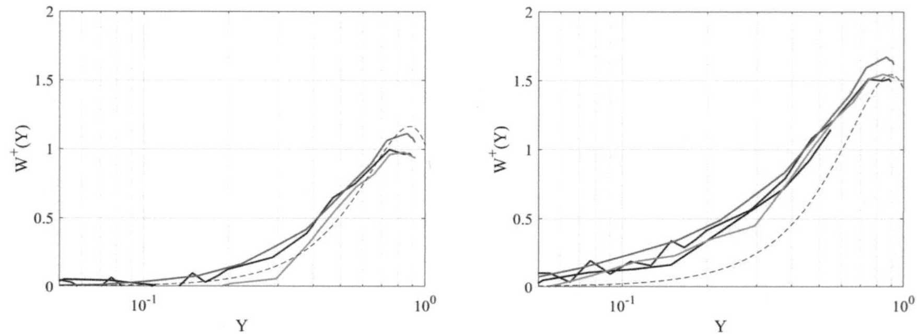

Fig. 2 Wake functions obtained form log-law subtraction ($\kappa = 0.39$, left; $\kappa = 0.42$, right). The wake functions are compared to Lewkowickz model

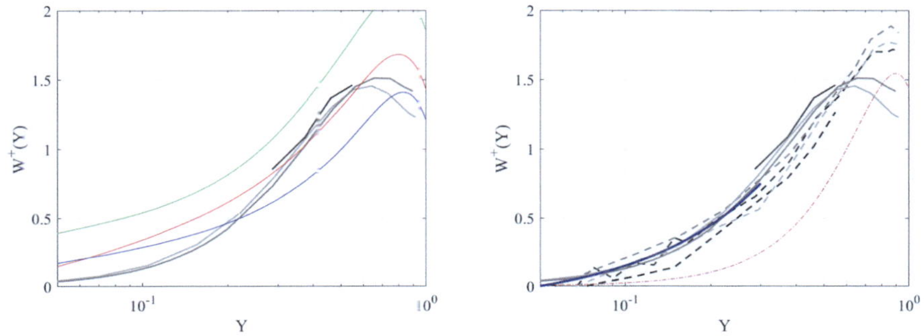

Fig. 3 Wake functions estimated from different pairs of Reynolds numbers. The reference Reynolds number is $Re_\tau = 39000$. The darker solid line corresponds to $Re_\tau = 22000$, the lighter gray line corresponds to $Re_\tau = 22000$. On the left, they are compared to the results in Fig. 13 by [8]. On the right, the wake functions are compare to the wake obtained form log-law subtraction ($\kappa = 0.42$, dashed gray lines) and to the Lewkowickz model. The thick solid line indicates a linear fit of the wake profiles up to $Y = 0.3$

4 Conclusions

A methodology to obtain the wake function without any *a priori* assumption on the log-law, originally proposed by [8], has been applied to velocity profiles obtained in a wide Reynolds number range in the Long Pipe facility in CICLoPE. Although the experimental errors can significantly affect the results, the wake functions obtained from three Reynolds-number pairs appear very consistent, showing Reynolds number invariance and similarity with results obtained at lower Reynolds number from DNS data. Unlike previously-reported wake models, the wake function appears to have a nearly-linear slope all the way to the wall. This is an indication that a Re-dependent term could contaminate the estimation of the the Von Kármán constant from individual velocity profiles.

Acknowledgements The work was carried out within the MOST–Sustainable Mobility National Research Center and received funding from the European Union Next-GenerationEU (PIANO NAZIONALE DI RIPRESA E RESILIENZA (PNRR)–MISSIONE 4 COMPONENTE 2, INVESTIMENTO 1.4–D.D. 1033 17/06/2022, CN00000023)

References

1. N. Furuichi, Y. Terao, Y. Wada, Y. Tsuji, Friction factor and mean velocity profile for pipe flow at high Reynolds numbers. Phys. Fluids **27**(9), 095108 (2015)
2. M.V. Zagarola, A.J. Smits, Mean-flow scaling of turbulent pipe flow. J. Fluid Mech. **373**, 33–79 (1998)

3. R. Örlü, T. Fiorini, A. Segalini, G. Bellani, A. Talamelli, P.H. Alfredsson, Reynolds stress scaling in pipe flow turbulence-first results from CICLoPE. Trans. R. Soc. A **375**, 20160187 (2017)
4. S. Pirozzoli, J. Romero, M. Fatica, R. Verzicco, P. Orlandi, Reynolds number trends in turbulent pipe flow: a DNS perspective. J. Fluid Mech. **926**, 1–20 (2021)
5. J. Tao, S. Rezaeiravesh, P. Schlatter, P.F. Hussain, Direct numerical simulations of turbulent pipe flow up to $Re_\tau \approx 5200$. J. Fluid Mech. **956**(A18), 1–18 (2023)
6. P.A. Monkewitz, K. Chauhan, H.M. Nagib, Self-consistent high-Reynolds-number asymptotics for zero-pressure-gradient turbulent boundary layers. Phys. Fluids **19**, 115101 (2007)
7. P. Luchini, Universality of the turbulent velocity profile. Phys. Rev. Lett. **118**, 224501 (2017)
8. P. Luchini, Structure and interpolation of the turbulent velocity profile in parallel flow. Eur. J. Mech. B. Fluids **71**, 15–34 (2018)
9. H.M. Nagib, P. Monkewitz, L. Mascotelli, T. Fiorini, G. Bellani, X. Zheng, A. Talamelli, Centerline Kármán "Constant" revisited and contrasted to log-layer karman constant at CICLoPE, in *10th International Symposium on Turbulence and Shear Flow Phenomena (TSFP10)* (Chicago, USA, 2017)
10. B.J. Cantwell, A universal velocity profile for smooth wall pipe flow. J. Fluid Mech. 834–874 (2019)
11. G. Bellani, A. Talamelli, The final design of the long pipe in CICLOPE, in *Progress in Turbulence VI. Springer Proceedings in Physics*, vol. 165. (Springer, 2016)
12. E.S. Zanoun, E. Öngüner, C. Egbers, G. Bellani, A. Talamelli, One-dimensional flow spectra and cumulative energy from two pipe facilities, in *Progress in Turbulence VIII* (2019), pp. 209–214
13. X. Zheng, G. Bellani, L. Mascotelli, R. Örlü, A. Ianiro, C.S. Vila, S. Discetti, J. Serpieri, M. Raiola, A. Talamelli, Y. Li, N. Jiang, Inter-scale interaction in pipe flows at high Reynolds numbers. Exp. Thermal Fluid Sci. **131**, 110529 (2022)
14. E.S. Zanoun, C. Egbers, R. Örlü, T. Fiorini, G. Bellani, A. Talamelli, Experimental evaluation of the mean momentum and kinetic energy balance equations in turbulent pipe flows at high Reynolds number. J. Turbulence **20**(15), 285–299 (2019)

Detailing History and Non-equilibrium Effects in Adverse Pressure Gradient Turbulent Boundary Layers

S. Romero, S. Zimmerman, J. Philip, and J. Klewicki

Abstract Utilizing the stress balance and a wall shear stress and pressure based velocity scale, u_{hyb}, introduced in Romero et al. (2022) non-equilibrium effects in adverse pressure gradient turbulent boundary layers are studied. Data from canonical flows, i.e. zero pressure gradient turbulent boundary layers, are used to provide context for this adverse pressure gradient turbulent boundary layer analysis. Adverse pressure gradient flows at varying Clauser pressure-gradient parameter β are studied. This parameter can be defined as the ratio of time-scales, $\beta = t_\Delta / t_{PG}$, characteristic of the turbulence and pressure gradient. Here, $t_\Delta = \Delta / u_\tau$, where $\Delta = (U_\infty / u_\tau)\delta^*$ is the Rotta-Clauser thickness, δ^* is the displacement thickness, and u_τ is the friction velocity. $t_{PG} = -(dU_\infty / dx)^{-1}$, where U_∞ is the free-stream velocity. For the adverse pressure gradient flow to achieve self-similarity, β, the ratio of two time-scales, has to be a constant. Therefore, under a non-constant β the flow takes time to reach self-similarity and history effects become apparent. From studying non-constant β cases, this study hypothesizes a link between the cross-over from equilibrium to non-equilibrium with a corresponding qualitative change to the internal leading order stress balance.

1 Introduction

The criteria of non-equilibrium boundary layers are studied using zero pressure gradient (ZPG) and adverse pressure gradient turbulent boundary layer (APG TBL)

S. Romero (✉) · J. Philip · J. Klewicki
Department of Mechanical Engineering, University of Melbourne, Parkville, Australia
e-mail: Sylvia.Romero789@gmail.com

J. Philip
e-mail: jimmyp@unimelb.edu.au

J. Klewicki
e-mail: klewicki@unimelb.edu.au

S. Zimmerman
Department of Mechanical Engineering, John Hopkins University, Baltimore, USA
e-mail: zimmerman@jhu.edu

R. Örlü et al. (eds.), *Progress in Turbulence X*, Springer Proceedings in Physics 404,
https://doi.org/10.1007/978-3-031-55924-2_11

data sets over a range of Reynolds number. Although the term "equilibrium" in TBLs takes on a variety of definitions these have been mainly set in terms of self-similarity or via properties of the stresses, e.g. [1, 2]. Self-similarity in terms of the stresses is a condition where the statistical profiles can be expressed in a self-preserving form, i.e., using local variable normalization. This is interrogated herein, as it holds the promise to embrace a more comprehensive framework for characterizing non-equilibrium and flow history effects.

1.1 Stress Balance and Hybrid Velocity Scale

The analysis begins with a consideration of the once-integrated mean momentum equation. The stress balance for the APG TBL is as follows:

$$\underbrace{\nu\frac{\partial U}{\partial y}}_{\text{VS}}+\underbrace{(-\overline{uv})}_{\text{RS}}+\underbrace{\int_0^y\left[-U\frac{\partial U}{\partial x}-V\frac{\partial U}{\partial y'}\right]dy'}_{\text{MIS}}+\underbrace{(-\frac{y}{\rho}\frac{dP}{dx})}_{\text{PS}}-\underbrace{\frac{\partial}{\partial x}\int_0^y\left[\bar{u^2}-\bar{v^2}\right]dy'}_{\text{negligible}}={u_\tau}^2, \quad (1)$$

where y' is a dummy variable of integration. Note that (1) contains Reynolds normal stress, $\overline{u^2}$ and $\overline{v^2}$, that will be shown to be negligible later. The viscous stress (VS) reaches unity at the wall ($y = 0$) and there the Reynolds stress (RS) is zero. Like in the ZPG case, the balance includes a non-constant mean inertia stress (MIS) term, and like the channel flow case there is a pressure stress (PS) term that is dependent on the wall-distance. In pipe and channel flow it is an analytical fact that $-\overline{uv}/u_\tau^2$ approaches a maximum of 1 as $Re \to \infty$. Strong analytical and empirical evidence in [3] shows that the Reynolds-averaged Navier-Stokes equation for a ZPG TBL can be reduced to a form similar to channel and pipe flow. Hence, in the boundary layer too, $-\overline{uv}/u_\tau^2$ approaches a maximum of 1 to within a second order correction as $Re \to \infty$. Unlike the ZPG case, for the APG case the magnitude of $-\overline{uv}/u_\tau^2$ is no longer bounded by 1. Previous observations in various APG TBL studies suggest that the friction velocity u_τ is not the characteristic velocity scale for APG TBLs, and these motivate consideration of more appropriate velocity scales. Theoretical arguments by the present authors [4] suggest that wall-distance scaling resulting from self-similar dynamics is consistent with both a single velocity scale leading to a log-law in mean velocity profile as well as a wall-distance varying velocity scale leading to a power-law mean velocity profile. Their analyses also show that wall-distance scaling observed in the wall-normal energy spectra remains virtually unchanged for measurements taken under an APG ramp as well as for measurements in a ZPG flow downstream of the APG, despite the increase in Reynolds number. This suggests that a wall-distance based scaling is appropriate for the APG cases.

Moving the PS term in (1) to the right hand side of the equation and neglecting the small normal turbulent stress terms, yields (2). The left hand side of (2) no longer contains a pressure-based term, and now matches the ZPG stress balance.

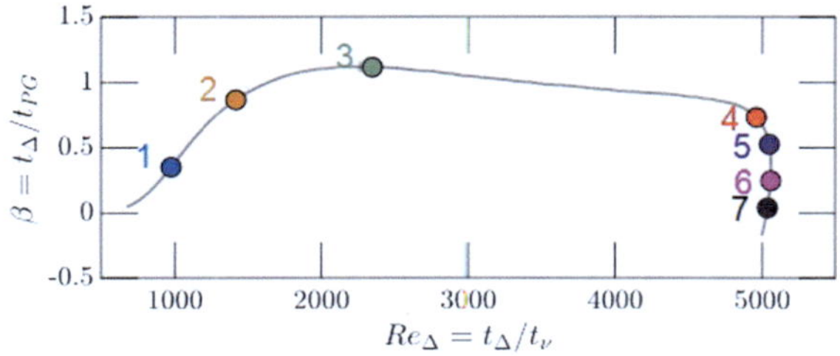

Fig. 1 $\beta = t_\Delta/t_{PG}$ versus $Re_\Delta = t_\Delta/t_\nu$, where $t_\nu = \nu/u_\tau^2$ and ν is the kinematic viscosity of the fluid. Markers represent the case studied in this paper. Data are from [6]

$$\underbrace{\nu\frac{\partial U}{\partial y}}_{\text{VS}} + \underbrace{(-\overline{uv})}_{\text{RS}} + \underbrace{\int_0^y \left[-U\frac{\partial U}{\partial x} - V\frac{\partial U}{\partial y'}\right] dy'}_{\text{MIS}} = u_\tau{}^2 + \underbrace{(\frac{y}{\rho}\frac{dP}{dx})}_{\text{PS}}, \tag{2}$$

Now in (2) the right hand side of the equation combines the wall-shear-stress velocity with a pressure stress-based velocity. This is the hybrid velocity scale denoted in (3). Note that the hybrid scale dependents on wall-distance. For the ZPG case the PS term is zero, and the velocity scale just becomes u_τ^2. In [5] this velocity scale,

$$u_{hyb}^2 \equiv u_\tau^2 + \frac{y}{\rho}\frac{dP}{dx} \tag{3}$$

was shown to exhibit relative success in scaling the Reynolds stress over a range of β and friction Reynolds number, $Re_\tau = u_\tau\delta/\nu$, conditions by significantly reducing case-to-case variations. The experimental data also revealed that at high Reynolds number APG profiles scaled by u_{hyb} show very little variation relative to one another even as β increases. However, the profiles under u_{hyb} scaling were seen to be of uniformly lower magnitude than the matched Reynolds number ZPG case. This speaks to the potential influence of the 'route' taken to a particular point in β-Re_τ parameter space on the apparent observed properties of the turbulence. This motivates examining the development of the stress balance for varying flow histories. To facilitate this aim, the remainder of the paper uses the Large-Eddy-Simulation (LES) data of Bobke et al. 2017 to observe these changes for increasing and decreasing β.

2 Results and Discussion

Seven locations from one of Bobke et al.'s data sets, marked in Fig. 1, are studied. At locations 1 and 2, β slowly increases with Reynolds number. Then at locations 3 and 4 β remains at a nominally fixed value of approximately 1 over a range of Reynolds number. At locations 5 through 7, however, β is now rapidly decreasing.

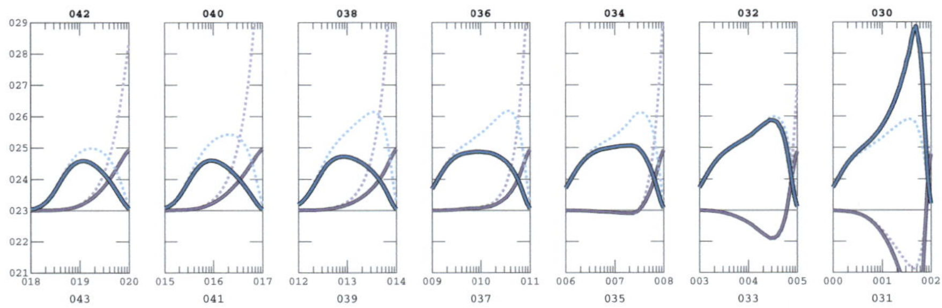

Fig. 2 Reynolds shear stress and mean inertia stress scaled by u_{hyb} and u_τ. The panels are numbered 1–7, and correspond to the locations indicated in Fig. 1. ▬: Reynolds shear stress scaled by u_{hyb}^2, ▬: mean inertia stress scaled by u_{hyb}^2. ▪▪▪▪▪: Reynolds shear stress scaled by ${u_\tau}^2$, ▪▪▪▪▪: mean inertia stress scaled by ${u_\tau}^2$. Data from [6]

The Reynolds shear stress scaled by u_{hyb}^2 and u_τ^2, and the mean inertia stress scaled by u_{hyb}^2 and u_τ^2 are plotted in Fig. 2. The wall-distance y is scaled by the boundary layer thickness δ. Although not shown here, when comparing Bobke et al.'s different β cases, the outer peaks of the viscous-scaled Reynolds stresses align well with a fixed y/δ position. In panels 1 through 4 the RS profiles scaled by u_{hyb}^2 remain below 1, while the profiles under u_τ^2 form an outer peak. In panels 5–7, however, β is now rapidly decreasing and the RS profiles under u_{hyb}^2 scaling form an outer peak that has increased to a value greater than 1. We now investigate why the hybrid velocity scaling suppresses the secondary peak for some cases and not for others.

The stress balance terms from (1) are shown in Fig. 3 to separate the effects of the velocity scale from the behaviors of the Reynolds shear stress and mean inerta stress. In this figure the terms are not normalized by a velocity scale and are given in the units reported by Bobke et al. In panels numbered 1–4 the changes in β are small, as seen in Fig. 1. At these locations the ${u_{hyb}}^2$ and the RS profiles merge closer together up to near the peak in the RS profile. This indicates that the VS and MIS terms are small in this region. In panels 5–7, however, β is now rapidly changing and the PS and MIS profiles exhibit a significant drop in the outer region and the local ${u_{hyb}}^2$ profile no longer behaves similarly to the local RS profile. Instead, the ${u_{hyb}}^2$ profile from panel 4 (plotted as red dashed lines) behaves similarly to the RS profile in panels 5–7. This observation suggests a reconsideration of what is implied by equilibrium. Namely, when the PS and RS terms locally balance, as in panels 1–4, the boundary layer can be considered to be in equilibrium according to this underlying structure with u_{hyb} as the characteristic velocity scale. Beyond this point (e.g., panels 5–7), where there is a qualitative change in the stress balance, the flow exhibits non-equilibrium behaviour and the u_{hyb} scaling breaks down. As such, in this study we explore the broader hypothesis that the cross-over from equilibrium to non-equilibrium, corresponding in the present scenario to where the MIS transitions from positive to negative, is dynamically characterized by a qualitative change to the internal leading order stress balance.

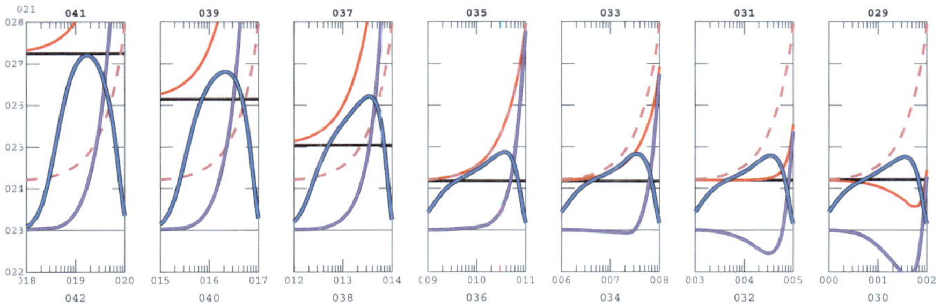

Fig. 3 Stress balance. The panels are numbered 1–7, and correspond to the locations indicated in Fig. 1. Lines ▬: local Reynolds shear stress, ▬: local mean inertia stress, ▬: local $u_{hyb}{}^2$, ▬: local $u_\tau{}^2$. ▬ ▪: Profile of $u_{hyb}{}^2$ from panel 4. Data from [6]

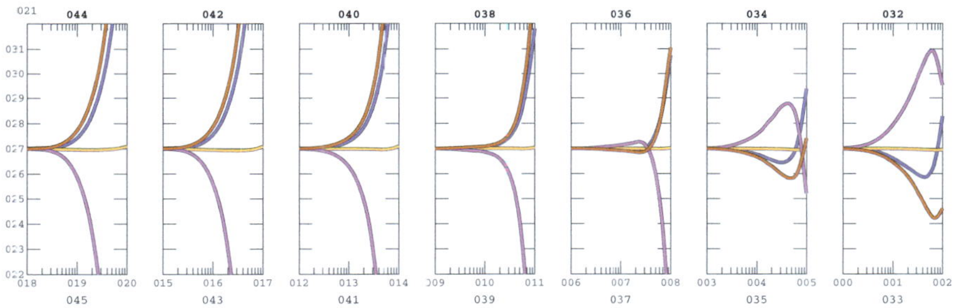

Fig. 4 Mean inertia separated into its smaller components. The panels are numbered 1–7, and correspond to the locations indicated in Fig. 1. Lines ▬: mean inertia stress, ▬: UV, ▬: momentum flux. ▬: negligible terms. Data from [6]

We further quantify this non-equilibrium effect by investigating the behavior of the mean inertia stress profile. In (4) the mean inertia stress is separated into its components:

$$\underbrace{\int_0^y \left[U\frac{\partial U}{\partial x} + V\frac{\partial U}{\partial y'}\right] dy'}_{\text{MIS}} = UV + \underbrace{\frac{\partial}{\partial x}\int_0^y \left[U^2\right] dy'}_{\text{momentum flux}}, \tag{4}$$

to better understand the sign change. These components are plotted in Fig. 4 including the normal stress shown in (1). In this Fig. V is negative at locations 5, 6, and 7, i.e., flow is wallward into the boundary layer, whereas in points 1 through 4 V is positive, that is, reflects the more typical case associated with boundary layer growth. Also in panels 5–7, the momentum flux is now positive, which means flow accelerates. Therefore, although the overall flow is under an APG condition ($\beta > 0$), the flow in the outer-region is essentially behaving like a favourable pressure gradient (FPG) this is where u_{hyb} breaks down, and the present hypothesis identifies as an apparent 'non-equilibrium' state coming into effect. It is noted that the behavior of the last location may be due to the formulation of the computation. The computation forces the flow back to the Blasius inflow profile, which could be causing an acceleration. This should not matter in the present interpretation, however, as it still constitutes a Navier-Stokes equation solution.

3 Conclusion

This present study explores a scenario where an APG TBL at nominally fixed β undergoes a rapid decrease in β. This induces a sign change in the mean inertia stress in the outer region. The transition of mean inertia stress from positive to negative leads to a qualitative change to the internal leading order stress balance. This qualitative change is hypothesized to indicate a cross-over from equilibrium to non-equilibrium flow state. Furthermore the hybrid velocity scaling works well under positive mean inertia stress. However, this velocity scale does not work when mean inertia stress becomes negative, e.g. rapidly decreasing β cases where the flow in the outer-region is essentially behaving like a FPG TBL. The broader implication (requiring future study) is that any qualitative change in the leading order stress balance presents a change from an equilibrium to non-equilibrium flow state.

References

1. A. Townsend, *The Structure of Turbulent Shear Flow*, vol. 15 (Cambridge University, 1956)
2. W. Devenport, T. Lowe, Equilibrium and non-equilibrium turbulent boundary layers. Prog. Aerosp. Sci. **131**, 100807 (2022)
3. C. Morrill-Winter, J. Philip, J. Klewicki, An invariant representation of mean inertia: theoretical basis for a log law in turbulent boundary layers. J. Fluid Mech. **813**, 594–617 (2017)
4. S. Romero, S. Zimmerman, J. Philip, J. Klewicki, Properties of the inertial sublayer in adverse pressure-gradient turbulent boundary layers. J. Fluid Mech. **937**, A30 (2022)
5. S. Romero, S. Zimmerman, J. Philip, J. Klewicki, Stress equation based scaling framework for adverse pressure gradient turbulent boundary layers. Int. J. Heat Fluid Flow **93**, 108885 (2022)
6. A. Bobke, R. Vinuesa, R. Örlü, P. Schlatter, History effects and near equilibrium in adverse-pressure-gradient turbulent boundary layers. J. Fluid Mech. **820**, 667–692 (2017)

Spectrum Analysis and Statistical Characteristics of Three Velocity Components in Pipe Flow at High Reynolds Number

M. Ono, N. Furuichi, and Y. Tsuji

Abstract This paper reports that the first result of a spectral analysis at High Reynolds Number Actual Flow Facility (Hi-Reff) up to $Re_\tau \approx 20000$ using a laser Doppler velocimetry (LDV). Pre-Multiplied Spectra (PMS) profile measured by LDV shows good agreement with the previous results measured by hot-wire anemometry. The wall-normal position dependence of the wavenumber of PMS peaks indicates that the dominant structure varies LSM, VLSM and LSM from the inner to outer region. Based on the relation between PMS and three velocity components turbulence intensity profiles, it is found that the dominant structure in the constant region of the wall-normal component is VLSM, on the other hand the dominant structure is changed from VLSM to LSM in the logarithmic regions of the streamwise and spanwise components.

1 Introduction

In order to clarify the universality of wall-bounded turbulence, the studies for statistics at high Reynolds numbers has increased in recent years. In past three decades, experimental facilities such as Superpipe, CICLoPE, and CoLaPipe for pipe flow have been constructed and experimental results of turbulence statistics at high Reynolds numbers over $Re_\tau = 10^4$ have been reported. We also reported the pipe friction factor, the mean velocity and the turbulence intensity profiles measured at "High Reynolds Number Actual Flow Facility (Hi-Reff)" [1–3]. However, most of the turbulence statistics reported in those previous works are only for the streamwise component.

M. Ono (✉) · N. Furuichi
National Institute of Advanced Industrial Science and Technology, 1497-1 Teragu, Tsukuba, Japan
e-mail: ono.marie@aist.go.jp

N. Furuichi
e-mail: furuichi.noriyuki@aist.go.jp

Y. Tsuji
Nagoya University, Furo-cho, Chikusa-ku, Nagoya, Japan
e-mail: c42406a@cc.nagoya-u.ac.jp

R. Örlü et al. (eds.), *Progress in Turbulence X*, Springer Proceedings in Physics 404,
https://doi.org/10.1007/978-3-031-55924-2_12

To understand the turbulence structure in more detail, not only for the streamwise component but also for the three velocity components, therefore the spanwise and wall-normal components are required. In our previous study, the three components turbulence intensity profiles were measured from $Re_\tau \approx 1000$ to 20000 using a laser Doppler velocimetry (LDV) and confirmed the logarithmic regions of the streamwise and spanwise component, and the constant region of the wall-normal component for $Re_\tau > 10000$ [4]. According to attached eddy hypothesis, the logarithmic and constant regions in the three components are expected to be coincide. In our results, the logarithmic regions of the streamwise and spanwise component are emerged in the same region ($0.055 < y/R < 0.25$), however, the constant region of the wall-normal component is emerged at closer region to the wall ($0.020 < y/R < 0.10$). To understand this inconsistency, the relation between spectrum of streamwise component and three components turbulence intensity profiles is investigated in this paper. First, we check the validity of the spectrum analysis for the high-Reynolds-number experimental data measured by LDV. Second, based on the comparison between the spectrum profile for streamwise component focusing large-scale motion (LSM) and very large-scale motion (VLSM), and turbulence intensity profiles for three velocity components, we discussed comprehensively about the logarithmic region.

2 Spectrum Analysis

In previous, most of spectral analysis for pipe flow at high Reynolds number have been conducted based on experimental data measured by hot-wire (HW). In this section, we explain the procedure of the spectral analysis for experimental data measured by LDV and validity of the spectrum data obtained. Since LDV detects the velocity of particles passing the measurement volume, the velocity data in time series are discrete. Therefore, it is necessary to reconstruct the equal time interval data from raw data to conduct the analysis by Fourier transform. In this experiment, the linear and the sample-and-hold (S+H) interpolations [5] were compared as reconstruction methods. The schematic of each interpolation is shown in Fig. 1.

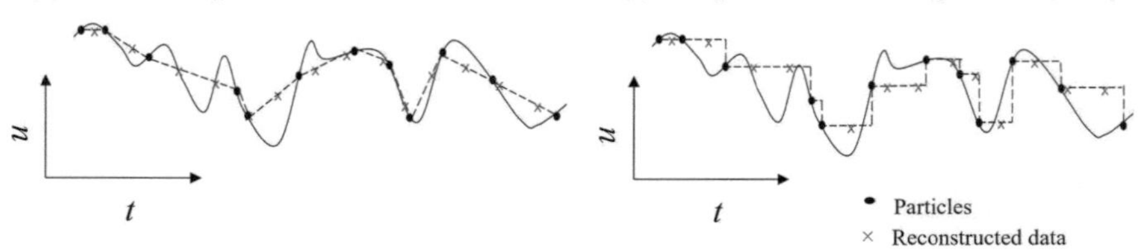

Fig. 1 The interpolation methods to reconstruct the equal time interval data for LDV. **a** Linear interpolation and **b** Sample-and-hold interpolation (S+H)

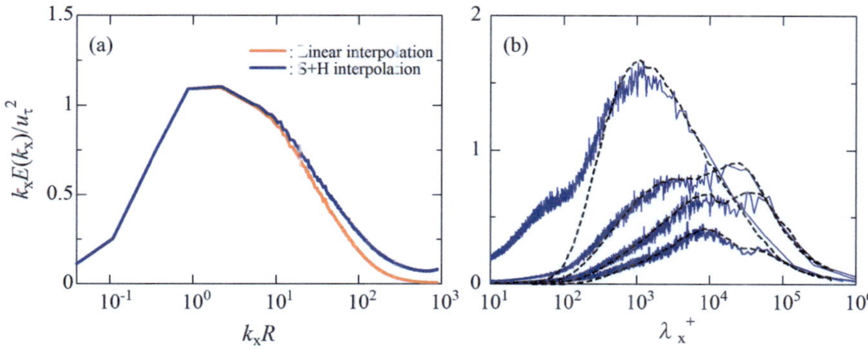

Fig. 2 **a** The comparison between the PMSs of $Re_\tau \approx 20000$ at $y/R = 0.05$ obtained from reconstructed data by the liner and the S+H interpolation methods. **b** The comparison of PMS distributions with previous data measured by hot-wire. The blue solid lines are present results for $y^+ \approx 15, 190, 870, 1890$, at $Re_\tau = 2960$. The broken lines are PMS reported by Monty et al. [7] for $y^+ \approx 15, 180, 800, 2000$ at $Re_\tau = 3005$

In Fig. 2a, PMS profiles obtained by the time-series data interpolated by two interpolation methods are shown. The difference between two interpolations is observed at high wavenumber region. As shown below, the integral value of the PMS is equal to the value of the turbulence intensity at that location.

$$u'^2 = \int_0^\infty E(k_x) dk_x = \int_0^\infty k_x E(k_x) d(\log k_x). \tag{1}$$

where, u'^2 is the variance of the streamwise velocity fluctuations and $E(k_x)$ is the power spectral density of the streamwise velocity fluctuation. $k_x = 2\pi f/U$ is the streamwise wavenumber, $\lambda_x = 2\pi/k_x$ is the streamwise wavelength. As same with the previous study [6], the integrated value of PMS were 5.4% lower than the turbulence intensity since the linear interpolation underestimates the energy in the high wavenumber region. On the other hand, the deviation between the integral value of PMS and the turbulence intensity was 0.08% in the S+H interpolation. Therefore, we reconstructed the equal time interval data using the S+H interpolation in this study.

Figure 2b shows the comparison between the present and the previous PMS results measured by HW (Monty et al. [7]), for four different measurement positions at $Re_\tau \approx 3000$. Except in the low wavelength region at $y^+ \approx 15$, both PMS profiles show nice agreement. The overestimation of PMS in the low wavelength region near the wall are due to the spatial resolution of the LDV. Contrast with the hot wire measurement, the turbulence intensity in LDV measurement is overestimated at near the wall due to the velocity gradient in the control volume. However, the spatial resolution affects only low wavelength at the near-wall region, and its effect is negligible in the logarithmic region, where is the subject of discussion in this study.

3 Results of Turbulence Intensity and Spectrum

Figure 3a shows the contour map of PMS at $Re_\tau \approx 20000$. In this contour map, the PMS peaks at each measurement location are indicated by open or solid triangles. Rosenberg et. al [8] reported that PMS peaks are scaled by wavenumber and the wall-normal position. They indicated that LSM peak locations are expressed by $k_x y = 0.4$ and $k_x R = 2.6$, and VLSM peaks are expressed by $k_x y = 0.045$ and $k_x R = 0.45$. The borders of two expressions for VLSM and LSM are same at $y/R = 0.1$. According to their scaling for PMS peak locations, the solid triangles in Fig. 3a indicate the VLSM derived peaks, the open triangles indicate the LSM derived peaks. From the present results, it was confirmed that the dominant structure changes from LSM, VLSM

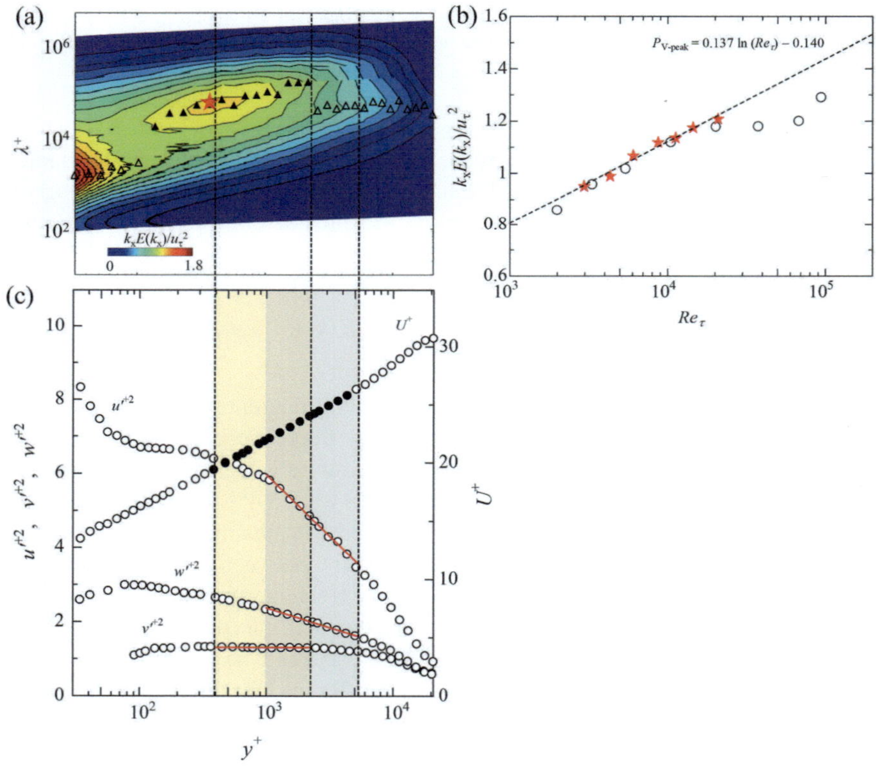

Fig. 3 **a** The contour map of PMS at $Re_\tau \approx 20000$. The open triangles (△) indicate LSM peaks and the solid triangles (▲) indicate VLSM peaks. The red star is the outer spectral peak. **b** The Reynolds number dependence of outer spectral peak. The open circles indicate previous results measured by HW [9]. **c** The mean velocity and turbulence intensity profiles for three velocity components at $Re_\tau \approx 20000$. The red lines are the logarithmic region at the streamwise/spanwise turbulence intensity profiles, and the constant region at the wall-normal turbulence intensity profile. The black circles indicate the logarithmic region at the mean velocity prolife

to LSM with the wall-normal position. Around the position where the dominant structure changes from LSM to VLSM / VLSM to LSM, the PMS profile shows clear bimodality.

The position where the VLSM has the highest PMS value in the pipe cross section, namely outer spectral peak, $P_{\mathrm{V-peak}}$ is indicated by a red star. The Reynolds number dependence of the outer spectral peak is shown in Fig. 3b. The outer spectral peak grows logarithmically with increasing Reynolds number, and it is expressed by $P_{\mathrm{V-peak}} = 0.137\ln(Re_\tau) - 0.140$. This result is very good agreement with the results measured by HW in previous study [9].

Figure 3c shows the mean velocity and turbulence intensity profiles for the three velocity components at $Re_\tau \approx 20000$. As mentioned in the introduction, the logarithmic regions of the streamwise and spanwise component in the turbulence intensity profiles are consistent, but the constant regions of the wall-normal component is not perfectly. Moreover, the logarithmic region in the mean velocity profile also is not consistent with the logarithmic regions in the streamwise turbulence intensity profiles. This result differs with previous studies [10]. However, the combined region, where is shaded area in Fig. 3c is consistent with the logarithmic region in mean velocity profile. To investigate the dominant structure in the logarithmic/constant region of the turbulence intensity profile, Fig. 3a, c are compared in the following. The starting-point of the logarithmic region of the mean velocity profile coincides with the location of the outer spectral peak as same with the result by Vallikivi et al. [9]. However, the logarithmic region of the mean velocity profile itself is not dominated by a single structure, as the inner side of the logarithmic region is dominated by VLSM and the outer side is dominated by LSM. For the turbulence intensity profiles, the dominant structure differs in the logarithmic/constant region: VLSM dominates the constant region of the wall-normal components, but the dominant structure changes from VLSM to LSM in the logarithmic regions of the streamwise and spanwise component. Thus, even in a region that is simply called a logarithmic region (or the constant region), the dominant structure which affects the characteristics of the statistics of each velocity component changes. As a future work, we will analyze the spectrum for the wall-normal and spanwise components, including in relation to the attached eddy hypothesis.

4 Summary and Future Works

The turbulence intensity profiles for three velocity components were measured using LDV up to $Re_\tau \approx 20000$ at Hi-Reff and the spectrum for the streamwise component was analyzed. The spectral results measured by LDV were nice agreement with the previous results measured by HW, and the validity of our experimental results was confirmed.

Although the logarithmic regions of the streamwise and spanwise components and the constant region of the wall-normal component do not perfectly coincide, their combined region coincides with the logarithmic region of the mean velocity profile. Based on the spectral analysis, it was observed that the constant region of the wall-normal component is dominated by VLSM, on the other hand, the dominant structure changes from VLSM to LSM in the logarithmic region of the streamwise and spanwise components. In the future, we plan to analyze also the wall-normal and spanwise spectrums to investigate the behavior of VLSM and LSM.

Acknowledgements This research was partially supported by JSPS KAKENHI Grant Numbers JP19KK0098, JP22K04943.

References

1. N. Furuichi, Y. Terao, Y. Wada, Y. Tsuji, Friction factor and mean velocity profile for pipe flow at high Reynolds numbers. Phys. Fluids **27**, 095108 (2015)
2. N. Furuichi, Y. Terao, Y. Wada, Y. Tsuji, Further experiments for mean velocity profile of pipe flow at high Reynolds number. Phys. Fluids **30**, 055101 (2018)
3. M. Ono, N. Furuichi, N. Kurihara, Y. Wada, Y. Tsuji, Reynolds number dependence of inner peak turbulence intensity in pipe flow. Phys. Fluids **34**, 045103 (2022)
4. M. Ono, N. Furuichi, Y. Tsuji, Reynolds number dependence of turbulent kinetic energy and energy balance of the 3-components turbulence intensity in pipe flow. J. Fluid Mech. (2023)
5. R. Adrian, C. Yao, Power spectra of fluid velocities measured by laser Doppler velocimetry. Exp. Fluids **5**, 17–28 (1986)
6. C. Chao, H. Leu, A fractal reconstruction method for LDV spectral analysis. Exp. Fluids **13**, 91–96 (1992)
7. J. Monty, N. Hutchins, H.C.H. Ng, I. Marusic, M.S. Chong, A comparison of turbulent pipe, channel and boundary layer flows. J. Fluid Mech. **632**, 431–442 (2009)
8. B. Rosenberg, M. Hultmark, M. Vallikivi, S.C.C. Bailey, A.J. Smits, Turbulence spectra in smooth- and rough-wall pipe flow at extreme Reynolds numbers. J. Fluid Mech. **731**, 46–63 (2013)
9. M. Vallikivi, M. Hultmark, A.J. Smits, Turbulent boundary layer statistics at very high Reynolds number. J. Fluid Mech. **779**, 371–389 (2015)
10. I. Marusic, J.P. Monty, M. Hultmark, A.J. Smits, On the logarithmic region in wall turbulence. J. Fluid Mech. **716**, R3 (2013)

Effects of Oscillating Spanwise Mean Pressure Gradient on Rotating Plane Couette Flow

O. Iida and T. Kanda

Abstract Effects of oscillating spanwise mean pressure gradient, i.e., OSMPG, on rotating plane Couette flow, especially on the mean velocity profile are investigated by DNSs of a spectral method. OSMPG disrupts vortex rolls at small rotation numbers, decreases skin friction, and induces linear mean velocity profiles at the channel centre. In contrast, the linear profile of zero absolute vorticity, typically observed at large rotation numbers, is found to be robust for OSMPG.

1 Introduction

Effects of system rotation have pivotal effects on turbulence in engineering flows, e.g., turbomachinery. One of the typical flow configurations for studying the effects of system rotation is a plane Couette flow [1]. At the small rotation numbers, mean velocity becomes the curve, and increases with an increasing distance from the bottom wall, while at the large rotation numbers, zero absolute vorticity, i.e., ZAV appears in the middle of the channel, where the absolute vorticity is zero, and the mean velocity becomes a linear profile. Previous studies indicate that S-shape mean velocity, i.e., SSMV, appears halfway toward the ZAV with an increase in rotation [2–5]. Especially, the work of Kawata and Alfredsson [3] discusses that the transport of the Reynolds shear stress into the channel centre is enhanced when the SSMV appears. In these previous studies, the rotation number $\mathrm{Ro_w} = 2\Omega\delta/\mathrm{U_w}$, defined by the rotation frequency 2Ω, channel half-width δ, and moving wall velocity U_w, is approximately $\mathrm{Ro_w} = 0.03$, when SSMV is observed. In contrast, in real flow related to turbomachinery with wall curvature, the oscillating spanwise mean pressure gradient, i.e., OSMPG, and resultant veering are inevitable. However, the research studies are few for the effects of OSMPG on the rotating plane Couette flows. Our interest

O. Iida (✉)
Nagoya Institute of Technology, Nagoya, Japan
e-mail: iida.oaki@nitech.ac.jp

T. Kanda
MHI Aerospace Systems Corp, Nagoya, Japan
e-mail: kandasyokuba@gmail.com

R. Örlü et al. (eds.), *Progress in Turbulence X*, Springer Proceedings in Physics 404,
https://doi.org/10.1007/978-3-031-55924-2_13

is in how robust each mean velocity profile, i.e., ZAV, and SSMV, is for the OSMPG and veering. As objectives, first, the effects of rotation on plane Couette flow and its mean velocity are investigated, and the mechanisms of SSMV are discussed. Then, the robustness of ZAV and SSMV is investigated under OSMPG.

2 Numerical Methods

In this study, DNSs of an incompressible turbulent plane Couette flow are performed with a spectral method. A flow is driven by moving walls with a constant velocity $\pm U_w$ in the streamwise directions; the upper and lower walls are forced to move in the $x-$ and $-x-$ directions, respectively. The wall-normal and spanwise directions are assumed to be the $y-$ and $z-$ directions, respectively. By assuming the anticyclonic system rotation in the spanwise direction, in which rotation frequency is defined as 2Ω, a flow is dynamically unstable in the entire regions of a channel. The Reynolds number $\mathrm{Re_w} = U_w\delta/\nu$ is set to be 200 to 1100, while $\mathrm{Ro_w}$ is set to be 0.01 to 0.7. The computational box of $15\pi\delta \times 2\delta \times 10\pi\delta$ is resolved by $256 \times 128 \times 164$ in the $x-$, $y-$, and $z-$ directions, respectively. The details of numerical methods are included in our previous study [6], while the detail of OSMPG is included in the subsection 3.2 of this paper. The statistics are calculated after the flow is considered to be fully developed, including the cases imposing OSMPG.

3 Results and Discussions

3.1 S-Shape Mean Velocity, SSMV

Table 1 represents the mean velocity gradient in the channel centre as the function of $\mathrm{Re_w}$ and $\mathrm{Ro_w}$. With an increase in $\mathrm{Ro_w}$, especially at $\mathrm{Re_w} = 1100$, the velocity profile shifts from a monotonically increasing curve to SSMV and finally becomes ZAV. At smaller $\mathrm{Re_w}$, SSMV cannot be observed.

The rotational effects on vortex rolls are discussed in Fig. 1, which shows isosurfaces of streamwise velocity fluctuations u', representing rolls, from the perspective of the upstream side of the channel at $\mathrm{Re_w} = 1100$. Rotation first attenuates turbulence, and generates rolls, as shown in Fig. 1a–c, then contrarily enhances turbulence, and disrupts rolls as shown in Fig. 1d. SSMV appears when the turbulence becomes weak, and not regular but irregular rolls appear as shown in Fig. 1b. In the cases where the rolls are discernible, as shown in Fig. 1b, c, low-speed fluids are transferred to the opposite side of the wall (upper wall) through the down gradient flux representing the red isosurfaces. Then, they are bounced back through the counter gradient flux representing the blue isosurfaces trapped in the channel centre in Fig. 1b representing the case of SSMV. Further increase in $\mathrm{Ro_w}$ attenuates turbulence, and

Table 1 The mean velocity gradient in the channel center as the function of $\mathrm{Re_w}$ and $\mathrm{Ro_w}$. Dark blue represents SSMV, while green represents ZAV. Light blue and red represent positive and negative absolute vorticity, respectively. ZAV at $\mathrm{Ro_w} = 0.5$ and 0.7 also represents the linear mean velocity profile in the channel centre

Re\Ro	0.01	0.02	0.03	0.04	0.05	0.07	0.085	0.1	0.2	0.3	0.5	0.7
200	0.238	–	0.07852	–	0.113	0.0596	0.0651	0.083	0.222	0.283	0.453	0.689
325	0.142	–	0.10363	–	0.054	0.0555	0.1071	0.119	–	0.342	0.463	–
500	0.083	–	0.04795	0.0354	0.026	0.0727	0.0937	0.106	–	0.260	–	–
700	0.019	–	0.02122	0.0159	0.021	0.0748	–	0.114	0.204	0.246	–	–
900	0.036	0.0255	0.00883	0.0063	0.026	0.0679	–	0.109	0.202	0.250	–	–
1100	0.035	0.0257	0.00352	-0.0164	0.031	0.0648	–	0.101	–	0.262	0.468	0.690

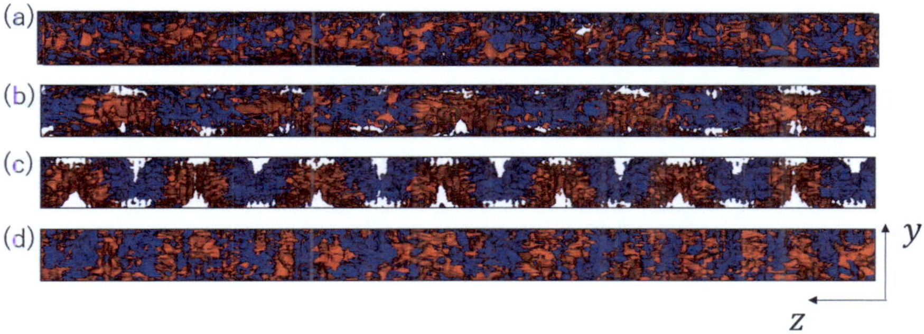

Fig. 1 Isosurfaces of streamwise velocity fluctuation $u'/u_\tau = -0.2$, where u_τ is the friction velocity at the wall; isosurfaces are viewed from the perspective of the upstream side of the channel of $\mathrm{Re_w} = 1100$. Red represents the regions of upward motion, and hence down gradient flux, while blue represents downward motion and counter gradient flux. **a** $\mathrm{Ro_w} = 0.01$, **b** $\mathrm{Ro_w} = 0.04$, **c** $\mathrm{Ro_w} = 0.05$, **d** $\mathrm{Ro_w} = 0.3$

shifts low-speed fluids back further to the lower wall through counter gradient flux as shown in Fig. 1c; blue isosurfaces approach further to the lower wall in comparison to Fig. 1b. So, in the case of SSMV, turbulent diffusion into the channel center is maximized, while the turbulent diffusion into the wall vicinity is minimized, which is quantitatively verified by the budget of turbulent kinetic energy [6].

3.2 *Effects of Oscillating spanwise Mean Pressure Gradient, OSMPG*

OSMPG is imposed as the periodic pulses of the opposite sign as shown in Fig. 2a. The amplitude of the mean spanwise gradient, i.e., $\mathrm{d}P/\mathrm{d}z$, normalized by u_τ of the case without OSMPG and δ, is set to be 0.5 or 2. The period of the pulse is set to

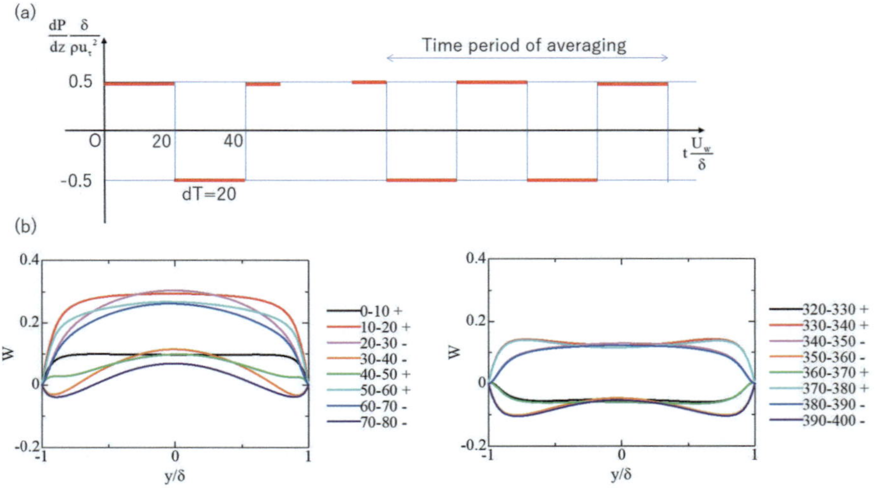

Fig. 2 **a** Time development of imposed mean pressure gradient, **b** Time development of induced mean spanwise velocity in Rew1100Ro0.04P0.5; left and right figures represent mean spanwise velocity W in the early and developed periods, respectively; W is averaged over the time indicated by the line captions, while the plus and minus signs of the captions represent the signs of the imposed pulse

be $20\delta/U_w$. As the nomenclature, the case where $\mathrm{Re_w} = 1100$, $\mathrm{Ro_w} = 0.04$, and OSMPG=0.5, is coined as Rew1100Ro0.04P0.5, while the case of the same $\mathrm{Re_w}$ and $\mathrm{Ro_w}$, and without OSMPG is coined as Rew1100Ro0.04. Figure 2b shows the time development of the mean spanwise velocity in Rew1100Ro0.04P0.5. The resultant spanwise mean velocity is synchronized with the OSMPG with the development of the time.

Figure 3 shows the development of the skin friction coefficient defined as $C_f = 2{u_\tau}^2/{U_w}^2$ after imposition of OSMPG. At the small $\mathrm{Ro_w}$, OSMPG decreases C_f; just after imposition of OSMPG=0.5, C_f significantly decreases, then it bounces back, though it remains small in comparison to C_f of no OSMPG. In contrast, with the imposition of OSMPG=2, no definite bounce back is observed and C_f remains small. The OSMPG effects on the vortex rolls are shown at $\mathrm{Ro_w} = 0.04$ in Fig. 4. Imposition of the OSMPG tends to suppress the spanwise meandering of rolls as shown in (b) and (e), and a further increase in OSMPG disrupts rolls as shown in (c) and (f); rolls tend to be fragmented as in the case of large rotation.

The effects of OSMPG on the mean velocity are remarkable as shown in Fig. 5a, b; SSMV, observed at $\mathrm{Re_w} = 1100$, disappears even in the case of small OSMPG. Further increase in OSMPG makes the mean velocity linear profile in all cases of $\mathrm{Re_w}$, which indicates that the linear mean velocity appears when the flow is turbulent and secondary flow effects are reduced. Finally, in Fig. 5c, d, OSMPG effects are investigated at large $\mathrm{Ro_m}$ where the flow is still turbulent. Because of the large rotation, the linear mean velocity profile and ZAV appear almost over the entire

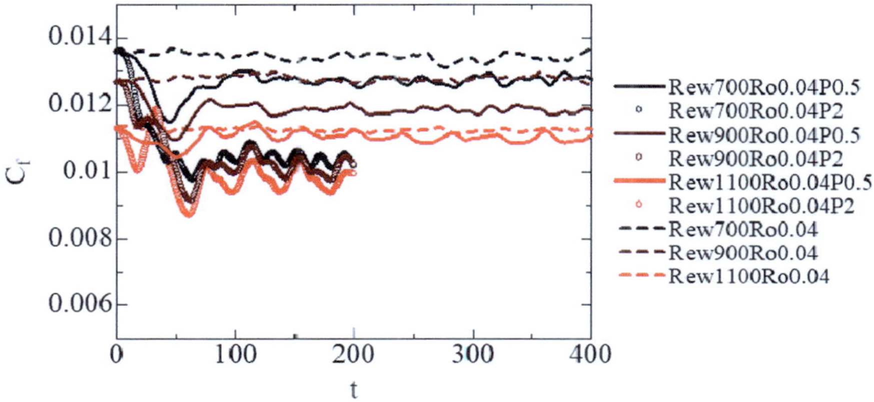

Fig. 3 Time development of streamwise skin friction coefficient C_f in the cases of $\mathrm{Ro_w} = 0.04$ and different $\mathrm{Re_w}$

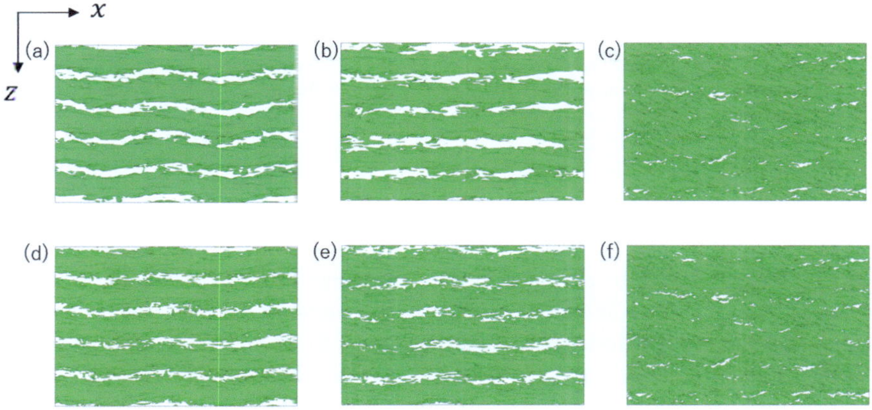

Fig. 4 Distributions of vertex rolls at $tU_w/\delta = 200$; isosurfaces of $u'/u_\tau = -0.2$; first and second rows represent the cases of $\mathrm{Re_w} = 700$ and 1100, respectively; left and right figures are cases of OSMPG=0, and 2, respectively, while center figures are cases of OSMPG=0.5. **a** Rew700Ro0.04, **b** Rew700Ro0.04P0.5, **c** Rew700Ro0.04P2, **d** Rew1100Ro0.04, **e** Rew1100Ro0.04P0.5, **f** Rew1100Ro0.04P2

region of the channel except the wall vicinity. The effects of OSMPG on ZAV are almost negligible, which must be because rolls are disrupted at large $\mathrm{Ro_w}$, and hence the effects of OSMPG are negligible

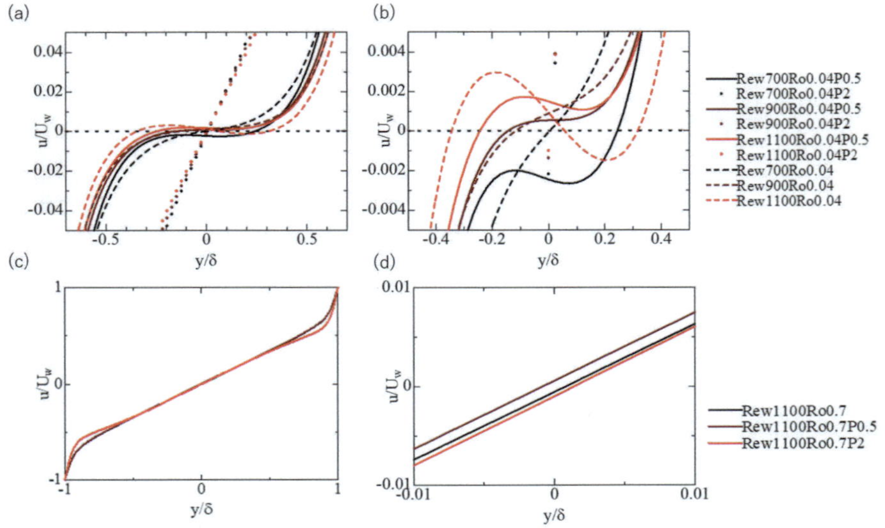

Fig. 5 Mean velocity profiles averaged over two periods of pulses after the flow is developed (see Fig. 2a); upper and lower figures represent cases of $Ro_w = 0.04$ and 0.7, respectively; left figures are close-ups of right figures

4 Conclusions

Effects of OSMPG are investigated on rotation plane Couette flow by the DNSs of a spectral method. The OSMPG disrupts the rolls at small Ro_w, decreases the skin friction coefficient, and induces a linear mean velocity profile at the channel centre. Linear mean velocity profile with ZAV at large Ro_w is robust for OSMPG, while S-shape mean velocity is vulnerable because the OSMPG disrupts the rolls.

References

1. K.H. Bech, H.I. Andersson, Turbulent plane Couette flow subject to strong system rotation. J. Fluid Mech. **347**, 289 (1997)
2. M. Salewski, B.B. Eckhardt, Turbulent states in plane Couette flow with rotation. Phys. Fluids 27:045109 (2015)
3. T. Kawata, P.H. Alfredsson, Turbulent rotating plane Couette flow: Reynolds and rotation number dependency of flow structure and momentum transport. Physical Review Fluids 1:034402 (2016)
4. J. Gai, Z. Xia, Q. Cai, S. Chen, Turbulent statistics and flow structures in spanwise-rotating turbulent plane Couette flows. Phys. Rev. Fluids 1:054401 (2016)
5. T. Kawata, P.H. Alfredsson, Scale interactions in turbulent rotating plane Couette flow: insight through the Reynolds stress transport. J. Fluid Mech. 879:255 (2019)
6. O. Iida, K. Kamiya, T. Kanda, Journal of Fluid Science and Technology, No 21-00239 (2022)

Assessment of Different Methods for Drag Penalty Predictions in Rough-Wall Boundary Layers

T. Medjnoun, M. A. Ferreira, R. Reinartz, B. Nugroho, J. Monty, N. Hutchins, and B. Ganapathisubramani

Abstract Accurate predictions of the drag penalty in rough-wall flows require careful characterisation of surface roughness to determine the equivalent sand-grain roughness height (k_s). The procedure involves measuring wall-shear stress (τ_w) using direct or indirect methods and analyzing velocity profiles. However, indirect methods often rely on assumptions whose validity cannot always be guaranteed. In this paper (partly based on the study [1]), wind tunnel measurements on a realistic rough surface scanned from a fouled ship hull are carried out to evaluate drag penalty predictions. Current data enabled the evaluation of k_s and associated wake parameters using several methods, which were then used for full-scale drag penalty predictions at high Reynolds numbers, with results showing the drag penalty could vary by nearly 15% among methods, highlighting the importance of exercising caution when employing such methods.

T. Medjnoun (✉) · B. Ganapathisubramani
Department of Aeronautics and Astronautics, University of Southampton, Burgess Road, Southampton, Hampshire SO17 1BJ, UK
e-mail: t.medjnoun@soton.ac.uk

M. A. Ferreira
Department of Process and Energy, Delft University of Technology, Delft, Netherlands

R. Reinartz
Department of Applied Physics, Eindhoven University of Technology, Eindhoven, Netherlands

B. Nugroho · J. Monty · N. Hutchins
Department of Mechanical Engineering, University of Melbourne, Grattan Street, Parkville, VIC 3010, Australia

R. Örlü et al. (eds.), *Progress in Turbulence X*, Springer Proceedings in Physics 404,
https://doi.org/10.1007/978-3-031-55924-2_14

1 Roughness Characterisation

The skin-friction coefficient for a zero-pressure gradient turbulent boundary layer can be expressed as,

$$\sqrt{\frac{2}{C_f}} = \frac{1}{\kappa}\ln(\delta^+) + B - \Delta U^+ + 2\frac{\Pi}{\kappa}, \tag{1}$$

with the superscript "+" denoting quantities non-dimensionalised by the friction velocity U_τ and kinematic viscosity ν. Assuming the universality of the Von-Kármán (κ and B) constants. Equation 1 can be rearranged to determine an expression for the roughness function:

$$\Delta U^+ = \overbrace{\sqrt{\frac{2}{C_f^S}} - \sqrt{\frac{2}{C_f^R}}}^{\text{I}} + \overbrace{\frac{1}{\kappa}\ln\left(\frac{\delta^{+R}}{\delta^{+S}}\right)}^{\text{II}} + \overbrace{2\frac{(\Pi^R - \Pi^S)}{\kappa}}^{\text{III}}, \tag{2}$$

with the superscripts S and R identifying quantities measured over smooth and rough wall flow conditions, respectively, $\delta^+ \equiv Re_\tau$ being the frictional Reynolds number while Π represents the wake strength parameter. Equation 2 denotes three main contributions in the roughness function; changes in (I) frictional drag, (II) frictional Reynolds numbers, and (III) the wake strength parameter. The function governing the wake in the outer flow is assumed to be universal (i.e. universal form), however, the wake strength (i.e. the amplitude of the function) can be different between surfaces. If $\Pi^R = \Pi^S$, then there is universal outer-layer similarity, and ΔU^+ can be determined by substituting the two different values of δ^+ in (2) on top of determining the term I. If also δ^{+R} and δ^{+S} are matched in the presence of universal outer-layer similarity, then ΔU^+ reduces to the difference in the frictional drag, and can be obtained from direct wall-shear stress measurements, without velocity information,

$$\Delta U^+ \equiv \Delta U^+_{DB} = \sqrt{\frac{2}{C_f^S}} - \sqrt{\frac{2}{C_f^R}} \tag{3}$$

When velocity and drag information are both available, ΔU^+ can be determined from the log-layer of the inner-scaled velocity profile. This is done by essentially subtracting the log-law from the measured inner-scaled velocity profile such that:

$$\Psi = U^+ - \left(\frac{1}{\kappa}\ln(\mathrm{y}^+ - \mathrm{d}^+) + \mathrm{B} - \Delta\mathrm{U}^+\right), \tag{4}$$

with d^+ being the zero-plane displacement, considered to be the height at which the mean drag acts. ΔU^+ (together with d^+) is subsequently determined by examining the optimum fit solution that returns the best plateau that occurs within the overlap region. The equivalent sand-grain roughness height k_s can then be determined by exposing the rough surface to different freestream speeds to obtain the variation of ΔU^+ as a function of k_s^+, using the expression:

$$\Delta U^+ \text{ or } \Delta U_{DB}^+ = \frac{1}{\kappa}\ln(k_s^+) - C \tag{5}$$

where, C is an additive constant, determined empirically to be equal to 3.5. Once $\Delta U^+/k_s$ are known, the wake strength parameter can be deduced through the modified-log function Ψ, which quantifies the maximum departure from the logarithmic behaviour; $\Pi = \kappa/2 \times \max(\Psi)$.

When drag cannot be directly measured, indirect methods must be employed. In this study, the first indirect method is the comprehensive shear-stress (CSS) method proposed by [2], which follows an iterative error minimisation procedure to determine the friction velocity U_τ using integral momentum equations - total shear stress balance together with the law-of-the-wall. The second method assumes the validity of the outer-layer similarity (OLS) hypothesis implying that both mean and turbulence intensity profiles are similar between the smooth- and rough-wall flows [3]. These different methods are summarised in Table 1.

2 Experimental Methods

Measurements are performed in an open-circuit suction-type wind tunnel at the University of Southampton (as shown in the schematics of Fig. 1). The turbulent boundary layer grows over a flat surface covered with biofouled roughness, with each tile consisting of some tubeworm and 8 barnacles. The barnacles are spaced 0.23δ mm and 0.46δ mm in both x-wise and z-wise horizontal directions, while the maximum roughness height is 6% of the boundary layer thickness δ. To determine the aerodynamic roughness lengthscale, wall-shear stress is directly measured using an in-house floating-element drag balance [4]. The balance is subjected to a series of nine free stream speeds (U_∞) ranging from 9 up to 27 ms^{-1}, with a total of five repetitions per velocity. Pre- and post-calibrations were conducted for each configuration without notable discrepancies. To examine the flow and its homogeneity, velocity fields are obtained from both planar and stereo particle image velocimetry (PIV) at a

Table 1 Methods used for the current biofouled rough-wall flow case study

Method	Wall-shear stress	Flow
Direct	Drag balance	PIV
Equation 3	Drag balance	Assumed
CSS	Log-law+TSS	PIV
OLS	Outer-layer similarity	PIV

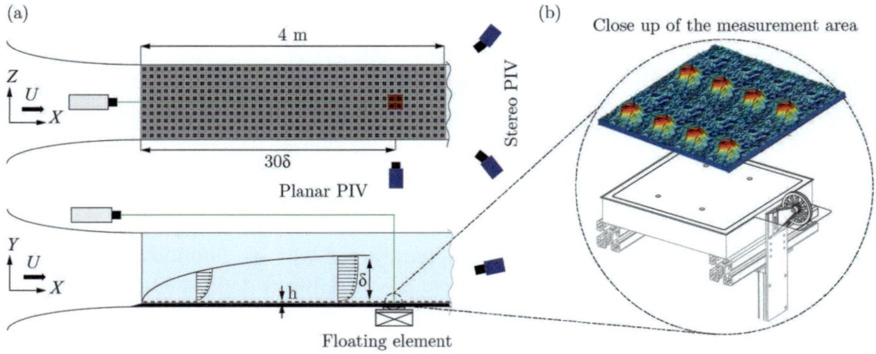

Fig. 1 **a** Schematics of the experimental setup of the planar-and stereo-PIV and **b** a close-up of the floating-element drag balance with the surface topography of a single repeating unit

similar location to that of the drag measurements (equivalent to 30δ downstream the inlet). All planar and stereoscopic PIV measurements are performed in the (x, y)- and (y, z)-planes, respectively, at five free stream speeds ranging from 10 up to 25 ms^{-1}.

3 Results: Lab-Scale

Results from direct drag measurements are shown in Fig. 2a and show that the skin-friction coefficient reaches a constant value, indicating the fully rough regime using the direct method shown by the invariance of C_f as a function of Re_x. However, estimates of the skin-friction coefficient from the OLS method, compared with the direct drag measurements underestimate C_f by close to 10% at the highest Reynolds number, as opposed to 6% when using the CSS method. The reason for this under-estimation is primarily due to a lack of collapse in both the mean and turbulence profiles between the rough and smooth surfaces (not shown for brevity). However, it is worth noting that for both indirect methods, C_f converges weakly towards those values, as the Reynolds number increases, albeit at different rates.

Now that the friction velocity is determined, the roughness function $\Delta U^+(k_s)$ is subsequently deduced by examining the Ψ profiles. Figure 2b shows the results from the different methods and reveals that (3) overestimates $\Delta U^+(k_s)$, while the CSS and OLS methods underpredict $\Delta U^+(k_s)$ as expected. The latter is expected and is a result of the lower estimated friction velocity, however, the former stems from the assumption of outer-layer similarity ($\Pi^R = \Pi^S$), in (3). This can be further explained by combining (2), (3) and (5), which results in the following expression:

$$\Pi^R = \Pi^S + \frac{1}{2}\ln\left(\frac{k_s}{k_{s\,DB}}\right) \tag{6}$$

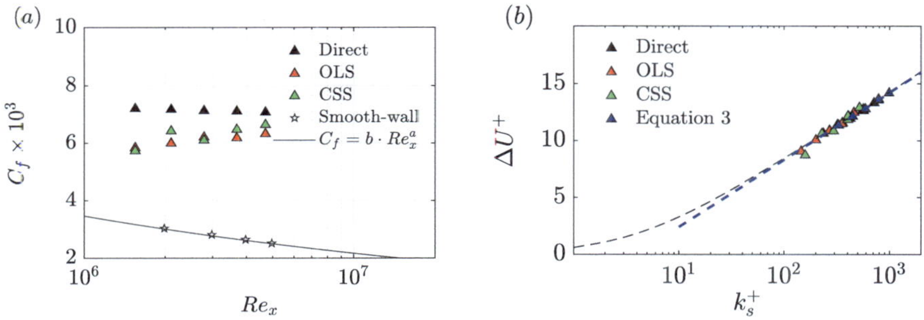

Fig. 2 **a** Variation of the skin-friction coefficient estimated using different methods for Config 1, with the smooth-wall data from [5], and the black solid line representing Schlichting's power-law. **b** Variation of the roughness function estimated the direct and indirect methods, with the solid black line representing the fully rough asymptote

Table 2 Comparison of the aerodynamic parameters results from the direct and indirect methods

Direct	OLS	CSS	Equation 3
$(k_s/\delta - \Pi^R)$	$(k_s/\delta - \Pi^S)$	$(k_s/\delta - \Pi^R)$	$(k_s/\delta - \Pi^S)$
(7.53% − 0.38)	(5.14% − 0.48)	(5.66% − 0.45)	(10.5% − 0.55)

Equation 6 shows that if $\Pi^R = \Pi^S$, then $k_s = k_{sDB}$. However, in the present study, $\Pi^R < \Pi^S$ (due to lack of collapse in the mean and turbulence profiles between smooth- and rough-wall flows), then $k_s < k_{sDB}$, which in turn means ΔU^+ measured will be smaller than ΔU^+_{DB} determined from (3). As also expected, both OLS and CSS methods underestimate the roughness functions. Results from using the different methods are collated in Table 2.

4 Results: Full-Scale Predictions

After determining the aerodynamic roughness parameters, a prediction of the spatially-averaged frictional drag C_F at high Reynolds numbers Re_x (with fixed U_∞ and ν but increasing x) can be made for each pair of (k_s, Π) results. The procedure proposed by [6] is employed, and it consists of an integral boundary layer evolution method based on numerical integration of the skin friction over a given length of the developing flow, assuming a velocity profile composed of a logarithmic and wake region (see [6] for more details).

Results of the full-scale prediction method are depicted in Fig. 3 and show that accurate prediction of skin friction remains possible using (k_s, Π) determined with Eq. 3, provided direct drag measurements at laboratory-scale data are available. Equation 3 results in prediction within 1% to those made by the direct measurements. However, when indirect drag estimates are used, C_F predictions at high Reynolds

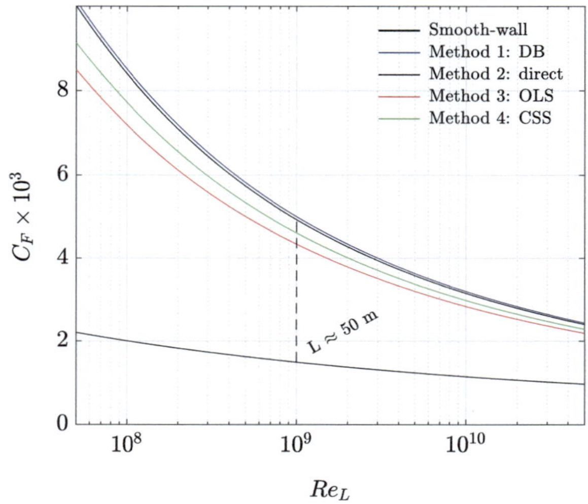

Fig. 3 Prediction of the averaged frictional drag C_F as a function of fetch Reynolds number Re_L at a unit Reynolds number $U_\infty/\nu \approx 2 \times 10^7$ with $U_\infty \approx 18$ ms^{-1} and $\nu \approx 9 \times 10^{-7}$ m^2 s^{-1}

numbers are systematically underestimated. At a Reynolds number of $Re_x \approx 10^9$ (equivalent to a Reynolds number of a 50 m long vessel at full-speed), assuming outer-layer similarity underestimated the frictional drag by over 12% while the comprehensive shear stress method resulted in approximately 6% difference, highlighting the importance of exercising caution when employing such methods. However, these differences tend to decrease as the Reynolds number increases, since most of the drag contributions stem from the changes in the log layer rather than changes in outer-layer wake.

References

1. T. Medjnoun, M.A. Ferreira, R. Reinartz, B. Nugroho, J. Monty, N. Hutchins, B. Ganapathisubramani, Assessment of aerodynamic roughness parameters of turbulent boundary layers over barnacle-covered surfaces. Exp. Fluids **64**(11), 169 (2023)
2. K.M. Womack, C. Meneveau, M.P. Schultz, Comprehensive shear stress analysis of turbulent boundary layer profiles. J. Fluid Mech. (2019)
3. A.A. Townsend, *The Structure of Turbulent Shear Flow*, 2nd edn. (Cambridge University Press, 1976)
4. M.A. Ferreira, E. Rodriguez-Lopez, B. Ganapathisubramani, An alternative floating element design for skin-friction measurement of turbulent wall flows. Exp. Fluids **59**, 155 (2018)
5. T. Medjnoun, C. Vanderwel, B. Ganapathisubramani, Characteristics of turbulent boundary layers over smooth surfaces with spanwise heterogeneities. J. Fluid Mech. **838**, 516–543 (2018)
6. J.P. Monty, E. Dogan, R. Hanson, A.J. Scardino, B. Ganapathisubramani, N. Hutchins, An assessment of the ship drag penalty arising from light calcareous tubeworm fouling. Biofouling **32**, 451–464 (2016)

A Method for Identifying the Freestream Turbulence/Turbulent Boundary Layer Interface

M. Asadi, P. A. Bullee, and R. J. Hearst

Abstract The interface between a turbulent boundary layer (TBL) and freestream turbulence (FST) is investigated by performing particle image velocimetry. Three different FST intensities are investigated together with a reference low-turbulence case. Utilizing vorticity distributions, a novel approach is introduced to delineate the boundary between the TBL and FST flows by identifying individual velocity thresholds for each instantaneous field. An increase in FST intensity results in a broader range of velocity thresholds. Conditionally averaged spanwise vorticity rapidly decays to zero on the FST side of the interface, while taking on a negative value within the boundary layer. The streamwise velocity exhibits a jump across the interface, whose magnitude increases with increasing FST intensity. The interface moves closer to the wall on average for increased FST intensities, leading to decreased intermittency values in the inner region of the TBL.

1 Introduction

Turbulent boundary layers (TBLs) are commonly observed in both industrial and environmental flows. The structure of canonical TBLs has been extensively studied. However, in practical scenarios, the wall-bounded flows are unavoidably impacted by freestream turbulence (FST). Numerous studies have explored the turbulence statistics of TBLs influenced by FST, but few have examined their instantaneous structure. To achieve this objective, identifying the instantaneous TBL and distinguishing it from the instantaneous freestream flow is crucial. Wu et al. [1] proposed that, similar to a turbulent/non-turbulent interface (TNTI) which demarcates the turbulent flow

M. Asadi (✉) · P. A. Bullee · R. J. Hearst
Norwegian University of Science and Technology, Trondheim, Norway
e-mail: masoud.as72@yahoo.com

P. A. Bullee
e-mail: pim.bullee@ntnu.no

R. J. Hearst
e-mail: jason.hearst@ntnu.no

R. Örlü et al. (eds.), *Progress in Turbulence X*, Springer Proceedings in Physics 404,
https://doi.org/10.1007/978-3-031-55924-2_15

from irrotational freestream, an interface marks the boundary of a TBL subjected to FST (henceforth denoted as the turbulent/turbulent interface, TTI). However, how to correctly identify the TTI is a topic of contemporary debate as researchers have used a variety of identification techniques.

Unlike previous TNTI studies (e.g., [2, 3]), Wu et al. [1] could not identify the TTI merely based on their direct numerical simulation (DNS) velocity data, i.e., by using a vorticity threshold. Instead, they used the normalized temperature as a passive scalar to mark the TTI. You and Zaki [4], however, refrained from utilizing a passive scalar, arguing that in the presence of FST, diffusion occurs on both sides of the interface. Instead, using a step-level approach in their DNS, they eliminated the viscous diffusion completely, albeit at the expense of disregarding the entrainment process. It is important to note that when vorticity is present in the freestream, the vortex stretching process operates on both sides of the TTI, resulting in a transport mechanism that is distinct from that for a passive scalar. Hence, it is essential to exercise caution while interpreting the interface indicated by the utilization of a passive scalar marker. Performing particle image velocimetry (PIV) measurements of a TBL subjected to FST, Hearst et al. [5] opted for a kinetic energy deficit threshold to separate the two flows in order to identify uniform momentum zones of the TBL.

In this study, a new method based on the vorticity signature is proposed to improve the TTI detection process for experimental data. The effect of FST intensities on the TTI characteristics is also studied.

2 Experimental Approach

The water channel facility at Norwegian University of Science and Technology was utilized to conduct the PIV experiments. An active grid ($M = 100$ mm, where M is the grid mesh length) was placed at the entrance of the recirculating free-surface water channel ($18M$ width $\times$ $110M$ length) to generate FST, which interacts with a TBL growing beneath it (see [6] for details of the active grid and water channel). The water channel was filled to a height of $H = 2.2M$ to conduct the FST experiments. Three different fully-random operational sequences were utilized to run the active grid. The rotational velocity, period, and acceleration of each grid bar were selected randomly from a top-hat distribution, where the mean rotational velocity (Ω) and the bandwidth (ω) were adjusted differently for each test case, resulting in three different FST cases referred to as ‘A’, ‘B’, and ‘C’ in the order of increasing turbulence intensity. Planar PIV measurements were performed in streamwise–wall-normal planes at a distance of $65M$ downstream of the grid. The active grid was then removed from the channel to achieve a reference low-turbulent case (REF), which was measured $95M$ downstream of the inlet for the friction Reynolds number to be comparable to that of the FST cases. Detailed parameters for each test case are listed in Table 1. The PIV system consisted of a 25-megapixel camera (LaVision Imager MX 25M) equipped with a 180 mm lens (Sigma f/2.8 EX DG), a dual-pulse Nd:YAG laser (Litron Nano L 200-15 PIV), and a programmable timing unit (LaVision PTU X). The field of view

Table 1 The freestream and boundary layer parameters together with the number of PIV fields and the interface location (y_i) statistics for the different test cases

Case	$\Omega \pm \omega$ (Hz)	U_∞ (ms^{-1})	u'_∞/U_∞ (%)	U_τ (mms^{-1})	δ (mm)	Re_θ	Re_τ	No. PIV fields	$\overline{y_i}$ (mm)	$\sigma(y_i)/\overline{y_i}$
REF	—	0.41	1.1	15.6	131	5543	2034	2000	111.7	0.17
A	1.5 ± 0.75	0.60	4.9	24.8	124	5029	2474	7000	70.5	0.49
B	0.5 ± 0.25	0.60	5.1	25.3	125	4709	2545	7000	62.1	0.55
C	0.05 ± 0.025	0.61	10.0	25.4	125	5093	2684	7000	55.5	0.61

(FOV) measured 200 mm in x and 200 mm in y (x is the streamwise and y is the wall-normal coordinate). LaVision DaVis 10.1 software was utilized to record the PIV images at 2 Hz and interrogate them (48 × 48 pixels window with 75% overlap).

3 Methodology and Results

The primary challenge in marking the TTI using a vorticity threshold lies in the existence of vorticity in the freestream zone [1, 4]. However, there is a crucial distinction between the vortical structures of a TBL and those of the FST. It is a characteristic feature of a TBL that it exhibits the presence of hairpin vortices, which are observed to form clusters [7]. As a consequence, the prograde vortices, which rotate in the same direction as the mean shear, dominate within the TBL. This phenomenon is demonstrated by the joint probability density function (p.d.f.) of the spanwise vorticity (ω_z) and y in Fig. 1a for case C, where the p.d.f. in the inner region is significantly skewed towards negative values. Homogeneous-isotropic FST, on the other hand, is made up of random vortical structures, which are formed by the turbulence-generating object and subsequently evolve downstream. As a result, a symmetric distribution of the spanwise vorticity about zero is witnessed in the region away from the wall within the freestream (Fig. 1a). Using the same phenomenon for instantaneous fields, we introduce a new method to identify the instantaneous TTI.

Figure 2 illustrates the stepwise procedure to identify the TTI between the instantaneous TBL and FST flow in a particular PIV field, as exemplified in Fig. 1b–e. The objective is to determine a suitable velocity threshold (u_{th}) for each instantaneous field, delineating the TTI by its continuous contour line (CCL). The process involves iteratively partitioning the field into a local FST area (LFST) and a region of interest ($\mathcal{R}$) beneath it, based on adjacent CCLs of constant velocity. To determine the identity of $\mathcal{R}$, i.e., whether it belongs to the instantaneous TBL flow or FST, the median value ($\mathcal{M}$) of ω_z within $\mathcal{R}$ ($\omega_{z,\mathcal{R}}$) as well as the Euclidean distance (D) between the p.d.f.s of $\omega_{z,\mathcal{R}}$ and $\omega_{z,\text{LFST}}$ (calculated using `pdist2` function in MATLAB) are evaluated. If D exceeds a threshold (D_{th}) and $\mathcal{M}$ is below a negative threshold ($\mathcal{M}_{th}$), $\mathcal{R}$ belongs to TBL. In this case, the iterative process concludes by identifying the upper CCL as the TTI. In effect, an individual u_{th} is identified for each particular

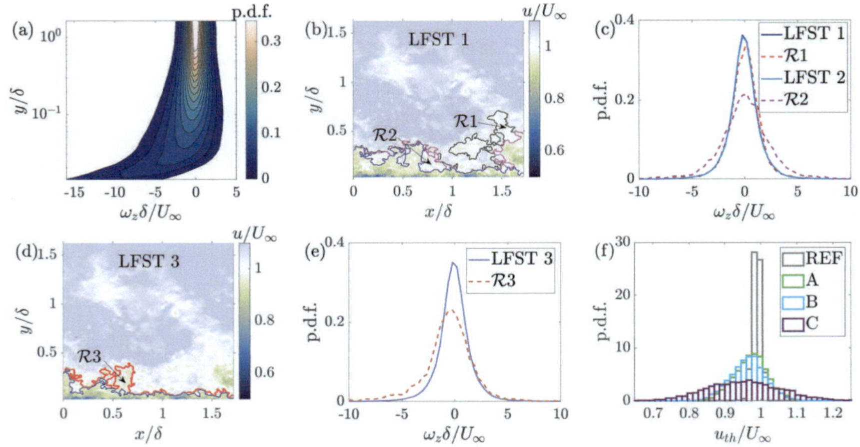

Fig. 1 **a** The distribution of vorticity at different wall-normal locations for case C. **b** and **d** A sample instantaneous streamwise velocity field of case C, with the regions of interest ($\mathcal{R}$) and local freestream (LFST) illustrated for the first and second iterations in **b** and the last iteration in **d**. Note that LFST 2 = LFST 1 $+\mathcal{R}1$. **c** and **e** the spanwise vorticity distributions in the different regions. $\mathcal{R}3$ satisfies both conditions (see Fig. 2) and the red contour line of $u_{th} = 0.985U_\infty$ is the detected interface. **f** Distributions of the identified velocity thresholds for all the cases

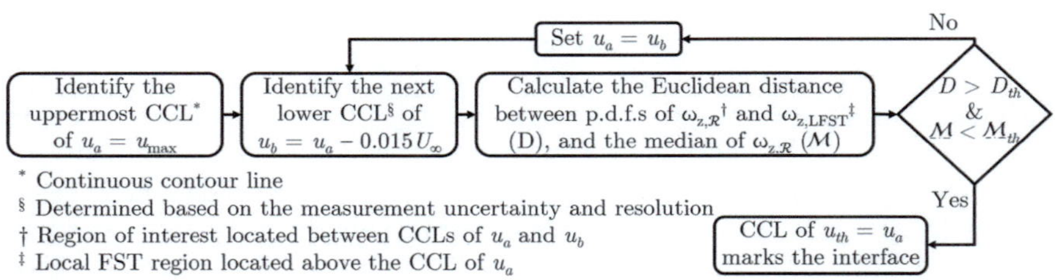

Fig. 2 The step-by-step diagram of the method

PIV field. Figure 1f depicts the p.d.f. of the identified u_{th} values for the different test cases. A significantly wider distribution of u_{th} is observed for FST cases compared to REF, dictating the necessity of selecting individual thresholds for each field instead of using a constant threshold value. In cases A, B, and C, no CCL was selected as the interface in 0.4, 1.2 and 4.8% of the PIV fields, respectively, suggesting that the discontinuity of the interface becomes more pronounced as the FST intensity increases.

The threshold values that effectively discriminate between the vorticity distributions inside the TBL and fluctuations of vorticity distributions within the FST region are evaluated for each case based on the variations of the vorticity distributions in the FST region. Through a thorough inspection, the threshold values are determined as $D_{th} = 3 \times \sigma(D_{i,\mathrm{FST}})$ and $\mathcal{M}_{th} = 3 \times \sigma(\mathcal{M}_{i,\mathrm{FST}})$, with $i = 1, 2, \cdots, N$, where N represents the total number of PIV fields. $D_{i,\mathrm{FST}}$ is the Euclidean distance between the ω_z p.d.f. in a portion of the local FST region of the i-th PIV field and the mean p.d.f. of ω_z in the same region that is averaged over all fields, while $\mathcal{M}_{i,\mathrm{FST}}$ is the median value of the spanwise vorticity within the FST region of the i-th PIV field. The former is evaluated within a region at the uppermost part of the PIV FOV ($\Delta y \approx 5\%$ of H) to ensure consistent data points for p.d.f. calculations, while the latter is calculated using the region above the uppermost CCL.

The conditionally averaged profiles are calculated by displacing the vertical coordinate to the interface location (y_i) and averaging the properties towards the boundary layer side ($y - y_i < 0$) and the freestream side ($y - y_i > 0$) of the interface. Figure 3a presents ω_z profiles conditionally averaged across the interface for all the cases. All four profiles show negligible mean spanwise vorticity content on the freestream side of the interface, while a negative value is observed on the boundary layer side. Thus, a step change is observed in the mean vorticity content when crossing the interface. The conditionally averaged streamwise velocity profiles are illustrated in Fig. 3b. For proper comparison, the profiles are first adjusted by subtracting u_{th} before averaging. An approximately uniform velocity profile within the FST is transformed to a uniformly sheared profile inside the TBL through a sharp jump on the interface. The magnitude of the jump increases with increasing FST intensities. The profiles of streamwise turbulence intensity zonally averaged inside the boundary layer (u'_{BL}/U_∞) and freestream (u'_{FS}/U_∞) are shown in Fig. 3c, indicating a uniform turbulence level inside the freestream irrespective of the wall-normal location, with

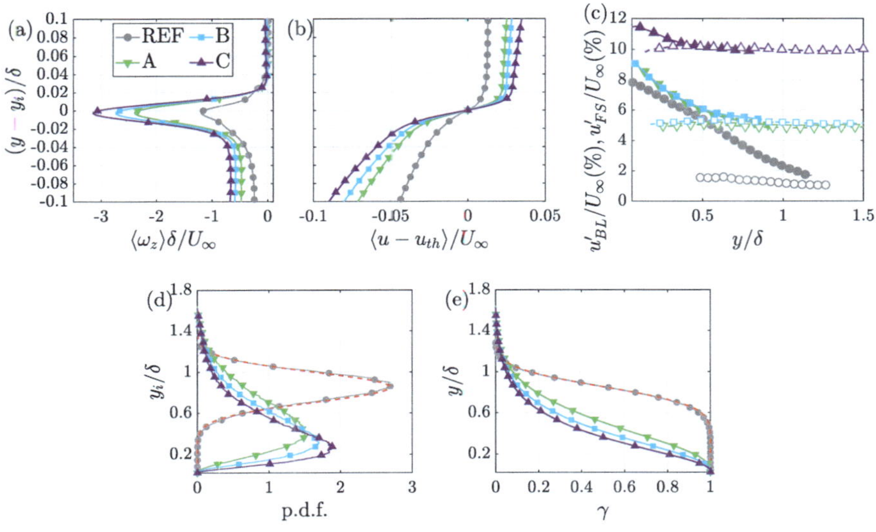

Fig. 3 **a** The spanwise vorticity and **b** the mean velocity profiles conditionally averaged across the interface. **c** The turbulence intensity profiles zonally averaged inside the boundary layer (the filled symbols) and freestream (the open symbols). **d** Distributions of the interface location, and **e** the intermittency profiles; red dashed lines show Gaussian distributions defined with REF's mean and standard deviation

values matching the FST intensities listed in Table 1. All the evidence presented in Fig. 3a–c provides a physical basis for the validity of the method in distinguishing between the TBL and FST flows.

Figure 3d and e illustrate the p.d.f.s of y_i and the intermittency (γ) profiles, respectively, for all the test cases. The results of case REF are well described by the Gaussian distribution (represented by the red dashed lines in Fig. 3d and e), showcasing the method's capability to identify the TNTI in the PIV fields of case REF. Conversely, the p.d.f.s of the FST cases no longer fit a Gaussian distribution and are markedly skewed towards the wall. Furthermore, with higher FST intensities, the peak in the p.d.f.s shifts closer to the wall, and the relative fluctuations in the interface location increase (see Table 1 for the mean and standard deviation of the interface location, denoted as $\overline{y_i}$ and $\sigma(y_i)$, respectively).

4 Conclusion

A novel method was introduced to identify the interface between a turbulent boundary layer (TBL) and freestream turbulence (FST). The method based upon the difference between the vorticital structures of the TBL and those of the FST, i.e., the fact that the spanwise vorticity distribution is skewed toward negative values in the TBL and thus different from the symmetric distribution of vorticity in the FST. The method was found to yield reliable results, with the mean spanwise vorticity exhibiting a negative value near the interface within the TBL while rapidly diminishing to zero on the FST side of the interface. This coincided with an abrupt increase in the streamwise velocity across the interface. Furthermore, the turbulence intensity zonally averaged inside the freestream possessed a uniform value regardless of the wall-normal location. When FST intensity was increased, the fluctuations of the interface location increased, and its average location moved closer to the wall. Furthermore, the distribution of the interface location deviated from a Gaussian distribution in the presence of FST.

Acknowledgements The funding for this study was provided by the Research Council of Norway through the WallMix project (project No. 288046).

References

1. X. Wu, J.M. Wallace, J.P. Hickey, Boundary layer turbulence and freestream turbulence interface, turbulent spot and freestream turbulence interface, laminar boundary layer and freestream turbulence interface. Phys. Fluids **31**, 045104 (2019)
2. J. Jiménez, S. Hoyas, M.P. Simens, Y. Mizuno, Turbulent boundary layers and channels at moderate Reynolds numbers. J. Fluid Mech. **657**, 335–360 (2010)
3. G. Borrell, J. Jiménez, Properties of the turbulent/non-turbulent interface in boundary layers. J. Fluid Mech. **801**, 554–596 (2016)

4. J. You, T.A. Zaki, Conditional statistics and flow structures in turbulent boundary layers buffeted by free-stream disturbances. J. Fluid Mech. **866**, 526–566 (2019)
5. R.J. Hearst, C.M. de Silva, E. Dogan, B. Ganapathisubramani, Uniform-momentum zones in a turbulent boundary layer subjected to freestream turbulence. J. Fluid Mech. **915**, A109 (2021)
6. Y. Jooss, L. Li, T. Bracchi, R.J. Hearst, Spatial development of a turbulent boundary layer subjected to freestream turbulence. J. Fluid Mech. **911**, A4 (2021)
7. X. Wu, P. Moin, Direct numerical simulation of turbulence in a nominally zero-pressure-gradient flat-plate boundary layer. J. Fluid Mech. **630**, 5–41 (2009)

Spectral Analysis of the Spatially Evolving Turbulent Channel Flow

E. Kannadasan, C. Atkinson, and J. Soria

Abstract The elementary structures of wall turbulence that carry most of the kinetic energy and momentum are typically referred to as energy-containing eddies (energy-eddies). Despite the general agreement that energy-eddies can sustain themselves at all relevant length scales [1, 2], their exact genesis and spatial evolution are still not well understood. In this study, energy-eddies at the inflow of a turbulent minimal-channel flow direct numerical simulation (DNS) are quenched and the spatial development of these eddies is studied at a friction Reynolds number $Re_\tau = 550$. Two synchronised DNSs are used in the current study: one is a fully resolved streamwise periodic channel flow, which is subsequently denoted as PCH-DNS, while the other is a fully resolved channel flow DNS with inflow-outflow boundary conditions, which will be denoted as IOCH-DNS. In the IOCH-DNS the inlet boundary condition is an inflow velocity field, which is a filtered version of the inflow of the PCH-DNS, with a convective outflow boundary condition applied at the domain exit [3]. The results demonstrate that energy-eddies are essential to maintain the eddies involved in the energy cascade. The results also show that the turbulent transport is the dominant transport term in the recovery of energy-eddies, and the energy-cascade eddies are primarily regenerated through wall-normal or scale-to-scale TKE transfer, rather than local production.

1 Introduction

Turbulence is a highly non-linear phenomenon characterized by the presence of multi-scale chaotic eddies, with wall-bounded turbulent flows exhibiting a cluster of recurring patterns known as energy-containing eddies (energy-eddies) [4]. These energy-eddies play a fundamental role in carrying energy and momentum within these flows, while the secondary eddies generated through the energy cascade are

E. Kannadasan (✉) · C. Atkinson · J. Soria
Laboratory for Turbulence Research in Aerospace and Combustion (LTRAC), Department of Mechanical and Aerospace Engineering, Monash University, Melbourne, Victoria 3800, Australia
e-mail: ezhilsabareesh.kannadasan@monash.edu

R. Örlü et al. (eds.), *Progress in Turbulence X*, Springer Proceedings in Physics 404,
https://doi.org/10.1007/978-3-031-55924-2_16

referred to as energy-cascade eddies [5]. Understanding the dynamics, origins, and interactions of energy-eddies is crucial for modeling and controlling industrial wall-bounded flows, yet it remains a significant challenge.

Townsend (1976) proposed a conceptual model for wall-bounded turbulence, suggesting that the size of energy-eddies spans a range of scales, from inner viscous to outer length scales, organized hierarchically in proportion to their distance from the wall [6]. In the near-wall region, consensus exists regarding the temporal self-sustaining cycle involving energy-eddies. This cycle involves interactions between streaks and vortices, where wall-normal ejections form streaks that subsequently break down, generating vortices [7–9]. This self-sustaining process extends to the logarithmic and outer regions of wall-bounded flows, even in the absence of near-wall energy-eddies, indicating that energy-eddies at different layers within the flow are self-sustaining independently [1, 2]. However, the mechanisms governing the genesis and spatial evolution of energy-eddies remain unknown. The focus of the present study is to remove the energy-eddies at the inflow of a turbulent channel and investigate the spectral energy cascade analysis of structures that emerge during the spatial evolution of energy-eddies.

2 Numerical Experiments

In this study, we investigate the spatial evolution of energy-containing eddies through direct numerical simulation (DNS) of incompressible turbulent flow in a plane channel at a friction Reynolds number (Re_τ) of 550. The channel has streamwise and spanwise domain sizes of 8π and π times the channel half-height (h), respectively. All quantities are non-dimensionalized using the friction velocity (u_τ) from fully developed channel flow. The grid spacings are $\delta x^+ = 6.65, \delta z^+ = 3.36$, in the streamwise and spanwise direction, respectively. The wall-normal grid spacing near the wall is $\delta y^+_{min} = 0.31$, reaching $\delta y^+_{max} = 2.87$ at the channel centerline, denoted in wall units relative to the viscous length scale (ν/u_τ).

Two synchronized DNS simulations are employed: one as a fully resolved streamwise periodic channel flow (PCH-DNS) and the other as a channel flow DNS with inflow-outflow boundary conditions (IOCH-DNS). In IOCH-DNS, the inlet boundary condition is a filtered version of PCH-DNS's inflow velocity field, with convective outflow boundary conditions applied at the exit. Statistics from IOCH-DNS are collected once it reaches a statistically stationary state, as initial transients wash out. A more detailed description of the synchronized dual-channel code and validation can be found in a previous work [12]. Energy-containing eddies are a direct result of turbulence production, manifest across multiple length scales, forming a complex interplay of scale interactions and energy cascades. The motions that originate specifically from the energy cascade process are referred to as energy-cascading eddies. The Fourier modes of the turbulent transport term resolves not only the energy-containing eddies but also those generated by energy cascade. The boundary that distinguishes the energy-containing eddies from the energy-cascading eddies needs to be identified to isolate the energy-eddies at the inflow of the IOCH-DNS. Cho et al. [5] investigated the spanwise spectra of the TKE and the turbulent transport of a fully developed channel flow. Their study suggested $\lambda_z = 3y$ as a reasonable boundary to separate the energy-eddies from the cascading eddies, where λ_z is the spanwise wavelength. Therefore, at a given spanwise Fourier mode $\lambda_{z,0}$, the fluctuations below the wall-normal profile with $\lambda_{z,0}/3$ is removed at the inflow of the IOCH-DNS, which represents that the energy-eddies are completely removed at the inflow. The detailed methodology of removing energy-eddies at the inflow of the DNS is discussed in detail in [3].

To analyze the turbulent kinetic energy (TKE) budget equation, we start with the equation:

$$\frac{\partial \langle e \rangle}{\partial t} = P - \epsilon - T_{turb} - T_p + T_\nu - C \tag{1}$$

Here, P, ϵ, T_{turb}, T_p, T_ν, and C represent turbulence production, viscous dissipation, turbulent transport, pressure transport, viscous transport, and convection terms, respectively.

In IOCH-DNS, due to spatial development in the streamwise direction, we perform a Fourier decomposition of velocity fluctuations in the spanwise direction:

$$u_i'(t, x, y, z) = \int \widehat{u_i'}(x, y, k_z) e^{ik_z z} dk_z \tag{2}$$

Here, $\hat{\cdot}$ denotes Fourier coefficients, and k_z is the spanwise wavenumber with $\lambda_z = (2\pi)/k_z$ representing the spanwise wavelength.

We then analyze the time rate of change of the TKE equation in Fourier space:

$$\frac{\overline{D\widehat{e}(k_z)}}{Dt} = \hat{P}(y, k_z) - \widehat{\epsilon}(y, k_z) - \widehat{T}_{turb}(y, k_z) - \widehat{T}_p(y, k_z) + \widehat{T}_\nu(y, k_z) \tag{3}$$

Here, $\hat{P}(y, k_z)$, $\widehat{\epsilon}(y, k_z)$, $\widehat{T}_{turb}(y, k_z)$, $\widehat{T}_p(y, k_z)$, and $\widehat{T}_\nu(y, k_z)$ are the spectral contributions of turbulence production, viscous dissipation, turbulent transport, pressure transport, and viscous transport, respectively.

3 Results and Discussion

To investigate energy transfer mechanisms in IOCH-DNS, the turbulent kinetic energy (TKE) budget in equation 1 is investigated. The TKE budget profiles along the wall-normal direction in both the IOCH-DNS and the fully developed PCH-DNS are shown in Fig. 1.

In IOCH-DNS, the production and dissipation profiles remain relatively consistent across the wall-normal distance, except in the buffer layer around $y^+ \simeq 15$, particularly for $x \leq 1.5$ h. As the flow progresses, TKE production and dissipation become

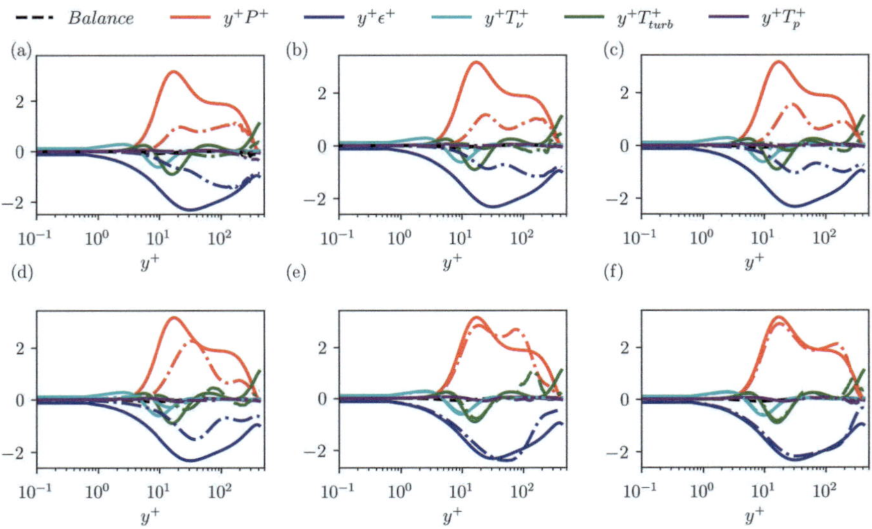

Fig. 1 Turbulent kinetic energy budget of the IOCH-DNS as a function of y at various streamwise locations: **(a)** $x = 0$ h, **(b)** $x = 1.5$ h, **(c)** $x = 3$ h, **(d)** $x = 6$ h, **(e)** $x = 12$ h, **(f)** $x = 24$ h. Solid line: PCH-DNS, dash-dotted line: IOCH-DNS

more pronounced in the buffer layer, while transport terms remain almost negligible for $x \leq 6$ h. This enhanced TKE production in the buffer layer may correspond to the formation of near-wall streaks, supported by amplified one-dimensional velocity spectra in the early stages of evolution, as discussed in [3]. Once near-wall streaks re-establish, the turbulent production term in the log-layer strengthens, as observed in Fig. 1e at $x = 12$ h. Notably, there is a significant energy influx through turbulent transport in the log-layer at $y^+ \simeq 150$ from the buffer layer at $x = 12$ h. These observations suggest either local energy production in the log-layer once near-wall streaks recover or energy transport from the buffer region to initiate the evolution of energy-eddies at the log-layer.

In the TKE budget discussed above, turbulent transport emerges as the dominant energy transport mechanism in the energy-eddy recovery. As a result, only the production, and turbulent transport spectra are examined in this paper.

Figure 2 shows how the pre-multiplied spanwise spectra of TKE production ($k_z^+ y^+ \hat{P^+}$) evolve along the streamwise direction in the IOCH-DNS. In the PCH-DNS, the TKE production peaks at $\lambda_z^+ \simeq 100$ and $y^+ \simeq 15$ (dashed black lines), indicating the energy source. In the IOCH-DNS, as the flow evolves, the strengthening of the $k_z^+ y^+ \hat{P^+}$ spectrum around $\lambda_z^+ \simeq 100$ (Fig. 2b, c, d) is observed which signifies the formation of near-wall streaks early in the evolution, consistent with the previous observations in [3].

Furthermore, the spectral density of TKE production above the cut-off wavelength decreases along the streamwise direction for $x \leq 6$ h, indicating the decay of the energy-cascade eddies. Beyond $x = 12$ h, the TKE production spectrum

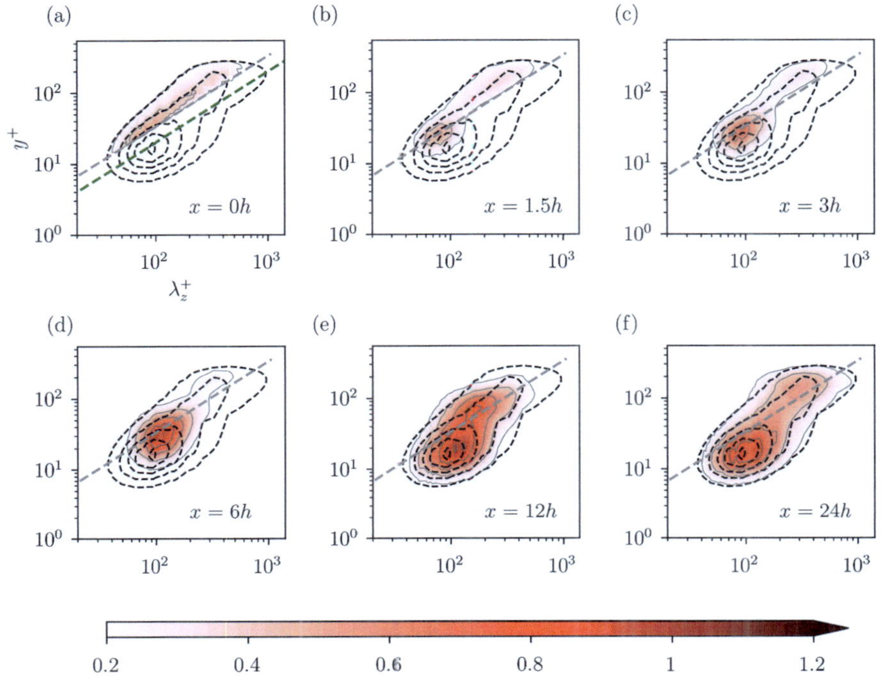

Fig. 2 The one-dimensional pre-multiplied production spectra of TKE ($k_z y^+ p^+$), as a function of y at various streamwise locations: **(a)** $x = 0\,\text{h}$, **(b)** $x = 1.5\,\text{h}$, **(c)** $x = 3\,\text{h}$, **(d)** $x = 6\,\text{h}$, **(e)** $x = 12\,\text{h}$, **(f)** $x = 24\,\text{h}$. The dashed black line represents the PCH-DNS data, with contour levels shown as [0.2, 0.4, 0.6, 0.8, 1, 1.2]. The red contours depict the IOCH-DNS data. Here, the dashed grey line is at $\lambda_z = 3y$ to indicate the spanwise cutoff wavelength, and dashed green line is at $\lambda_z = 5y$

above the ridge aligns with that of the PCH-DNS, indicating that the production of TKE in the region of energy-cascade eddies resumes once the near-wall streaks are re-established. This re-establishment is characterized by the recovery of the production spectrum at the typical wall-normal location of $y^+ = 15$ and spanwise lengths scale of $\lambda_z^+ \simeq 100$, exhibiting the same spectral density as observed in the corresponding PCH-DNS spectrum.

The pre-multiplied spanwise turbulent transport spectra $k_z^+ y^+ \hat{T}_{turb}^+$ exhibit both positive and negative regions, as shown in Fig. 3. This behavior arises because the integral of transport terms over the domain is zero due to energy conservation [5].

In the PCH-DNS, the negative turbulent transport, depicted as blue dashed line contours, aligns with the ridge $\lambda_z = 5y$ and corresponds to the positive TKE production spectra. The positive turbulent transport region above the ridge $\lambda_z = 3y$, indicated by red dashed line contours, signifies the classical energy cascade, transferring energy from large to small scales. There's also a weak positive turbulent transport region near the wall for $y^+ < 10$ in the PCH-DNS, attributed to the inverse energy cascade from small to large scales [5].

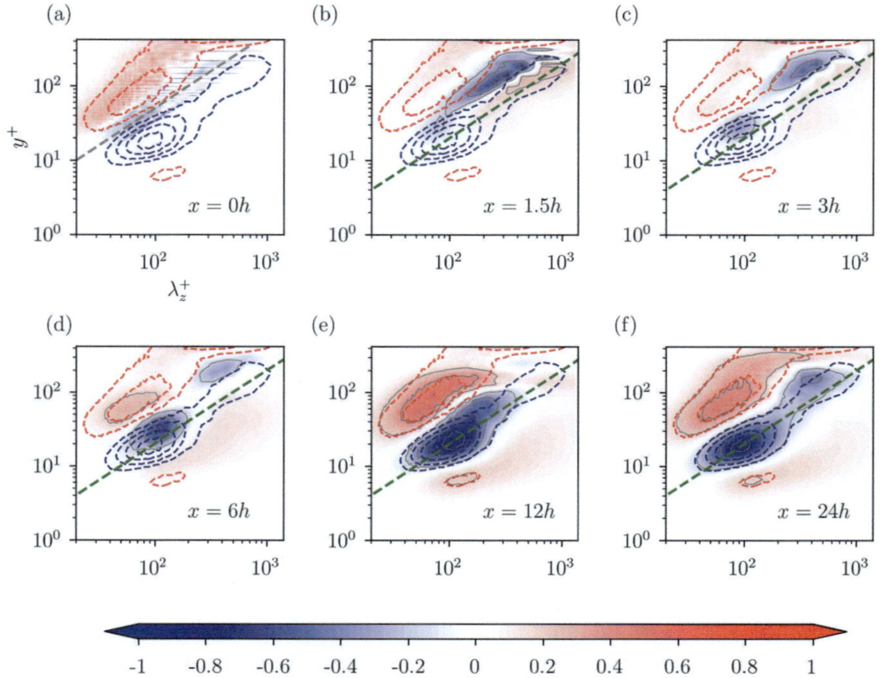

Fig. 3 The pre-multiplied spanwise spectra of turbulent transport ($k_z^+ \, y^+ \hat{T}_{turb}^+$), as a function of y at various streamwise locations: **(a)** $x = 0\,\mathrm{h}$, **(b)** $x = 1.5\,\mathrm{h}$, **(c)** $x = 3\,\mathrm{h}$, **(d)** $x = 6\,\mathrm{h}$, **(e)** $x = 12\,\mathrm{h}$, **(f)** $x = 24\,\mathrm{h}$. PCH-DNS (dashed red lines and dashed blue lines), IOCH-DNS (contour enclosed in grey line). Here, the dashed grey line is at $\lambda_z = 3y$ to indicate the cutoff spanwise wavelength, and the dashed green line is at $\lambda_z = 5y$

In the IOCH-DNS, the region displaying positive turbulent transport above the $\lambda_z = 3y$ ridge remains relatively dormant until reaching $x = 6\,\mathrm{h}$. This suggests a limited transfer of TKE from energy-eddies to energy-cascade eddies during this phase. This observation aligns with the findings in our previous study [3], where observed a decline in energy above the ridge for $x < 6\,\mathrm{h}$ was observed, as shown by the $k_z\phi_{vv}$ and $k_z\phi_{ww}$ spectra.

The positive turbulent transport region becomes active at $x = 6\,\mathrm{h}$ where the energy at the near-wall streaks location ($\lambda_z^+ = 100$) recovers. However the production spectra at $x = 6\,\mathrm{h}$ shows that there is no active TKE production above the ridge $\lambda_z = 3y$. This suggests that the production of energy-cascade eddies primarily results from wall-normal or scale-to-scale TKE transfer, facilitated through turbulent transport from the vicinity of the near-wall streaks, rather than localized TKE generation above $\lambda_z = 3y$.

4 Conclusion

In this study, the spatial evolution of energy-eddies in turbulent channel flow through detailed spectral analysis of turbulent kinetic energy (TKE) terms is investigated. The results highlight the crucial role of the turbulent transport term in this spatial evolution. Notably, our findings indicate that the regeneration of energy-cascade eddies predominantly occurs through wall-normal or scale-to-scale TKE transfer, rather than local TKE production. This insight significantly enhances our comprehension of energy transfer mechanisms within the spatial evolution of energy-eddies in wall-bounded turbulent flows. Furthermore, it is worth noting that further scale interaction analysis may unveil even more intricate mechanisms of energy transfer in the spatial evolution of energy-eddies.

Acknowledgements The authors acknowledge funding from the Australian Research Council through a Discovery Grant, as well as computational resources from the Pawsey Supercomputing Centre and the National Computational Infrastructure (NCI) via computational grants awarded by the National Computational Merit Allocation Scheme (NCMAS), funded by the Australian Government. Ezhilsabareesh Kannadasan acknowledges support from a Monash Graduate Scholarship (MGS) and a Monash International Tuition Scholarship (MITS).

References

1. Y. Hwang, C. Cossu, Self-sustained process at large scales in turbulent channel flow. Phys. Rev. Lett. **105**(4), 044505 (2010)
2. O. Flores et al., The three-dimensional structure of momentum transfer in turbulent boundary layers. J. Fluid Mech. **564**, 165–233 (2006)
3. E. Kannadasan, C. Atkinson, J. Soria, Spectral analysis of the evolution of energy-containing eddies. J. Fluid Mech. **955**, R1 (2023)
4. L.F. Richardson, *Weather Prediction by Numerical Process* (Cambridge University Press, 1922)
5. M. Cho, Y. Hwang, H. Choi, Scale interactions and spectral energy transfer in turbulent channel flow. J. Fluid Mech. **854**, 474–504 (2018)
6. A.E. Perry, M.S. Chong, On the mechanism of wall turbulence. J. Fluid Mech. **119**, 173–217 (1982)
7. J. Jimenez et al., The structure of intense vorticity in isotropic turbulence. J. Fluid Mech. **225**, 65–90 (1991)
8. J.M. Hamilton et al., Regeneration mechanisms of near-wall turbulence structures. J. Fluid Mech. **287**, 317–348 (1995)
9. F. Waleffe, On a self-sustaining process in shear flows. Phys. Fluids **9**(4), 883–900 (1997)
10. A.A. Townsend, *The Structure of Turbulent Shear Flow* (1976)
11. J. Jiménez, A. Pinelli, The autonomous cycle of near-wall turbulence. J. Fluid Mech. **389**, 335–359 (1999)
12. K. Ezhilsabareesh, C. Atkinson, A. Lozano-Duran, P.J. Schimd, J. Jiménez, J. Soria, Effect of limited near-wall inlet data on the direct numerical simulation of turbulent channel flow. J. Phys. Conf. Ser. **1522**(1), 012019 (2020)

Free Shear Flows

The Near- and Intermediate-Wakes of Cylinders Under the Influence of Freestream Turbulence

L. Li and R. J. Hearst

Abstract A circular and a square cylinder were subjected to four different freestream turbulence (FST) inflows to investigate their impact on the spatial evolution of the wake. These turbulence intensities spanned from 1.1 to 13.8%, while the Reynolds number based on the cylinder diameter D was $Re_D \approx 3.3 \times 10^4$. The wake flow fields were captured using an expansive 2-D planar particle image velocimetry (PIV) set-up, offering a physical field of view extending nearly 1 m in length, or $20D$ downstream of the cylinder. Elevated FST triggers an earlier onset of apparent self-similar behaviour in both the wake centreline velocity deficit U_s and the wake half-width ℓ. Moreover, the rise in FST leads to a reduction in the average number of coherent vortices linked to wake vortex shedding, underscoring how the introduction of freestream turbulence disperses the spatial propagation of these vortices.

1 Introduction

Cylindrical objects with various cross-sectional shapes are commonly found in both natural and urban environments. Therefore, comprehending the flow field around these objects holds significant scientific and practical implications. Vickery's early work [1] demonstrated that elevated freestream turbulence (FST) can decrease the root-mean-square fluctuations of the lift coefficient of a square cylinder by 50%. Cheung and Melbourne [2] further illustrated that the drag coefficient can be significantly affected by FST at the same Reynolds number based on the cylinder diameter D, attributed to FST triggering an earlier onset of the supercritical flow regime. Moreover, Surry [3] discovered that even within the subcritical regime, FST has the tendency to artificially elevate the Reynolds number. These studies achieved turbulence intensities nearing 20%. On a smaller scale, Kwok [4] and Norberg and Sundén [5] demonstrated that even minor turbulence on the order of 1% can lead

L. Li · R. J. Hearst (✉)
Department of Energy and Processing Engineering, Norwegian University of Science and Technology, Kolbjørn Hejes Vei 2, 7034 Trondheim, Norway
e-mail: jason.hearst@ntnu.no

R. Örlü et al. (eds.), *Progress in Turbulence X*, Springer Proceedings in Physics 404,
https://doi.org/10.1007/978-3-031-55924-2_17

to noteworthy alterations in the boundary layer and pressure distributions around a cylinder. Blackburn and Melbourne [6] observed that FST can induce a transition in the fluctuation energy of the lift coefficient from a Strouhal number of 0.2, linked to vortex shedding, towards lower frequencies. Kankanwadi and Buxton [7] observed that the entrainment mass flux across the wake-freestream interface is significantly influenced by the freestream turbulence level, to the extent that it can reverse the flux direction in extreme scenarios.

The majority of the previously mentioned studies concentrated on the aerodynamics of the cylinder itself under the impact of FST, rather than examining the spatial development of the wake downstream. Upon reviewing numerous studies on cylinder flow [8–10], it is evident that a systematic examination of the effects of varying FST on different cylinder geometries is presently lacking. Eames et al. [11] suggest that the centreline wake velocity deficit U_s and the wake half-width ℓ diverge from canonical self-similar behaviour under conditions of high FST. Nevertheless, their investigation exhibited noticeable inflow inhomogeneity, and the Reynolds number was relatively low ($\mathcal{O}(10^3)$). The objective of this study is to bridge this research gap by methodically examining the spatial evolution of the wake behind different cylinder geometries, subjected to varying levels of homogeneous and isotropic freestream turbulence.

2 Experimental Set-Up

The experiments were performed in the large closed-loop wind tunnel at the Norwegian University of Science and Technology, featuring a test section with dimensions $2.71\ \text{m} \times 1.80\ \text{m} \times 11.15\ \text{m}$. For the present study, the freestream velocity was held at $U_\infty \approx 11\ \text{ms}^{-1}$. The background freestream turbulence intensity for the empty test section during the experiments was $\langle u_1'^2 \rangle^{1/2}/U_\infty = 1.1\%$. Subsequent enhancements in flow conditioning have led to a reduction of this value to around 0.25%.

To generate various freestream turbulence inflows, an active grid was positioned close to the inlet of the test section. The grid comprises 90 independently actuated rods arranged in a bi-planar layout, with square wings affixed to each rod in a space-filling pattern. The grid has a mesh length of $M = 0.1$ m. Motor actuation protocols were communicated via serial ports from MATLAB. Three distinct actuation protocols were employed, and when combined with a reference case with the grid removed, this resulted in a total of four inflow cases. More details on the active grid can be found in Kildal et al. [12]. Table 1 details the relevant turbulence characteristics of the inflow conditions. It is worth noting that both the root mean square (RMS) variations in U_1 and the isotropy ratios suggest a state of nearly homogeneous and isotropic inflows.

The study involved testing two cylinder cross-sectional geometries: circular and square. Both the circular cylinder diameter and the side length of the square cylinder are $D = 0.05$ m. The Reynolds number based on the cylinder diameter is $Re_D \approx 3.3 \times 10^4$. The cylinders span the full width of the test section and are anchored at a

Table 1 Summary of the inflow turbulence characteristics

Case name	U_∞ [ms^{-1}]	u'_1/U_∞ [%]	RMS_{U_1}/U_∞ [%]	$\overline{u'_1/u'_2}$	L_{11}/D	η/D [10^{-3}]	Re_λ	Re_D
REF	10.5	1.1	2.5	0.79	5.12	13.41	70	32593
FST8	10.6	7.5	1.1	1.10	4.33	4.84	544	32664
FST12	10.6	12.1	2.5	1.01	5.68	3.56	755	32613
FST14	10.8	13.8	1.7	1.10	7.86	3.44	958	33130

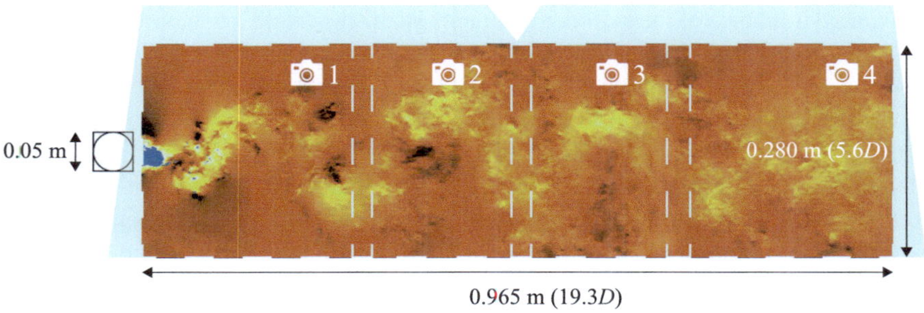

Fig. 1 Schematic of the experimental set-up showing the PIV lasers and camera fields of view. The size and position of the cylinders relative to the field of view are to scale

position $64M$ downstream from the active grid, as well as 0.75 m above the tunnel floor, using four struts. The central 50% of the cylinder span is left unobstructed to the freestream. The overall blockage ratio is estimated to be 5.4%, falling below the 6% threshold recommended by West and Apelt [13] for the application of blockage corrections.

Instantaneous wake flow fields were captured using planar particle image velocimetry (PIV). To create a single illuminated light sheet behind the cylinders, two dual-pulse Litron Nano L Nd-YAG lasers (532 nm, 200 mJ) were positioned in sequence. Additionally, four cameras positioned in sequence recorded a continuous instantaneous flow field, with dimensions of 0.965 m $\times$ 0.280 m or $19.3D \times 5.6D$. The camera system comprised two LaVision Imager MX 25 cameras equipped with Zeiss 100 mm lenses flanking two LaVision Imager LX 16 cameras featuring Nikon 200 mm lenses. Figure 1 shows a schematic of the PIV set up with an example instantaneous velocity field showing the field of view. In each test case, a total of 2000 image pairs were captured at a sampling rate of 1.5 Hz. Vector fields were computed using DaVis 10.1 through multiple passes. The final pass employed a 48×48 window with 75% overlap, resulting in a vector spacing of 0.72 mm. Post-calculation stitching of the vector fields was conducted using MATLAB.

3 Wake Spatial Development

Figure 2 illustrates the wake centreline velocity deficit U_s and the wake half-width ℓ, both normalized by U_∞ and D, respectively. Observing the near wake ($x_1/D < 3$), we note that turbulence intensity does not appear to impact the length and intensity of the recirculation region immediately behind the circular cylinder. Conversely, for the square cylinder, elevated FST leads to the elongation of the recirculation region and an elevation in the peak reverse flow velocity. This observation underscores the substantial influence of the cylinder geometry on the flow field's behaviour in the near wake, particularly under distinct inflow conditions. As we move to the intermediate wake ($3 < x_1/D < 20$), both U_s and ℓ exhibit power law behaviour with respect to x_1. The orange dotted lines in Fig. 2 correspond to $U_s/U_\infty \propto (x_1/D)^{-0.5}$ and $\ell/D \propto (x_1/D)^{+0.5}$, indicative of the analytical power law behaviour within the self-similar region of a planar wake. Notably, as FST increases, the rates of change for both U_s and ℓ align more closely with the analytical solutions. This suggests that elevated FST initiates an earlier onset of self-similar behaviour in the first-order statistics.

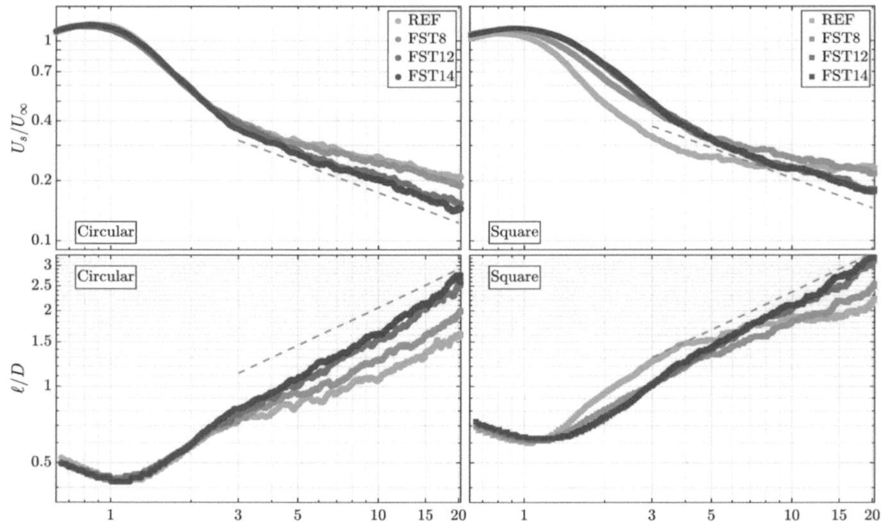

Fig. 2 Spatial evolution of the normalised wake centreline velocity deficit U_s/U_∞ (top row) and the normalised wake half-width ℓ/D (bottom row) for all the test cases. The orange dotted lines show the analytical power law relations in the self-similar region

4 Vortex Spatial Development

To identify vortices in each instantaneous velocity field, the vortex detection method of Graftieaux et al. [14] is employed. In essence, the process involves calculating the summation of the cross product of the velocity fluctuation vectors with a spatial kernel containing the position vectors around a central point P. When a vortex is centred around P, the summation yields a high value. Performing this calculation for all points in the instantaneous velocity fluctuation field generates a mapping where regions around a vortex yield high values. Vortices can subsequently be detected by establishing a threshold value on the mapping. The practical computation of these mappings is facilitated by the efficient algorithm developed by Zigunov [15], which relies on 2-D convolution.

Figure 3 illustrates the average vortex count across all 2000 instantaneous velocity fluctuation fields in all cases. In this study, the spatial window size was fixed at 107×107 vectors, providing optimal outcomes for identifying larger coherent vortices while excluding the smaller ones linked to turbulence. A sensitivity analysis indicated that window sizes exceeding 53×53 exhibited the same mean trend for both cylinder geometries. It is evident that the average number of vortices steadily declines as FST increases, and both cylinders demonstrate comparable values for the tested inflow cases. The impact of elevated FST on the spatial evolution of coherent vortices in the cylinder wakes is evident. This leads to an early disintegration of coherence, ultimately resulting in a reduced number of vortices.

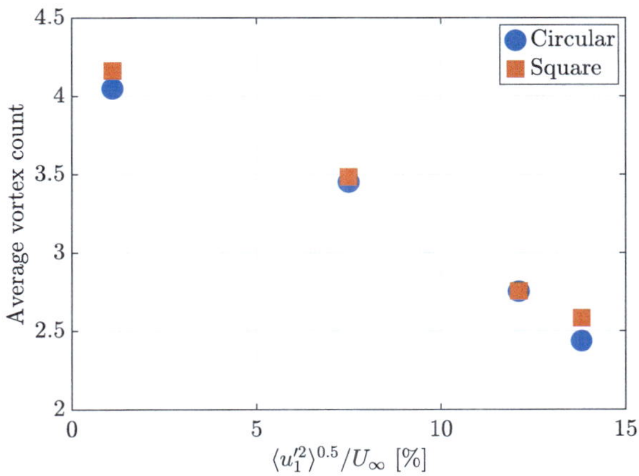

Fig. 3 Average number of detected vortices relative to the freestream turbulence intensity for both cylinder geometries

5 Conclusions

This study examines the impact of freestream turbulence on the wake development of both a circular and a square cylinder. Using PIV at a moderately high Reynolds number of $Re_D \approx 3.3 \times 10^4$, a large instantaneous velocity field was acquired, extending up to 1 m or $20D$ downstream of the cylinder.

Within the near wake ($x_1/D < 3$), the behaviour of the recirculation region to varying freestream turbulence strongly depends on the cylinder geometry. This indicates that cylinder geometry significantly influences the statistics of the near wake. Transitioning to the intermediate wake ($3 < x_1/D < 20$), elevated freestream turbulence leads to an earlier onset of self-similar behaviour in the spatial evolution of both the centreline velocity deficit U_s and the wake half-width ℓ. Additionally, an increase in freestream turbulence contributes to the dispersion of coherent vortices in the wake, resulting in a reduction in the number of vortices as FST levels rise.

Acknowledgements Funded by the European Union (ERC, GLITR, 101041000). Views and opinions expressed are however those of the authors only and do not necessarily reflect those of the European Union or the European Research Council Executive Agency. Neither the European Union nor the granting authority can be held responsible for them.

References

1. B.J. Vickery, Fluctuating lift and drag on a long cylinder of square cross-section in a smooth and in a turbulent stream. J. Fluid Mech. **25**(3), 481–494 (1966)
2. J.C.K. Cheung, W.H. Melbourne, Turbulence effects on some aerodynamic parameters of a circular cylinder at supercritical numbers. J. Wind Eng. Ind. Aerod. **14**(1–3), 399–410 (1983)
3. D. Surry, Some effects of intense turbulence on the aerodynamics of a circular cylinder at subcritical Reynolds number. J. Fluid Mech. **52**(3), 543–563 (1972)
4. K.C.S. Kwok, Turbulence effect on flow around circular cylinder. J. Eng. Mech. **112**(11), 1181–1197 (1986)
5. C. Norberg, B. Sundén, Turbulence and Reynolds number effects on the flow and fluid forces on a single cylinder in cross flow. J. Fluid. Struct. **1**(3), 337–357 (1987)
6. H.M. Blackburn, W.H. Melbourne, The effect of free-stream turbulence on sectional lift forces on a circular cylinder. J. Fluid Mech. **306**, 267–292 (1996)
7. K.S. Kankanwadi, O.R.H. Buxton, Turbulent entrainment into a cylinder wake from a turbulent background. J. Fluid Mech. **905**, A35 (2020)
8. C.H.K. Williamson, Vortex dynamics in the cylinder wake. Annu. Rev. Fluid Mech. **28**(1), 477–539 (1996)
9. J.F. Derakhshandeh, Md.M. Alam, A review of bluff body wakes. Ocean Eng. **182**, 475–488 (2019)
10. B. Forouzi Feshalami, S. He, F. Scarano, L. Gan, C. Morton, A review of experiments on stationary bluff body wakes. Phys. Fluids **34**(1), 011301 (2022)
11. I. Eames, C. Jonsson, P.B. Johnson, The growth of a cylinder wake in turbulent flow. J. Turbul. **12**, N39 (2011)
12. O. Kildal, L. Li, R.J. Hearst, Ø.W. Petersen, O. Øiseth, On the use of an active turbulence grid in wind tunnel testing of bridge decks. J. Wind Eng. Ind. Aerod. **233**, 105331 (2023)

13. G.S. West, C.J. Apelt, The effects of tunnel blockage and aspect ratio on the mean flow past a circular cylinder with Reynolds numbers between 10^4 and 10^5. J. Fluid Mech. **114**, 361–377 (1982)
14. L. Graftieaux, M. Michard, N. Grosjean, Combining PIV, POD and vortex identification algorithms for the study of unsteady turbulent swirling flows. Meas. Sci. Technol. **12**(9), 1422 (2001)
15. F. Zigunov, 2D Vortex Core Tracking - Gamma 1 - Super Fast (2023)

Characteristics of the Turbulent/Turbulent Interface Geometry in a Cylinder Wake

J. Chen and O. R. H. Buxton

Abstract The spatial evolution of the geometrical properties of the turbulent/turbulent interface (TTI) of a cylinder wake is examined in comparison to a turbulent/non-turbulent interface (TNTI). The cylinder is positioned at different downstream distances from a grid. Various grids are employed to generate background turbulence with varying turbulence intensities and integral length scales. A planar laser induced fluorescence experiment was carried out to capture the interface between the wake and the ambient flow, spanning from the near to the relatively far field, at a Reynods number of 3800 based on the inlet velocity and the cylinder diameter. A transition region of the interfaces' spreading towards the ambient flow is identified, beyond which the interfaces propagate at a slower rate than upstream and their mean positions scale with the local wake half-width. It is also found that the locations of both the TNTI and TTI have non-Gaussian probability density functions (PDFs) in the near wake because of the influence of the large-scale coherent motions. The length scale in the background turbulence significantly influences the TTI fractal dimension in the near wake, whilst the turbulence intensity only becomes important for the fractal dimension farther downstream.

1 Introduction

The spreading of turbulence into previously irrotational fluid depends on viscous diffusion of vorticity across a well-defined thin layer, usually referred to as a turbulent/non-turbulent interface (TNTI), which bounds the turbulent region and separates it from the outer, non-turbulent regions. The properties of the interface are directly related to the turbulent entrainment and mixing and have been studied extensively (see the review of [1]). However, for most industrial and environmental

J. Chen · O. R. H. Buxton (✉)
Imperial College London, London, UK
e-mail: o.buxton@imperial.ac.uk

J. Chen
e-mail: jiangang.chen@imperial.ac.uk

R. Örlü et al. (eds.), *Progress in Turbulence X*, Springer Proceedings in Physics 404,
https://doi.org/10.1007/978-3-031-55924-2_18

flows in practice, the background flow is also turbulent, a typical example of which is the wake of a wind turbine developing in the atmospheric turbulent boundary layer or the turbulent wake of other upstream wind turbines [2]. Compared to TNTIs, our knowledge of turbulent/turbulent interfaces (TTIs) remains limited, notwithstanding its prevalence in the physical world.

In their recent studies of TTIs, Kankanwadi and Buxton [4, 5] observed that both the turbulence intensity and the integral length scale in the ambient flow correlate to enhanced entrainment in the presence of the large-scale coherent vortices in the near wake of a cylinder, whilst a contrasting result was observed in the far field in which background turbulence was observed to suppress the entrainment rate relative to a non-turbulent background, with the turbulence intensity of the background playing the dominant role. An immediate question raised by these observations is "how does the behavior of the TTI, produced in a cylinder wake spatially evolve from near to far field with the coherent motions diminishing?". This is of central interest to the present study.

2 Experimental Setup and Methodology

The experiment was carried out in the same water flume as in [4] (Fig. 1a). A cylinder with a diameter of $d = 0.01$ m was vertically mounted in the middle of the flume test section. The incoming velocity of the flow was $U_1 = 0.38$ m/s,

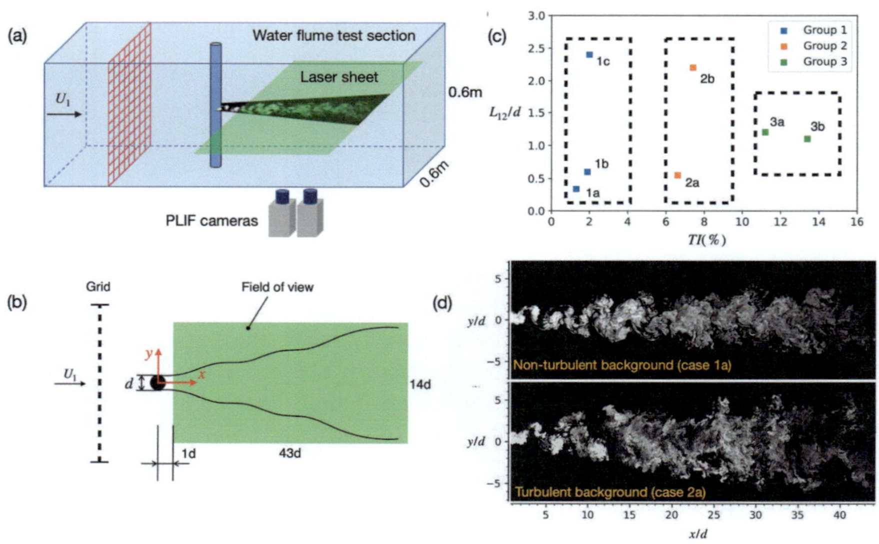

Fig. 1 **a**, **b** Conceptual sketch of the experimental setup and some denotations. **c** Parameter space (TI, L_{12}) of the background flow in the middle of the field of view at $x/d = 20$. **d** A comparison of the measured PLIF images of the wake without (the upper image) and with (the lower image) turbulence in the background

corresponding to a Reynolds number $Re(U_1, d)$ of about 3800. The turbulent wake was marked with a high-Schmidt-number fluorescent dye Rhodamine-6G and a planar laser induced fluorescence (PLIF) experiment was carried out to capture the interface between the wake and the ambient flow using two cameras (Phantom V641) with a field view of $14d \times 43d$ (Fig. 1b). 2000 images were captured for each measurement case and the measurement spatial resolution was 0.1 mm/pix. Upstream of the cylinder, four different grids (including two regular and two fractal grids, see [4] for details of the grids), were used to generate the background turbulence. The cylinder was placed at distances downstream of various grids, such that the parameter space (turbulence intensity TI, integral length scale L_{12}) was explored as widely as possible. We explored seven cases of (TI, L_{12}), and divided them into three groups according to their turbulence intensity. The distribution of (TI, L_{12}) without the cylinder at $x/d = 20$, i.e. the middle of the field of view, is displayed in Fig. 1c. The variation of TI and L_{12} with respect to their values at $x/d = 20$ within the field of view is about 2% and $0.1d$ for the cases of regular grids (cases 2a, 3a, 3b in Fig. 1c); for the cases of fractal grids (cases 1b, 1c and 2b), the variations are about 1% for TI and $0.5d$ for L_{12}, which does not affect their relative positions in figure 1c. A TNTI case (case 1a) with no grid upstream, is also included for comparison against the result of TTIs. Two typical measured PLIF images with non-turbulent and turbulent backgrounds are shown in Fig. 1d.

To identify the interfaces of the wake from the PLIF images, we employed the same methodology as [7] whose experimental configuration is similar to our present study. The detailed process of determining the interface can be found in [6]. A typical example of the detected interface (case 1a) on both sides of the wake is given in Fig. 2. All the TTI cases were determined using the same method with case-dependent threshold values.

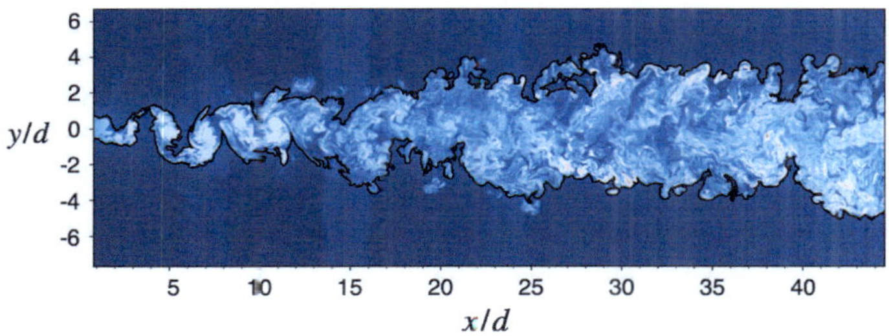

Fig. 2 Interface lines determined by selecting the longest continuous contours on both sides of the wake (case 1a)

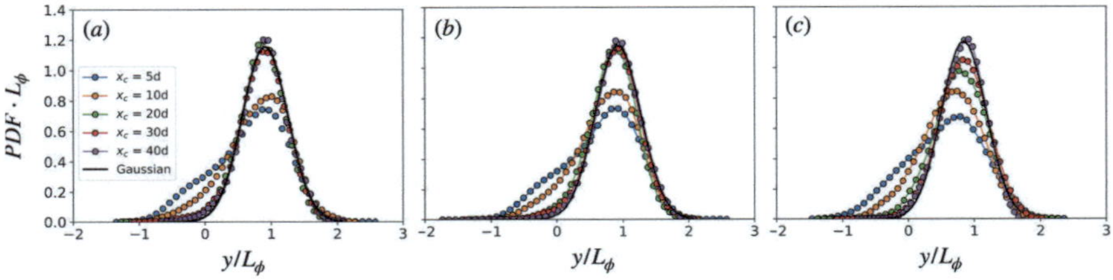

Fig. 3 Streamwise development of the PDFs of both TNTI and TTI position. **a** TNTI case 1a, **b** TTI case 2a, **c** TTI case 3a

3 PDFs of TNTI and TTI Positions

With detected interfaces from the near to the far field, and examined the downstream evolution of the PDFs of both TNTI and TTI. Figure 3 shows the PDFs at five streamwise locations (x/d = 5, 10, 20, 30, and 40) for both the TNTI (case 1a, Fig. 3a) and two representative TTI cases with higher background turbulence intensities (case 2a and case 3a respectively in Fig. 3b, c). The PDFs for both TNTIs and TTIs noticeably diverge from a Gaussian distribution at positions closer to the cylinder. As x/d increases to $x/d = 40$, the TNTI PDF gradually approaches Gaussianity, while the TTI PDFs always show discernible departure from the Gaussian distribution (Fig. 3b, c). It is intriguing to observe that the most likely locations of both TTIs and the TNTI scale approximately with the local wake half-width L_ϕ which is estimated from the mean profile of the light intensity $\overline{\phi}(x, y)$ of the PLIF images at the corresponding x position.

Figure 4 provides additional confirmation that the average position of the interface $\overline{y}$ for both the TNTI case and all the TTI cases exhibit scaling with the local wake-half width L_ϕ, after a critical position at $x/d \approx 15$. This critical position is corresponding to an apparent turning point of the slope of $\overline{y}$. After this position, $\overline{y}$ grows noticeably more slowly than farther upstream, for both TNTI and TTI cases (Fig. 4a). This implies that the entrainment rate upstream of $x/d \approx 15$ is faster than after this position. It is conjectured that the transition position might be contingent upon the dynamics of the near-wake coherent vortices which have been reported to be important for near-wake large-scale engulfment [5] and decay significantly from $x/d = 10$ to 20 (Fig. 1d).

4 Fractal Dimensions of the TNTI and TTI

Apart from the PDFs of the interface position, the geometrical properties of the interface can also be characterized by its ‘roughness’, which is reflected by its fractal dimension. The length of a fractal ‘line’ L_I which follows a power law with increased resolution scale r, is given by $L_I(r) \sim r^{1-D}$ where D is the fractal dimension. To

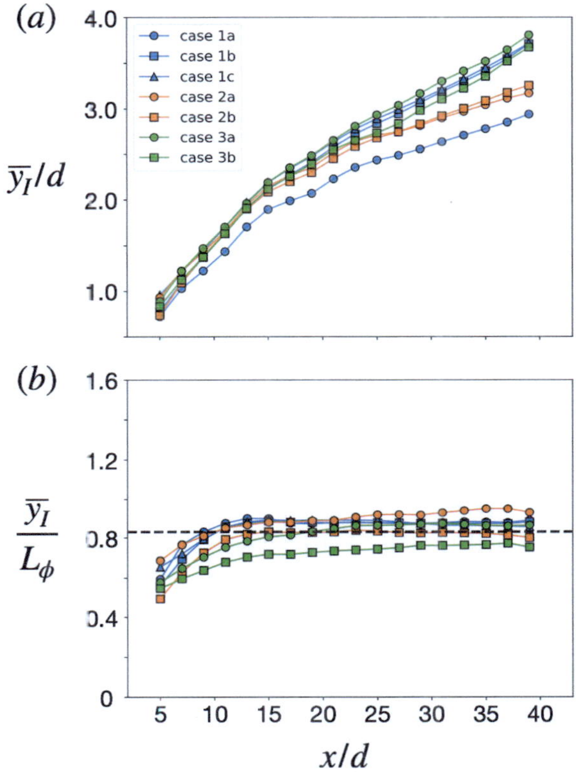

Fig. 4 Streamwise distribution of the mean interface position **a** $\overline{y_I}/d$ and **b** $\overline{y_I}/L_\phi$ of all cases

assess the streamwise evolution of the fractal dimension of the interfaces, we utilized the 'filtering method' in [6]. Figure 5a shows the scaling of the mean length of the filtered interface of the TNTI case $L_i(\Delta_f)$ with respect to the scale of the filter Δ_f at various streamwise positions. In the figure, x_c/d located between 4 to 40 is the centre position of the examination window with span of $8d$. A clearly defined scaling range exists between $\Delta_f/d = 0.2$ and 1 for all the examined positions. Applying this analysis to all the measured seven cases, it is found that the background turbulence intensity (Fig. 5b) and integral length scale (Fig. 5c) have an effective influence on the fractal dimension of the interface. In various flow regions, different parameters take on dominance: in the near wake, the integral length scale is the more crucial parameter; however, as the flow progresses downstream and the coherent vortices degrade significantly, the impact of the integral length scale diminishes, giving way to the gradual predominance of turbulence intensity.

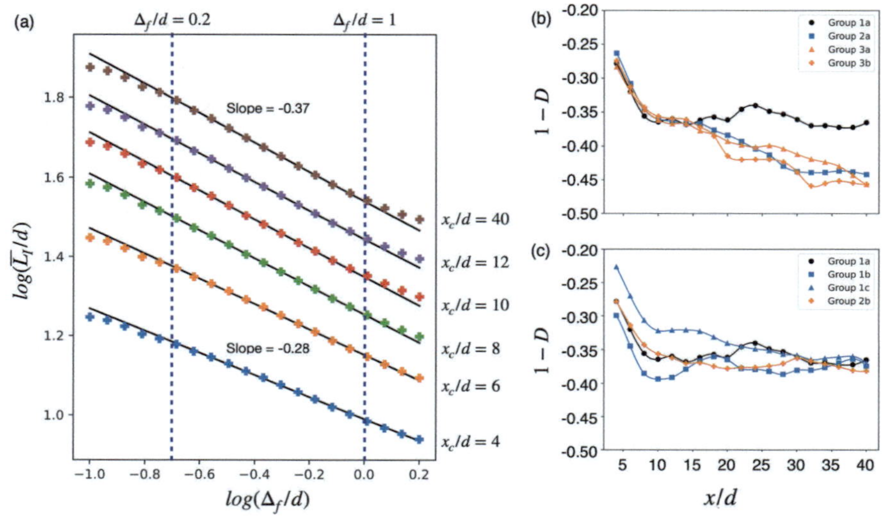

Fig. 5 **a** Scaling of the interface length of the TNTI at different streamwise x/d positions. **b** Effect of turbulence intensity, and **c** effect of integral length scale on the streamwise distribution of fractal dimensions of the TNTI and all TTI cases

5 Conclusion

We examined the spatial evolution of the geometry of TTIs in a turbulent wake. A transition region is noted in the mean interface spreading outwards at $x/d \approx 15$, after which the interfaces exhibit a noticeably diminished spreading rate. It is conjectured that the different spreading rates before and after this transition region are associated with the dynamics of the large-scale coherent vortices. After this region, the mean position of the interfaces, including both TNTI and all TTI cases, display a reasonable scaling with the wake half-width. It is also observed that the positions of both the TNTI and TTI have non-Gaussian PDFs in the near wake, displaying evident negative skewness attributed to the impact of the large-scale coherent motions. The turbulence intensity leads to a higher fractal dimension of the interface in the far wake, while the influence of the integral length scale is more pronounced in the near-wake region.

Acknowledgements The authors would like to acknowledge the Engineering and Physical Sciences Research Council for funding the work under grant no. EP/V006436/1.

References

1. C.B. da Silva, J.C. Hunt, I. Eames, J. Westerweel, Interfacial layers between regions of different turbulence intensity. Annu. Rev. Fluid Mech. **46**, 567–90 (2014)
2. F. Porté-Agel, M. Bastankhah, S. Shamsoddin, Wind-turbine and wind-farm flows: a review. Bound.-Layer Meteorol. **174**, 1–59 (2020)
3. K.F. Kohan, S.J. Gaskin, On the scalar turbulent/turbulent interface of axisymmetric jets. J. Fluid Mech. **950**, A32 (2022)
4. K.S. Kankanwadi, O.R.H. Buxton, Turbulent entrainment into a cylinder wake from a turbulent background. J. Fluid Mech. **905**, A35 (2020)
5. K.S. Kankanwadi, O.R.H. Buxton, Influence of freestream turbulence on the near-field growth of a turbulent cylinder wake: turbulent entrainment and wake meandering. Phys. Rev. Fluids **8**(3), 034603 (2023)
6. J.G. Chen, O.R.H. Buxton, Spatial evolution of the turbulent/turbulent interface geometry in a cylinder wake. J. Fluid Mech. **969**, A4 (2023)
7. R.R. Prasad, K.R. Sreenivasan, Scalar interfaces in digital images of turbulent flows. Exp. Fluids **7**(4), 259–264 (1989)

Origins of Dual Hairpin-Horseshoe Vortex Arrangement in the Wake of Oscillating Foils

S. Verma, K. Seitz, and A. Hemmati

Abstract Three dimensional wake evolution of an oscillating foil in coupled heaving and pitching motion is evaluated numerically at range of chord-based Strouhal number ($0.32 \leq St_c \leq 0.56$), while the phase-offset ($\phi°$) is fixed at $\phi = 90°$. The increase in St_c reveals growth of a supplementary horseshoe-like system in the wake. These large scale structures evolve along with the secondary hairpin arrangement that has been already identified in wakes. The horseshoe-like system forms on account of bending strains of the primary leading edge vortex (LEV) in the presence of a developing trailing edge vortex (TEV). The bending is further accelerated at higher St_c, which coincides with increasing circulation strength of developing TEVs. This promoted emergence of the horseshoe-like system early on in the wake.

1 Introduction

Vortex instabilities can significantly impact the functioning of bio-inspired robotic swimmers and micro vehicles [1]. For oscillating foils, particularly, the spanwise instability of LEV fosters the development of additional streamwise vortex structures. This eventually leads to a three-dimensional wake [6, 11]. Understanding the mechanisms that elucidate the spatio-temporal dynamics of these instabilities is important, given their direct influence on the forces and rolling moments generated on lifting surfaces, such as wings [5].

The impact of varying kinematics on the two-dimensional wake and the propulsive efficiency of oscillating foils are substantial [3, 7]. Verma and Hemmati [11] offered an intricate account of the evolution characteristics of secondary instabilities and

S. Verma · K. Seitz · A. Hemmati (✉)
116 St & 85 Ave, Edmonton, AB T6G 2R3, Canada
e-mail: arman.hemmati@ualberta.ca

S. Verma
e-mail: suyash@ualberta.ca

K. Seitz
e-mail: kdseitz@ualberta.ca

R. Örlü et al. (eds.), *Progress in Turbulence X*, Springer Proceedings in Physics 404,
https://doi.org/10.1007/978-3-031-55924-2_19

intermediate structures at ϕ of 90° and St_c of 0.32. However, the effects of diverse kinematic configurations, specifically St_c, are not thoroughly explored. This study extends the discussion of Verma and Hemmati [11] by examining the influence of increasing St_c on the growth of secondary structures.

Verma and Hemmati [6] quantitatively established a potential connection between spanwise undulations of LEVs and the elliptic instability of vortex pairs in foils exhibiting coupled pitching and heaving motions. These connections were in correspondence with low and high reduced frequencies. The former showed subtle corrugations on shed vortex structures, while the latter displayed more prominent hairpin-horseshoe vortex formations in the wake. Visbal [2] have further highlighted the significant effects of varying reduced frequencies on LEV dynamics. In a recent study, Son et al. [10] examined LEV instabilities in heaving oscillations for both high aspect ratio wings and infinite span foils. The investigation revealed alterations in the strength of primary LEV and TEV, which coincides with changes in the onset mechanisms of instability, with increasing St_c. This study advances the wake characteristics of secondary vortex structures and its association with St_c.

2 Problem Description

The flow around an infinitely span (2D) foil is numerically examined using DNS at a range of chord-based ($St_c = fc/U_\infty = 0.32 - 0.56$) and amplitude-based Strouhal numbers ($0.05 \leq St_A \leq 0.4$). The cross section of the foil resembles a teardrop foil, which was used in recent experimental and computational investigations [3, 6, 8, 9, 11]. The Reynolds number is $Re = U_\infty c/\nu = 8000$. The kinematics of the foil is prescribed by a coupled heaving and pitching motion. The motion profiles of heave (h) and pitch (θ) are represented as $h(t) = h_o \sin(2\pi f t)$ and $\theta(t) = \theta_o \sin(2\pi f t + \phi)$, respectively. Here, pitching has a phase advancement (or offset) of ϕ relative to heaving. We utilized the Overset Grid Assembly (OGA) method in OpenFOAM to perform combined pitching and heaving motions. OGA is based on a finite volume solver, which has been well validated for studying wake and vortex dynamics [4]. Second-order accurate temporal (backward Euler) and spatial (central-difference) schemes were employed to discretize the governing equations.

The computational domain included a stationary background grid and an overlapping grid that modeled the foil motion. The background grid extended 32c, 25c and πc in the streamwise, cross-stream and spanwise directions, respectively. A uniform velocity with zero pressure gradient was assigned at the inlet boundary, while Neumann outflow condition was prescribed at the outlet boundary. The domain sides were assigned a periodic boundary condition. All remaining edges were assigned a zero gradient condition for both velocity and pressure, whereas a no-slip wall condition was ensured for the moving foil. More details on the validation for the numerical solver can be found in Verma and Hemmati [6].

3 Results and Discussion

Verma and Hemmati [11] presented insights on the evolution of secondary hairpin-like structures on account of deforming secondary LEV at $St_c = 0.32$ and $\phi = 90°$. These streamwise coherent structures transform to paired-rib configuration on account of continuous straining induced by the primary LEV. Verma et al. [12] extended the analysis to increasing St_c and ϕ recently, establishing the association of kinematics and growth mechanisms of secondary hairpin-like structures. Here, we identify and briefly discuss the presence of a supplementary horseshoe-like system that coexists with the hairpin-like arrangement identified previously [11, 12].

Figure 1 provides the temporal evolution of the wake at $\phi = 90°$ and $St_c = 0.40$, within an oscillation cycle. The vortical structures are identified using $Q^+ = Qc^2/U_\infty^2 = 0.4$, where Q represents the second invariant of velocity gradient tensor, c is the chord length of foil and U_∞ is the freestream velocity. At $t^+ = 0.5$ in Fig. 1a, growth of hairpin structures is evident at $X^+ \approx X/c = 1$. These hairpins are attributed to the core vorticity outflux from secondary LEV [12]. These structures further outgrow to form paired rib configuration on account of continued straining [6]. The presence of horseshoe-type structures is also noticed downstream of $X^+ \approx 3$, which are visually larger in size than secondary hairpin-like structures observed at

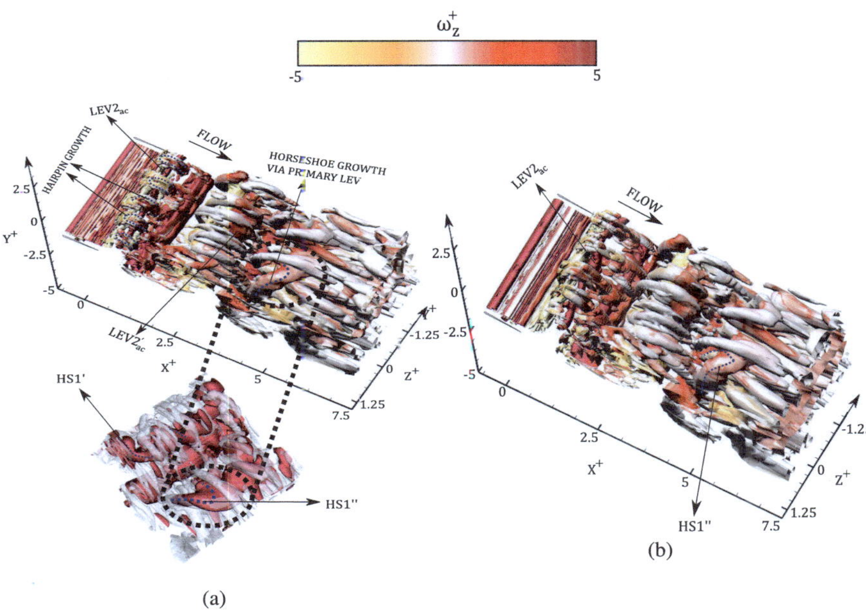

Fig. 1 Wake evolution at $\phi = 90°$ and $St_c = 0.40$. The oscillation time corresponds to **a** $t^+ = 0.5$ and **b** $t^+ = 0.75$. The vortical structures are represented using iso-surfaces of $Q^*(= 0.05)$, which are colored based on $|\omega_z^+| = |\omega_z c/U_\infty| = 5$

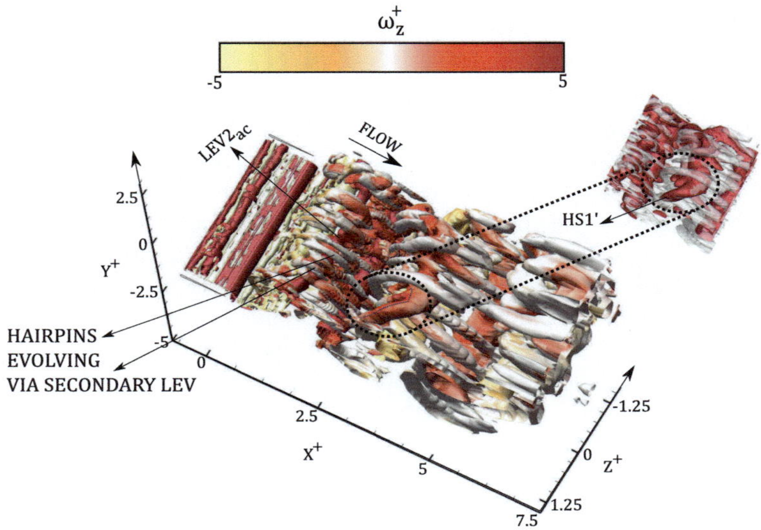

Fig. 2 Wake evolution at $\phi = 90°$ and $St_c = 0.40$. The oscillation time corresponds to $t^+ = 1$. The vortical structures are represented using iso-surfaces of $Q^*(= 0.05)$, which are colored based on $|\omega_z^+ = \omega_z c/U_\infty| = 5$

$X^+ = 1$. These are also identified as $HS1'$ and $HS1''$ in Fig. 1a. Temporal wake evolution reveals that the primary LEV undergoes bending as it advects downstream, and consequently forms $HS1'$ and $HS1''$. Figures 1b and 2 depicts the evolution stages of $HS1'$ and $HS1''$ at $t^+ = 0.75$ and 1, respectively, where these structures co-exist with the first system of secondary hairpin-like vortices.

As St_c increase beyond 0.40, an accelerated bending of primary LEV further lead to an early emergence of horseshoe-like structures. This is qualitatively confirmed in Fig. 3a–d, presenting near wake ($X^+ < 4$) evolution at a range of St_c from 0.32 to 0.56, and $\phi = 90°$. The time instant corresponds to $t^+ = 0.75$. At $St_c = 0.32$, $LEV1_{ac}$ and $TEV1_c$ are marked along with secondary hairpin-like structure $HP1$ in Fig. 3a. $LEV1_c$ does not undergo prominent bending in spanwise direction and hence fails to reveal any horseshoe-like system in the wake. At $St_c = 0.40$, bending of the primary LEV and the resulting horseshoe-like formation became evident at $X^+ \geq 2.5$ (see Fig. 2). At $St_c > 0.40$, however, bending and the resulting horseshoe-like formation of primary LEV coincide within a streamwise distance of $X^+ \approx$ 1.875–2.5.

In order to understand the accelerated bending of the primary LEV and its consequent transformation to the supplementary horseshoe-like system, the variation of circulation ($\Gamma^+ = \Gamma/U_\infty c$) magnitudes for LEV and TEV are now assessed. These can qualitatively reveal if the kinematics at higher St_c coincide with an increased strain field near the foil trailing edge. Verma et al. [12] described the coincident growth of secondary hairpin-like structures and strengthening of the LEV at increasing St_c towards 0.56. Here, Γ^+_{TEV} at a given St_c is estimated to highlight the contribution of TEVs in promoting enhanced straining and bending of LEVs. Figure

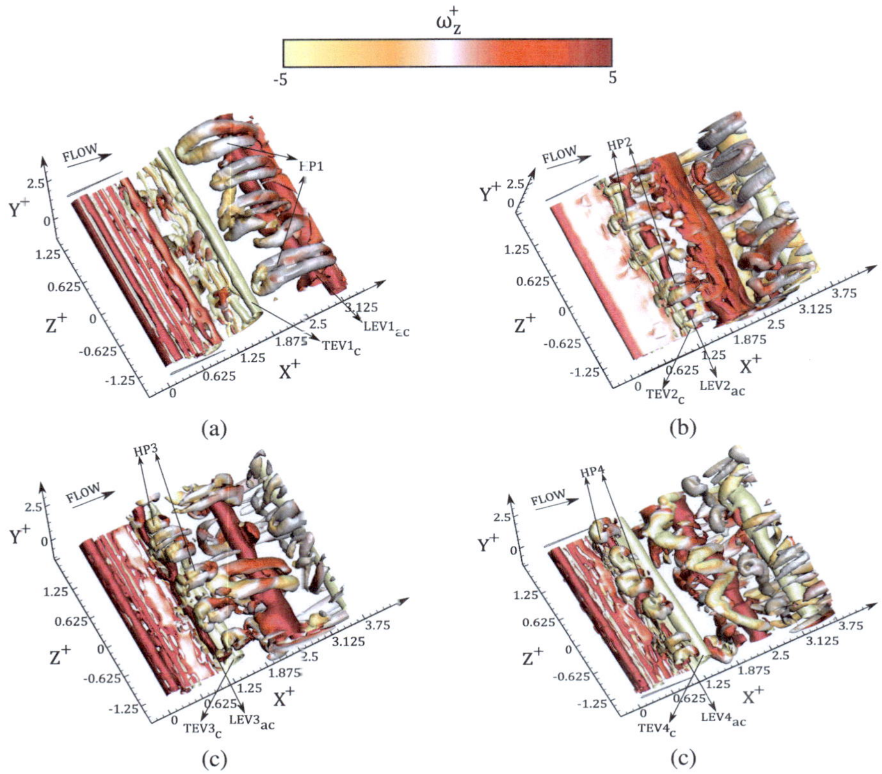

Fig. 3 Wake evolution at $\phi = 90°$, and **a** $St_c = 0.32$, **b** $St_c = 0.40$, **c** $St_c = 0.48$, **d** $St_c = 0.56$. The oscillation time corresponds to $t^+ = 0.75$. The vortical structures are represented using iso-surfaces of $Q^*(= 0.05)$, which are colored based on $|\omega_z^+| = |\omega_z c/U_\infty| = 5$

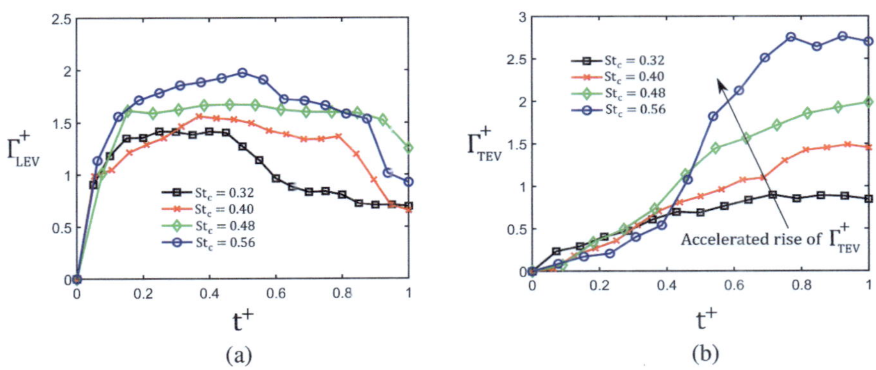

Fig. 4 Variation of Γ^+ for the **a** LEV and **b** TEV structures identified in Fig. 3a–d, at $\phi = 90°$ and $0.32 \leq St_c \leq 0.56$

4a demonstrates the increase in peak Γ^+_{LEV} with increasing St_c. The trends of Γ^+_{TEV} also depicts an accelerated increase. This suggests that the relative strengthening of TEVs compared to LEVs, ahead of $t^+ = 0.8$, promote a stronger straining on the primary LEV. This coincides with an early emergence of the horseshoe-like system in the wake.

4 Conclusions

Dual hairpin-horseshoe like system is identified in the wake of an oscillating foil in a coupled heaving and pitching motion. At $\phi = 90°$, increasing St_c coincides with an accelerated bending of the primary LEV, while it sheds and advects downstream from the trailing edge. Consequently, early transformation of the primary LEV to multiple horseshoe-like spanwise pairs coincides with increasing St_c. The decreasing circulation strength of LEVs relative to TEVs suggests that a stronger TEV contributes to the enhanced bending strain on the primary LEV. This, therefore, contributes to an early emergence of the horseshoe-like system in the wake.

References

1. J.M. Ortega, R.L. Bristol, O. Savas, Experimental study of the instability of unequal-strength counter-rotating vortex pairs. J. Fluid Mech. **474**(474), 35–84 (2003)
2. M.R. Visbal, High-fidelity simulation of transitional flows past a plunging airfoil. AIAA J. (2009)
3. T. Van Buren, D. Floryan, A.J. Smits, Scaling and performance of simultaneously heaving and pitching foils. AIAA **57**(9), 3666–3677 (2018)
4. S. Verma, A. Hemmati, Performance of overset mesh in modeling the wake of sharp-edge bodies. Computation **8**(3), 66 (2020)
5. N. Chiereghin, S. Bull, D.J. Cleaver, I. Gursul, Three-dimensionality of leading-edge vortices on high aspect ratio plunging wings. Phys. Rev. Fluids **5**(6), 064701 (2020)
6. S. Verma, A. Hemmati, Evolution of wake structures behind oscillating hydrofoils with combined heaving and pitching motion. J. Fluid Mech. **927**, A23 (2021)
7. S. Verma, A. Hemmati, Characterization of bifurcated dual vortex streets in the wake of an oscillating foil. J. Fluid Mech. **945**, A7 (2022)
8. S. Verma, B.R.S. Freeman, A. Hemmati, Effects of Reynolds number and average angle of attack on the laminar scaling of oscillating foils. Phys. Fluids **34**(3), 031905 (2022)
9. S. Verma, A. Hemmati, Route to transition in propulsive performance of oscillating foil. Phys. Rev. E **105**(4), 045102 (2022)
10. O. Son, A.K. Gao, I. Gursul, C.D. Cantwell, Z. Wang, S.J. Sherwin, Leading-edge vortex dynamics on plunging airfoils and wings. J. Fluid Mech. **940**, A28 (2022)
11. S. Verma, A. Hemmati, Influence of kinematics on the growth of secondary wake structures behind oscillating foils. Int. J. Heat Fluid Flow **102**, 109146 (2023)
12. S. Verma, M.S.U. Khalid, A. Hemmati, On the association of kinematics, spanwise instability and growth of secondary vortex structures in the wake of oscillating foils. Proc. R. Soc. A **479**(2276) (2023)

Transfer Mechanism of a Passive Scalar in Grid Turbulence with Mean Scalar Gradient

M. Wang, T. Yurikusa, K. Iwano, Y. Sakai, and Y. Ito

Abstract We perform DNSs of grid turbulence with mean scalar gradient and conduct scale-by-scale analysis to reveal the interscale transfer of passive scalar. The scale-by-scale scalar (SBSS) equation, which can be deemed as an extended equation of the Karman–Howarth–Monin–Hill (KHMH) equation, is derived and employed in the DNS data to evaluate the interscale transfer between two points. The results reveal that the inverse cascade phenomenon of the scalar appears in the upstream regions. Further analysis reveals that the existence of fluid mass with an unmixed scalar plays an essential role in the inverse cascade phenomenon.

1 Introduction

Scalar diffusion and mixing in turbulence are seen in a wide variety of industrial and environmental flows. Therefore, a fundamental understanding of scalar transfer in turbulence is of great importance. In particular, when a scalar is introduced to the flow field under a non-premixed condition, its transport becomes a multi-scale

M. Wang · T. Yurikusa · Y. Ito (✉)
Graduate School of Engineering, Nagoya University, Nagoya, Japan
e-mail: yito@nagoya-u.jp

M. Wang
e-mail: cybrexrem@gmail.com

T. Yurikusa
e-mail: yurikusa-takuya@khi.co.jp

K. Iwano
Department of Mechanical Systems Engineering, Okayama University of Science, Okayama, Japan
e-mail: iwano@nagoya-u.jp

Y. Sakai
Global Engagement Center, Nagoya University, Nagoya, Japan
e-mail: ysakai@mech.nagoya-u.ac.jp

R. Örlü et al. (eds.), *Progress in Turbulence X*, Springer Proceedings in Physics 404,
https://doi.org/10.1007/978-3-031-55924-2_20

process dominated by the velocity field, which necessitates the multi-scale analysis of the scalar field.

In this study, we perform DNSs of grid turbulence with mean scalar gradient and conduct scale-by-scale analysis to reveal the interscale transfer of passive scalar. We derive and analyze the transport equation of the scalar structure function without using any assumptions. The primary goal of this study is to refine the understanding of interscale scalar transfer in given flows by revealing the associated structures and placing particular emphasis on the inverse cascade phenomenon.

2 Numerical Details

Figure 1a shows a schematic view of the computational domain as a rectangular box with dimensions $L_x \times L_y \times L_z = 32M \times 6M \times 6M$, where M is the mesh size of the grid as defined in Fig. 1b. Here, x, y, and z represent streamwise, vertical, and spanwise directions, respectively. A uniform flow of $U_0 = 1$ was supplied from the inlet. The initial concentration of a passive scalar was set to $C = C_0 = 1$ for the upper-half layer of the stream and $C = 0$ for the lower-half. The Reynolds number based on M and U_0 was set to $Re_M = 5000$. Figure 2 shows a color contour map of the instantaneous scalar for $Re_M = 5000$. The scalar distribution in the y direction is confirmed to be stepwise at the inlet and initially sharp.

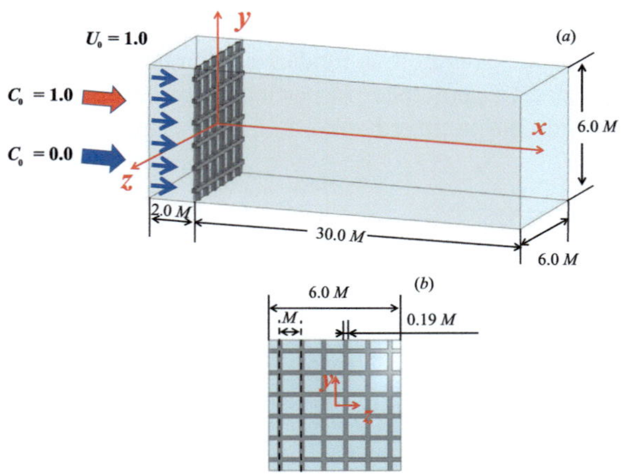

Fig. 1 Schematic view of **a** the computational domain and **b** the turbulence-generating grid

3 Scale-by-Scale Analysis

3.1 The Scale-by-Scale Scalar (SBSS) Equation

The scale-by-scale scalar (SBSS) equation was derived [1–4] following the routine of the generalized KHMH equation, as shown in (1).

$$\underbrace{\frac{\partial\,[\delta c]^2}{\partial t}}_{4A_{t,c}}+\underbrace{\left(\frac{\langle U_i\rangle+\langle U_i'\rangle}{2}\right)\frac{\partial\,[\delta c]^2}{\partial X_i}}_{4A_c}+\underbrace{\frac{\partial \delta u_i\,[\delta c]^2}{\partial r_i}}_{4\Pi_c}+\underbrace{\frac{\partial\langle\delta U_i\rangle\,[\delta c]^2}{\partial r_i}}_{4\Pi_{U,c}}$$
$$=\underbrace{-2\delta c\delta u_i\frac{\partial\delta\langle C\rangle}{\partial r_i}-\left(u_i+u_i'\right)\delta c\frac{\partial\delta\langle C\rangle}{\partial X_i}}_{4P_c}\underbrace{-\frac{\partial}{\partial X_i}\left(\frac{\left(u_i+u_i'\right)[\delta c]^2}{2}\right)}_{4T_c} \tag{1}$$
$$+\kappa[\underbrace{2\frac{\partial^2}{\partial r_i^2}}_{4D_\kappa}+\underbrace{\frac{1}{2}\frac{\partial^2}{\partial X_i^2}}_{4D_{X,\kappa}}]\,[\delta c]^2\underbrace{-2\kappa\left[\left(\frac{\partial c}{\partial x_i}\right)^2+\left(\frac{\partial c'}{\partial x_i'}\right)^2\right]}_{4\chi}$$

The superscript $'$ indicates the values at x', the symbol $\langle\cdots\rangle$ indicates time average. Further, $\delta\langle U_i\rangle=\langle U_i\rangle-\langle U_i'\rangle$, $\delta\langle C\rangle=\langle C\rangle-\langle C'\rangle$, $\delta u_i=u_i-u_i'$, $\delta c=c-c'$, $X_i=(x_i+x_i')/2$, and $r_i=x_i-x_i'$. For clarity, $\langle U_i\rangle$ is the k-th component of the mean velocity vector; $\langle C\rangle$ is the mean scalar; u_i is the k-th component of the fluctuation velocity vector, and c is the scalar fluctuation. Additionally, A_c in (1) is the advection of $[\delta c]^2$ by the mean flow; Π_c is the nonlinear transfer term; $\Pi_{U,c}$ is the linear transfer term; P_c is the turbulent production term; T_c is the turbulent transport term; D_κ is the molecular diffusion term in the r space; $D_{X,\kappa}$ is the molecular diffusion term along X; χ is the scalar fluctuation dissipation, and $A_{t,c}$ is the time dependence term. Owing to statistical steadiness, $A_{t,c}$ becomes 0.

Figure 3 reveals the results for SBSS where the superscript a indicates circumferentially averaged. The molecular diffusion $\langle D_\kappa\rangle^a$ is dominant and balances with the dissipation $\langle\chi\rangle^a$ near $r/\lambda=0$, regardless of the streamwise position and Reynolds number. This is because the Schmidt number in the present study is 1, and scalar transport in this region is dominated by viscous dissipation, as in the case of turbulent energy. Due to the existence of the mean scalar gradient, the production term

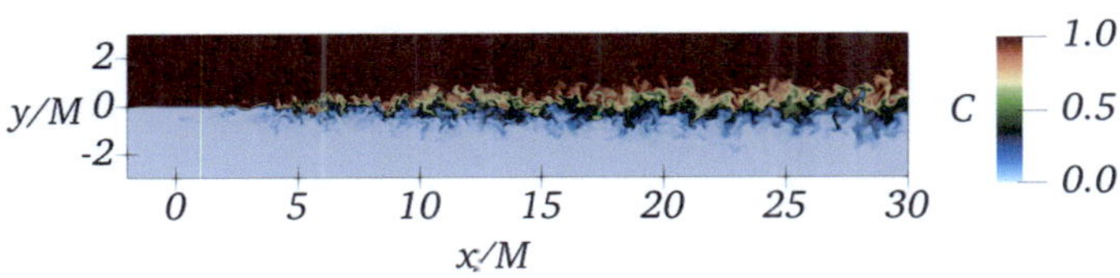

Fig. 2 Contour map of the instantaneous scalar concentration for $Re_M=5000$

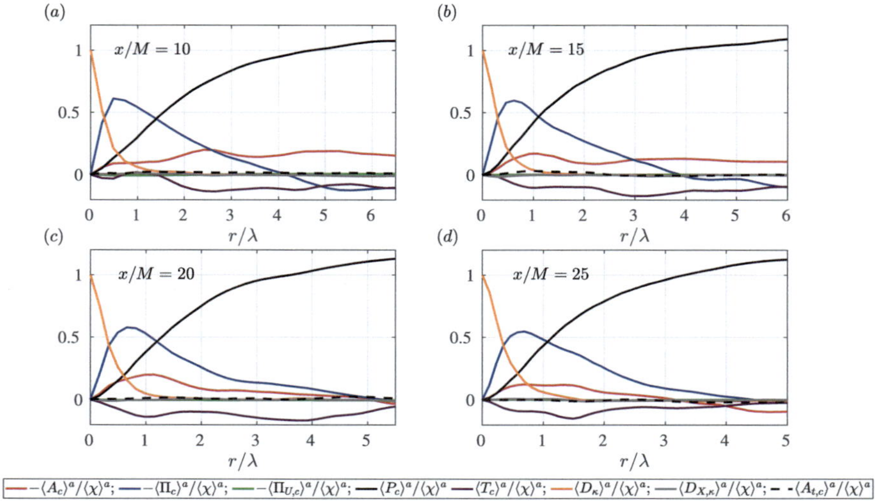

Fig. 3 Circumferentially-averaged terms in the SBSS equation normalized by the averaged dissipation rate $\langle \chi \rangle^a$ at **a** $x/M = 10$, **b** $x/M = 15$, **c** $x/M = 20$, and **d** $x/M = 25$

$\langle P_c \rangle^a$ becomes large with increasing r. The turbulent transport term $\langle T_c \rangle^a$ exhibits negative values and roughly balances with the advection term $\langle A_c \rangle^a$. Considering the similarity of the governing equation with the velocity field, this balance is interpreted to demonstrate that a small fluid mass, with small scalar fluctuation, is transported by a large eddy due to sweeping motion, as reported in the previous studies of box turbulence by DNS that the advection tends to cancel the turbulent transport term by significant organization from to large scales motion to the small scales motion, implying a connection between large and small scales in turbulence. Meanwhile, the non-linear transfer term $\langle \Pi_c \rangle^a$ reaches a peak at around $r/\lambda = 1$ and decreases with increasing r in all cases. In other words, the inverse cascade is caused by non-linear scalar transport.

3.2 Decomposition of the Nonlinear Transfer Term

To reveal the key factor causing the inverse cascade for the scalar, Π_c is decomposed into three directions at $x/M = 10$:

$$-\Pi_c = -\frac{\partial}{\partial r_i}\delta u_i\,[\delta c]^2\,, = -\underbrace{\frac{\partial}{\partial r_x}\delta u\,[\delta c]^2}_{\Pi_{c,x}} - \underbrace{\frac{\partial}{\partial r_y}\delta v\,[\delta c]^2}_{\Pi_{c,y}} - \underbrace{\frac{\partial}{\partial r_z}\delta w\,[\delta c]^2}_{\Pi_{c,z}}. \tag{2}$$

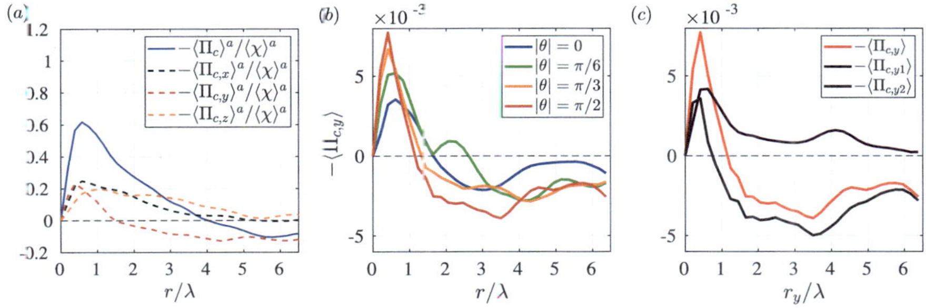

Fig. 4 **a** Decomposition results of the nonlinear transfer term for three directions, **b** the angle dependence of the nonlinear transfer term, and **c** the further decomposition in y direction

$$-\Pi_{c,y} = \underbrace{-\frac{1}{2}[\delta c]^2\left(\frac{\partial v'}{\partial y'} + \frac{\partial v}{\partial y}\right)}_{\Pi_{c,y1}} \underbrace{-\,\delta v\delta c\left(\frac{\partial c'}{\partial y'} + \frac{\partial c}{\partial y}\right)}_{\Pi_{c,y2}}, \quad (3)$$

The decomposition of the nonlinear transfer term Π_c (Fig. 4a) reveals that the y component is dominant in the nonlinear transfer, and the results are most significant in y where the angle $\theta = \pi/2$ as Fig. 4b. It is also clarified that this phenomenon occurs in the direction of the mean scalar gradient(the same as the vertical y–axis). Further decomposition in this direction as shown in Fig. 4c exhibits that the component $\Pi_{c,y2} = \delta v\delta c\left(\partial c/\partial y + \partial c'/\partial y'\right)$ plays a dominant role in this process. Here, $\Pi_{c,y2}$ consists of δv, δc, and $\left(\partial c'/\partial y' + \partial c/\partial y\right)$, and all terms can take both positive and negative values. Therefore, to identify the couplings of the terms causing the negative flux of the non-linear term, the probability of each coupling was calculated.

Figure 5a shows the probability of couplings between terms consisting of $\Pi_{c,y2}$, $P(\omega_i)$, where ω_i is the event of each coupling. The first, second, and third signs of each coupling at the axis of abscissa in the figure are the signs for δv, δc, and $\left(\partial c'/\partial y' + \partial c/\partial y\right)$, respectively. It is revealed that when the separation distance is large($r_y/\lambda = 6$), where the interscale scalar transfer is inverse (small to large scales), the direction of the scalar flux is determined by the vertical velocity fluctuation difference between two points $\left(\partial c/\partial y + \partial c'/\partial y'\right)$ since the mutual sign of the events of coupling $(+ - -)$ and $(- + -)$ is the third one. The conditional average results based on the event of coupling $(+ - -)$ (the most significant event of coupling for inverse cascade at large scale) reveal that a negative scalar gradient always appears with the occurrence of the inverse cascade as shown in Fig. 5b.

This can be explained as follows: in the off-central region of $\left|r_y/\lambda\right| > 2$, this fluid mass diffuses following the mean scalar gradient, and according to the statistical results shown previously in Fig. 5a, which caused a negative scalar gradient near the fluid mass as shown in Fig. 5b. This phenomenon occurs frequently and associated with the inverse cascade often occurs. It also implies that turbulent vertical diffusion is large in conditions where the inverse cascade is observed. In contrast, in the central region $\left|r_y/\lambda\right| < 2$, although a similar scalar distribution that scalar concentration

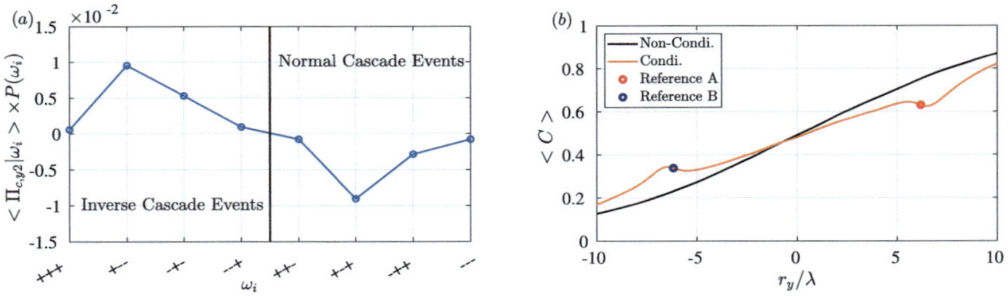

Fig. 5 **a** Conditional average of the component of nonlinear transfer term $\Pi_{c,y2}$ and **b** separation dependence of conditional average scalar concentration conditioned on the event $(+--)$ at $r_y/\lambda = 6$

locally decreases near the reference point is exhibited, the scalar mixing is more active, and scalar fluctuation is mainly dominated by intensive small-scale motions in the central region. Therefore, the sign of $\Pi_{c,y2}$ is determined by δv, and the inverse cascade phenomenon is not observed.

4 Conclusion

Inter-scale scalar transport analyses are conducted on grid turbulence with a mean scalar gradient. A two-point correlation analysis utilizing the SBSS equation is carried out. The findings demonstrate that the nonlinear transfer term, representing the velocity fluctuation difference in the radial direction and the scalar fluctuation difference at two points, exhibits negative values at larger scales (indicating an inverse cascade) in upstream locations. Further analysis reveals that this phenomenon is primarily attributed to the nonlinear scalar transfer term aligned with the mean scalar gradient (specifically, in the vertical (y) direction in this study). The gradient of the scalar fluctuations determines the direction of inter-scale scalar transfer at larger scales. Moreover, conditional averages results based on the component of $\Pi_{c,y2}$ suggests that the presence of unmixed scalar in fluid mass plays a crucial role in the occurrence of the inverse cascade phenomenon, with these conditions exhibiting a significant turbulent scalar flux.

Acknowledgements Computer simulations were mainly carried out by the supercomputers at Institute of Fluid Science, Tohoku University. Part of the work was carried out under the Collaborative Research Project of the Institute of Fluid Science, Tohoku University. Part of the study was supported by Kakenhi (No. 20K04264, No. 21K03851, No.22K03937).

References

1. A.M. Yaglom, On the local structure of a temperature field in a turbulent flow. Dokl akad nauk sssr **69**, 743–746 (1949)
2. R.J. Hill, Structure-function equations for scalars. Phys. Fluids **14**, 1745–1756 (2002)
3. J.C. Vassilicos, Dissipation in turbulent flows. Annu. Rev. Fluid Mech. **47**, 95–144 (2015)
4. M. Wang, T. Yurikusa, K. Iwano, Y. Sakai, Y. Ito, Y. Zhou, Y. Hattori, Scale-by-scale analysis of interscale scalar transfer in grid turbulence with mean scalar gradient. Phys. Fluids **35**, 045153 (2023)

Influence of Free-Stream Turbulence on Dynamic Strain Measurements of a Cantilevered Cylinder in a Turbulent Cross-Flow

F. J. G. de Oliveira, Z. S. Khodaei, and O. R. H. Buxton

Abstract The relationship between different "flavours" of free-stream turbulence and the developed fluctuating strain within/deflections of a circular cylinder is analysed. Regardless of the influence of the free-stream turbulence, bluff bodies at sufficiently high Reynolds number produce turbulent wakes in which multiple turbulent length/time scales exist concurrently and interact with the body's structure, imprinting these multi-scale physics onto the structural response. The present analysis aims to provide more insight on the effect of different free-stream characteristics on the experienced fluctuating loads acting on cantilevered structures, imposed by the conditions both upstream and downstream of the body. A set of experiments is carried out in a water channel, combining simultaneous particle image velocimetry (PIV) and strain measurements, using Rayleigh backscattering sensors (RBS), along four representative polar position sensing lines, to characterise the cylinder's response before and after the shear-layer separation. The cylinder is mounted as a cantilever beam, supported at one end with 95% of its body submerged and exposed to the influence of the flow at Re = 25000, whilst being subjected to the rising tip vortices developing at its free end. The presence of free-stream turbulence is shown to be more relevant to the standard deviation of the strain development along the upstream face of the wake-generating body with little change observed on the downstream face of the body when comparing turbulent/non-turbulent cross-flow.

F. J. G. de Oliveira (✉) · Z. S. Khodaei · O. R. H. Buxton
Imperial College London, London, UK
e-mail: f.oliveira22@imperial.ac.uk

Z. S. Khodaei
e-mail: z.sharif-khodaei@imperial.ac.uk

O. R. H. Buxton
e-mail: o.buxton@imperial.ac.uk

R. Örlü et al. (eds.), *Progress in Turbulence X*, Springer Proceedings in Physics 404,
https://doi.org/10.1007/978-3-031-55924-2_21

1 Introduction

The presence of free-stream turbulence has a direct and indirect impact on the fluctuating loads of bluff-bodies exposed to cross-flows; the latter by modifying the vortex-shedding process within the bluff body wake [1], whereas free-stream turbulence has also been identified as the main driver of drag variations in previous works [2]. In addition, wind turbines and other structural representations in engineering are often described as cantilevered structures, subjected to 3D flow events at the free-end which originate tip vortices, producing an evolving shedding process along their spanwise direction [3]. For the current analysis a cylinder with aspect ratio $L/D = 10$ is mounted as a cantilever with the fixing above the free-surface of the water. The submerged portion of the cylinder is exposed to cross-flow, allowing for the development of two spanwise cellular regions: one dominated by the alternating regular von and the other by tip vortices shed from the free end of the cylinder, inducing downwash into the wake [3].

Recent developments in structural health monitoring techniques using RBS [4] allow the use of fibre optics as a way to acquire fine details of the structural response of the cylinder to different flow conditions, with a fine spatial sensing distribution thanks to optical frequency-domain reflectometer (OFDR) technology, enabling the extraction of the imprinted flow effects on the structure. This chapter aims to summarise the ongoing work being conducted to better understand the effect of free-stream turbulence parameters such as integral length scale $\mathcal{L}$ and turbulence intensity TI on the developing strain over a cantilevered cylinder, by coupling PIV and strain measurements. This will provide correlations between developing flow phenomena to structural response regimes, with novel temporally and spatially correlated measurements of combined high-speed PIV distributed strain sensing over the surface of a wake-generating body.

2 Experimental Methodologies

The current set of experiments was conducted in the water flume, of the hydrodynamics laboratory of the Department of Aeronautics at Imperial College London. During the experiments, the flume had an operating cross-section of $56 \times 60\,\text{cm}^2$. An acrylic tube with diameter $D = 50$ mm, 2.5 mm wall thickness and 524 mm length, was vertically assembled in the mid section of the flume, with a 95% body submersion (wet length of ≈ 490 mm). A regular grid was placed upstream of the cylinder to generate background turbulence, to explore the effects of TI and $\mathcal{L}$ in the free-stream flow, generating two test cases characterised by $\{TI(\%), \mathcal{L}/D\}$, denoted by NoGrid with $\{2.2, 0.08\}$ and TurbGrid $\{8.5, 0.13\}$. The incoming free-stream velocity of the flow, without the presence of the turbulence generating grid is 0.49m/s. The Reynolds number based on this velocity and D is ≈ 25000. The cylinder was set as a cantilever and filled with water to neglect buoyancy effects, and the tip of the cylinder was left free to allow the development tip vortices and respective induced downwash [3].

Simultaneous measurements of strain using RBS and 2D-PIV were conducted, to further analyse how free-stream turbulence and captured flow events influence the evolution of flow-induced strain. The RBS sensors were embedded on the cylinder's surface to accurately represent the structural response of the cylinder. Figure 1a represents a schematic of the fibre optic sensors layout used, as well as the PIV fields of view (FOV), overlayed with the fibre sensors path. A Phantom v641 was used to obtain a FOV of $7D \times 5D$ (respectively in the stream-wise and spanwise direction) in two separate experiments capturing the upstream (FOV A1) and downstream (FOV A2) regions of the cylinder. The measurements of the acquired strain using the RBS sensors were coupled with PIV, both set at a frequency of acquisition of 50 Hz for a total time (T) of 54s ($\approx$ 100 vortex shedding cycles). The resulting images were processed, employing a multi-pass, fast Fourier transform to estimate the average particle displacement per interrogation window. The final interrogation area consisted of 16×16px, resulting in a spatial resolution of 2.4mm for both FOV A1 and A2. Figure 1b corresponds to a schematic of the cross-plane of the cylinder, where the fibre paths $[u_1, d_1, d_2, u_2]$ are defined, characterised by constant polar angles of respectively $\theta_f = [45, 135, 225, 315]°$, covering the spanwise length. In the Reynolds number regime for these experiments, flow separation has been seen to occur at $\theta = [101, 257]°$ [5]. The location of the sensors was set to capture the localised induced strain upstream and downstream of the shear-layer separation location to be able to analyse the direct effect of free-stream turbulence and wake generation on the body's structural response. The sensing network of RBS possesses

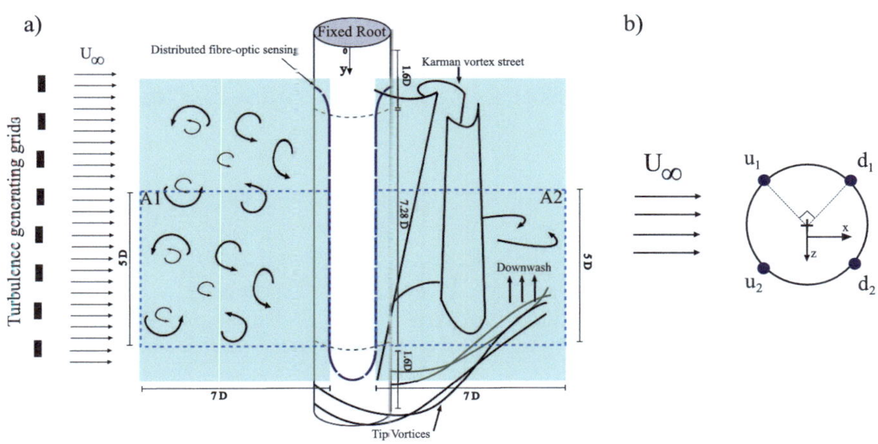

Fig. 1 Experimental layout schematic: **a** fibre optic path, PIV fields of view (FOV) captured and main flow events over the cylinder; **b** fibre sensing path localisation in a plane orthogonal to the spanwise direction of the cylinder

a spatial resolution of 2.6 mm where each sensing line consists of 140 sensing points, along 7.28 D (364 mm) in the spanwise direction of the cylinder.

3 Flow Induced Strain Measurements

A PIV snapshot of the velocity magnitude for the two test cases is represented in Fig. 2. The downwash induced by the 3D events at the tip of the cylinder is more noticeable in the wake of the cylinder, especially in case (a), following the documented 'Lee Model' [3], where the wake is advected towards the root of the cylinder, due to the interaction between the von Kàrmàn vortex street and the rising tip vortices.

The presence of free-stream turbulence has been associated with an increased vibration of cantilevered bluff bodies [6]. Distributed sensing networks allow for a better understanding of the evolution of strain along the spanwise direction, and the fingerprint of the several flow phenomena acting on the structure. A Reynolds decomposition is applied to the strain signal to extract the mean ($\overline{\varepsilon}$) from the fluctuating (ε') strain field. The mean strain field along the sensing lines for the two experimental conditions shows little variation, suggesting a similar time-averaged impact of the flow on the structure for both conditions. The addition of free-stream turbulence

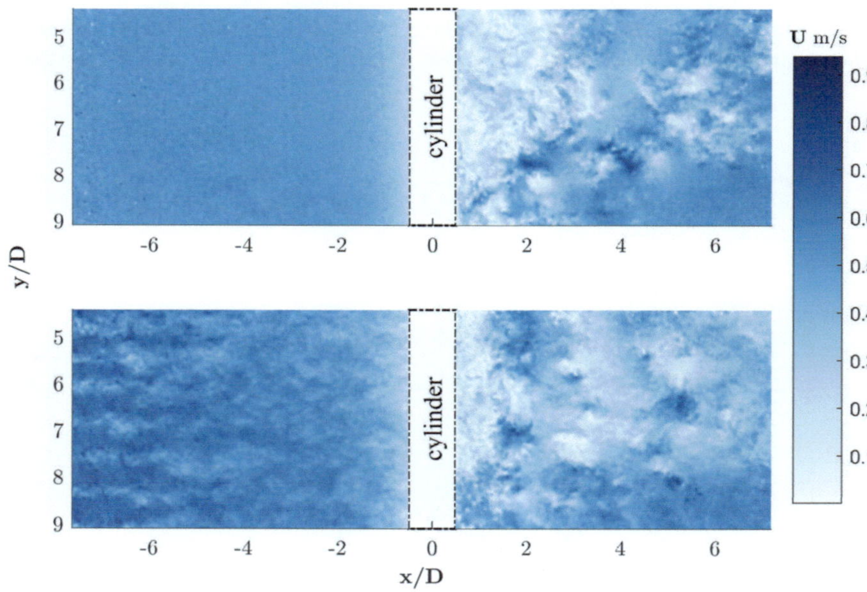

Fig. 2 PIV snapshots from the flow fields of velocity magnitude **U**: top—NoGrid, bottom—TurbGrid. Both upstream flow events generated by the turbulence generating grids, and downstream events generated by the turbulent wake of the cylinder are captured in the FOV A1 and A2 respectively

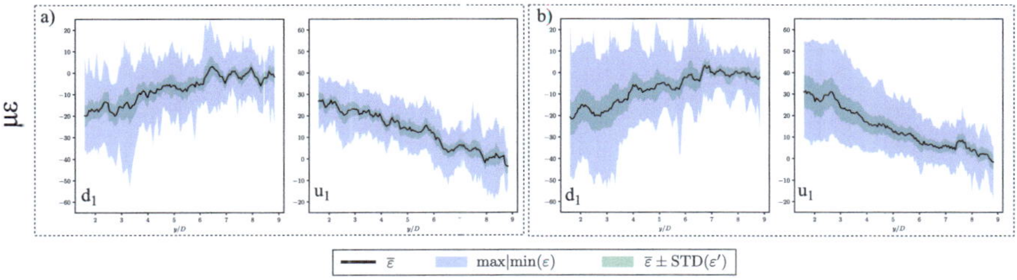

Fig. 3 Spanwise distribution of mean strain and maximum, minimum and standard deviation range of fluctuating strain in the cylinder for u_1, and d_1, for the prescribed free-stream conditions

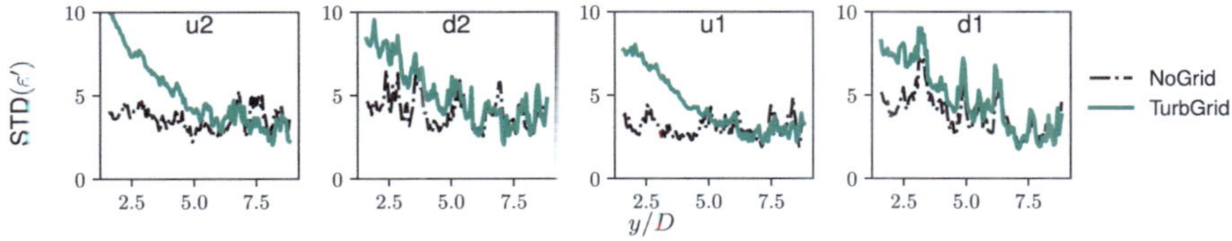

Fig. 4 Spanwise STD evolution of ε' along the 4 distributed sensing lines, for the two prescribed free-stream conditions

enhanced the magnitude of fluctuating strain, especially near the root (fixing region of the cylinder) where the bending moment is maximum. The distributions of strain along two representative sensing lines downstream and upstream of the cylinder are shown in Fig. 3. Furthermore, the introduction of free-stream turbulence accounts for an increase in the range of maximum and minimum values of strain, and increased standard deviation, consequently contributing to the increased gradual accumulation of fatigue damage over time. In addition, we can further distinguish the effect of the wake in the region downstream of the cylinder (d_1), characterised by its increasingly unsteady behaviour, representative of the cylinder's turbulent wake.

The evolution of the standard deviation of ε' along the four distributed sensing lines for the two experimental conditions is presented in Fig. 4. The evolution of $STD(\varepsilon')$ within u_1 and u_2 between the strain for NoGrid and TurbGrid shows great discrepancy, generated by the direct influence of the fluctuating loads induced by the fluctuating velocity field upstream of the cylinder. Nonetheless, d_1 and d_2 present a similar distribution of $STD(\varepsilon')$, suggesting that the induced free-stream turbulence has a bigger impact on the developing strain upstream than downstream, despite the near-wake changes induced by free-stream turbulence [1].

As strain and flow measurements are available concurrently, as a result of being triggered at the same time during the same experimental conditions, an analysis between the fibre optic data and flow measurements frequency content may be performed to analyse how flow structures interact with the wake-generating body. The spanwise evolution of the influence of the near wake region on the cylinder is then assessed by evaluating the spectra of ε', in sensing line u_2, and the stream-wise velocity fluctuations u' spectra captured by FOV A2, at $x/D = 1.5$ in Fig. 5. Vor-

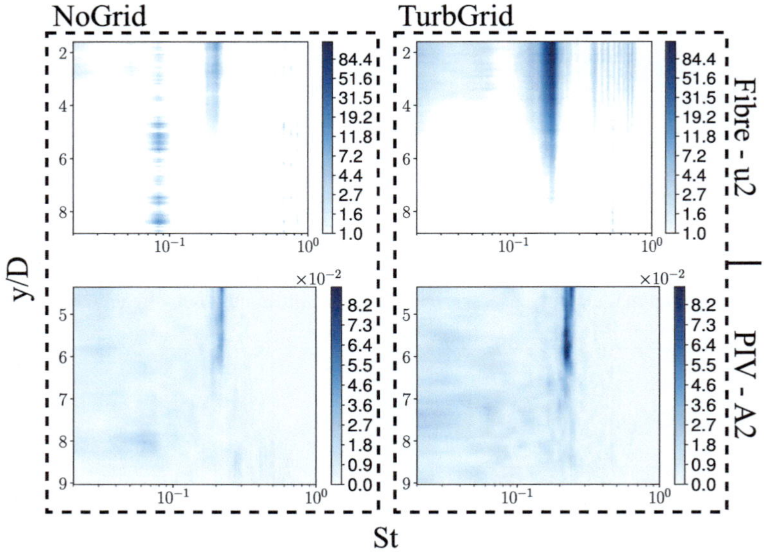

Fig. 5 On the top, spanwise fluctuating strain spectra ε', obtained with sensing line u_2 for case NoGrid—left—and TurbGrid—right. On the bottom, spanwise evolution of fluctuating stream-wise velocity spectra at location $x = 1.5D$, with the referential origin located at the centre of the cylinder, for case NoGrid—left—and TurbGrid—right

tex shedding at $St = (fD)/U_\infty \approx 0.2$ is clearly evident in the strain signal of the cylinder, providing an accurate depiction of the evolution of this phenomena along the cylinder's spanwise direction, increasing towards the root of the cylinder where the von Kàrmàn vortex street remains fairly unchanged, away from 3D effects at the tip [3]. The near-root region corresponds to the location of maximum strain. This is a result of integrating spanwise flow events occurring along the cylinder, which collectively contribute to the cumulative root bending moment. Furthermore, the presence of free-stream turbulence increases the intensity of the spectra on the identified vortex shedding band both in the strain and flow fields. The presence of tip vortices characterised by low frequencies—$St = [0.06, 0.09]$ [3] is identified in the flow field, appearing in isolated regions along the spanwise direction of the cylinder—$y/D \in [7.8, 8.3]$, as they evolve towards the root, whilst being reflected in the strain signal over the cylinder exciting the structure in $St = [0.08, 0.09]$, where cells of low St are also identifiable.

Acknowledgements The authors wish to acknowledge financial support given by EPSRC through grant no. EP/V006436/1.

References

1. K.S. Kankanwadi, O.R.H. Buxton, Influence of freestream turbulence on the near-field growth of a turbulent cylinder wake: turbulent entrainment and wake meandering. Phys. Rev. Fluids **8**(3) (2023)
2. P.W. Bearman, T. Morel, Effect of free stream turbulence on the flow around bluff bodies. Prog. Aerosp. Sci. **20** (1983)
3. R. Porteous, D.J. Moreau, C.J. Doolan, A review of flow-induced noise from finite wall-mounted cylinders. J. Fluids Struct. **51**, 240–254 (2014)
4. C. Xu, Z. Sharif Khodaei, Shape sensing with Rayleigh backscattering fiber optic sensor. Sensors (Basel) **20**(14) (2020)
5. M.M. Cicolin, O.R.H. Buxton, G.R.S. Assi, P.W. Bearman, The role of separation on the forces acting on a circular cylinder with a control rod. J. Fluid Mech. **915** (2021)
6. P. Wang, C.W. Wong, Y. Zhou, Turbulent intensity effect on axial-flow-induced cylinder vibration in the presence of a neighboring cylinder. J. Fluids Struct. **85** (2019)

Scaling of Turbulence Moments in Compressible Axisymmetric Jets

K. Y.-N. Hinh, R. J. Martinuzzi, and C. T. Johansen

Abstract Self-preservation of the axisymmetric jet is revisited to derive scaling laws for compressibility effects. While compressibility acts to attenuate turbulence levels, there is no consensus in the proposed scaling of the Reynolds stresses and other turbulence moments. Whereas earlier studies rely on empirical evidence, similarity analysis describes compressibility effects through the spreading rate and finds different scaling depending on the direction and order of the velocity moment. The scaling laws for second and third velocity moments are then validated with experimental results, providing new insights on Reynolds stress anisotropy.

1 Introduction

In compressible turbulent shear flows, an inverse relationship between the convective Mach number, M_c, and turbulence levels is observed. The effect of increasing M_c is observed through attenuation of the Reynolds shear stress and the growth rate of the shear layer. For jets, using the half-width, b, to scale the shear layer thickness, the attenuation can be characterized by the compressible spreading rate to the incompressible limit. With b' denoting the streamwise gradient, an empirical function, $\Phi(M_c)$, is then defined as,

$$b' = b_0' \Phi(M_c). \tag{1}$$

where the subscript, 0, denotes the incompressible case. The convective Mach number characterizes the propagation of disturbances in the shear layer and is

K. Y.-N. Hinh (✉) · R. J. Martinuzzi · C. T. Johansen
University of Calgary, Calgary, Canada
e-mail: kenneth.hinh@ucalgary.ca

R. J. Martinuzzi
e-mail: rmartinu@ucalgary.ca

C. T. Johansen
e-mail: johansen@ucalgary.ca

R. Örlü et al. (eds.), *Progress in Turbulence X*, Springer Proceedings in Physics 404,
https://doi.org/10.1007/978-3-031-55924-2_22

approximated in the free jet as $M_c = 0.5U_m/a$, with U_m as the centerline velocity and a as the speed of sound.

It is shown that $\Phi(M_c)$ scales the spreading rate and Reynolds shear stress [1, 2], but there remains disagreement in both experimental and numerical studies on whether the Reynolds normal stresses are affected by compressibility [1, 3–5]. Previously, compressible shear flows are found to be self-preserving [2, 3]. Here, similarity analysis is employed to determine scaling laws for second and third velocity moments to clarify observations from previous studies. From the scaling laws, the anisotropy due to compressibility can be determined to provide insights on the modelling of the pressure-strain which is a key mechanism in regulating the development of compressible shear layers [1]. The scaling laws are then validated against planar particle image velocimetry (PIV) measurements of a Mach 1.25 axisymmetric turbulent free jet.

2 Self-preservation of Reynolds' Equations

Conventional Reynolds averaging is used where an overline indicates a temporal mean. The density-velocity correlations are small compared to the other terms, as per Morkovin's hypothesis allowing this simplification up to Mach = 1.5 for free shear flows [6]. The Reynolds-Averaged Navier-Stokes equations expanded in general tensor notation, are then,

$$\nabla_i\left(\rho U^i\right) = 0, \tag{2}$$

$$\rho U^j \nabla_j U_i = \nabla_i \overline{p} - \nabla_j\left(\rho \overline{u_i u^j}\right) + \nabla_j \overline{\tau}_i^j, \tag{3}$$

where: i, j, k are reserved for indices; ∇_k is the covariant derivative, U_i is the mean velocity, u_i the fluctuating velocity, $\overline{p}$ the mean pressure, ρ the mean density and $\overline{\tau_i^k}$ the mean stress tensor.

For the axisymmetric jet a cylindrical coordinate system, (x, r, θ), with corresponding physical velocity components, (u, v, w), is chosen. Introducing scaling functions to define the similarity variables:

$$\eta = r/b, \tag{4}$$

$$U = U_m(x, M_c) f(\eta), \tag{5}$$

$$\rho = \rho_m(x, M_c) g(\eta), \tag{6}$$

$$\overline{u_i u_j} = R_{ij}(x, M_c) h_{ij}(\eta), \tag{7}$$

where $f(\eta)$, $h_{ij}(\eta)$ and $g(\eta)$ are radial shape functions of η, defined by the jet-half width, b. The tensor $\overline{u_i u_j}$ now expresses physical components. The scaling functions U_m and R_{ij} represent the streamwise variation in the flow and are functions of x and M_c. For the current analysis, the effect of compressibility on turbulence is isolated from mean density gradient effects. Substituting into the momentum equations with continuity,

$$gf^2 + \frac{f'}{\eta}\int_0^\eta g\tilde{f}\eta d\eta = \frac{R_{12}}{b'U_m^2}\frac{1}{\eta}\frac{d(gh_{12}\eta)}{d\eta} - \frac{R_{11}}{U_m^2}\eta\frac{d(gh_{11})}{d\eta} +$$
$$\frac{R'_{22}b}{U_m^2 b'}\left[\int_0^\eta \frac{1}{\hat{\eta}}\frac{d(\hat{g}\hat{h}_{22}\hat{\eta})}{d\hat{\eta}}\right]d\hat{\eta} - \frac{R_{22}}{U_m^2}\frac{d}{d\eta}\left[\eta\int_0^\eta \frac{1}{\hat{\eta}}\frac{d(\hat{g}\hat{h}_{22}\hat{\eta})}{d\hat{\eta}}\right]d\hat{\eta}, \quad (8)$$

which from the first term on the right had side shows the similarity requirements for the Reynolds shear stress, $R_{12} \propto b'U_m^2$, which allows a larger class of flows to be self-preserving [7] extending the concept to the compressible jet [3]. Note, R_{22} appears through the substitution of pressure with the radial momentum equation. In (8), the normal Reynolds stresses do not scale with b' and appear as streamwise gradients where their scaling is indirect. Investigation on the similarity conditions of the Reynolds stress transport (RST) equations can clarify the scaling of the other second velocity moments. Written in tensor notation,

$$\nabla_k\left(\rho U^k\overline{u_i u_j}\right) = -\nabla_k\left[\overline{\rho u_i u_j u^k} - \overline{u_i\left(\tau_j^k\right)''} - \overline{u_j\left(\tau_i^k\right)''}\right]$$
$$-\left(\overline{\rho u_j u^k}\right)\nabla_k U_i - \left(\overline{\rho u_i u^k}\right)\nabla_k U_j + \overline{u_i\nabla_j p''} + \overline{u_j\nabla_i p''} + \epsilon_{ij}, \quad (9)$$

with p'' the fluctuating pressure and $\left(\tau_i^k\right)''$ the fluctuating stress tensor. The viscous terms are separated into dissipation, ϵ_{ij} and diffusion terms, where the latter is negligible. With $\pi_{ij}(\eta)$, $t_{ijk}(\eta)$ and $e_{ij}(\eta)$ as shape functions, and $\Pi_{ij}(x, M_c)$, $T_{ijk}(x, M_c)$ and $E_{ij}(x, M_c)$ as scaling functions, the following correlations can also be written in similarity variables,

$$\overline{u_i\nabla_j p''} = \Pi_{ij}(x, M_c)\pi_{ij}(\eta), \quad (10)$$
$$\overline{u_i u_j u_k} = T_{ijk}(x, M_c)t_{ijk}(\eta), \quad (11)$$
$$\epsilon_{ij} = E_{ij}(x, M_c)e_{ij}(\eta). \quad (12)$$

For the transport of $\overline{uv}$ for example, with no swirl and second-order streamwise gradient terms neglected, the remaining terms scale with:

$$\underbrace{\frac{\partial\rho U\overline{uv}}{\partial x} + \frac{1}{r}\frac{\partial r\rho V\overline{uv}}{\partial r}}_{\propto \frac{U_m R_{12} b'}{b}} = -\Bigg(\underbrace{\frac{1}{r}\frac{\partial r\rho\overline{uv^2}}{\partial r}}_{\propto \frac{T_{122}}{b}} - \underbrace{\frac{\rho\overline{uw^2}}{r}}_{\propto \frac{T_{133}}{b}} + \cancel{\frac{\partial\rho\overline{vu^2}}{\partial x}}\Bigg)$$
$$-\Bigg(\underbrace{\overline{uv}\frac{\partial\rho V}{\partial r}}_{\propto \frac{U_m R_{12} b'}{b}} + \underbrace{\overline{v^2}\frac{\partial\rho U}{\partial r}}_{\propto \frac{U_m R_{11}}{b}} + \cancel{\overline{u^2}\frac{\partial\rho V}{\partial x}} + \cancel{\overline{uv}\frac{\partial\rho U}{\partial x}}\Bigg) + \underbrace{\Bigg(\overline{u\frac{\partial p''}{\partial r}} + \cancel{\overline{v\frac{\partial p''}{\partial x}}}\Bigg)}_{\propto \Pi_{12}}. \quad (13)$$

From the scaling relationship, $R_{12} \propto b'U_m^2$, from (8), the remaining scaling functions can be found for the RST equations and are tabulated in Table 1. Noting other

Table 1 Scaling of double and triple-velocity correlations for each transport equation

Equation	$R_{ij}/U_m^2 \propto$						$T_{ijk}/U_m^3 \propto$								
	11	12	13	22	23	33	111	112	113	122	123	133	222	223	333
U	–	b'	–	–	–	–	–	–	–	–	–	–	–	–	–
$\overline{uv}$	–	b'	–	b'^2	–	–	–	–	–	b'^2	b'^2	–	–	–	–
$\overline{u^2}$	Const.	b'	–	–	–	–	Const.	b'	–	–	–	–	–	–	–
$\overline{v^2}$	–	–	–	b'^2	–	–	–	–	–	–	–	–	b'^3	b'^3	–
$\overline{w^2}$	–	–	–	–	–	b'^2	–	–	b'	–	–	–	–	b'^3	–
$\overline{uw}$	–	–	b'	–	–	–	–	–	–	–	–	b'^2	–	b'^3	–
$\overline{vw}$	–	–	–	–	b'^2	–	–	–	–	–	b'^2	–	–	–	b'^3

than $\overline{u^3}$, the streamwise gradient terms must be neglected, as the radial-gradient terms scale separately for the same turbulence moment.

3 Experimental Setup

A schematic of the Mach 1.25 axisymmetric jet is shown in Fig. 1. Data was collected using planar two-component particle image velocimetry (PIV). The flow was perfectly expanded. Both the ambient air and the jet were seeded with $\sim$1 μm Di-Ethyl-Hexyl-Sebacat, yielding a Stokes number of $\sim$0.1 for this flow satisfying response relaxation requirements. The illumination source was a dual cavity Nd:YLF laser (*Photonics* DM20-527-DM) with energy of 20 mJ per pulse. 2600 image pairs were collected and processed using a multi-pass strategy resulting in a vector field with 12×12 pixel window size with 75% overlap, providing approximately 50 vectors across the jet exit diameter. Further details can be found in Hinh et al. [3].

Fig. 1 Schematic of free jet. Not to scale

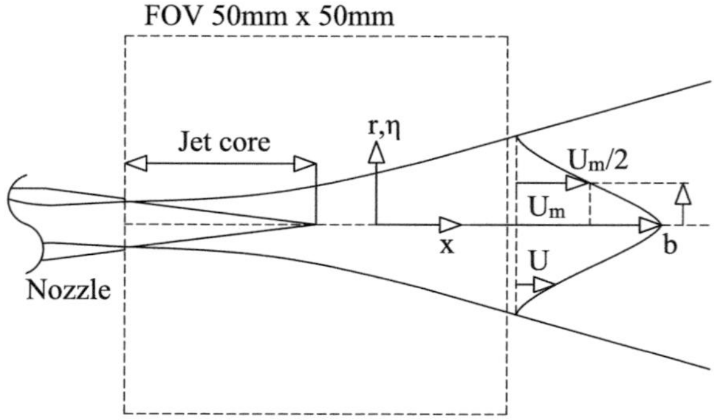

4 Discussion

The effect of compressibility on turbulence can be characterized by the influence of $\Phi(M_c)$ on b' shown in Fig. 2. A large scatter in the behavior of $\Phi(M_c)$ is observed in the literature, but the compressible jet agrees well with the curve for the planar mixing layer from Dimotakis [8]. Although the spreading rate is no longer constant the relationship $U_m \propto b^{-1}$ still holds through the momentum conservation of the jet, indicating that the velocity decay is also reduced by $\Phi(M_c)$. The mean velocity profile is plotted in Fig. 3, and scales with the centerline velocity, U_m, showing a Gaussian profile and is not effected by compressibility.

To date, the effect of M_c on turbulence moments relied on empirical observation lacking a physical basis for the difference the Reynolds stress' attenuation. Similarity analysis provides scaling relations of turbulence moments in a self-preserving flow. In the compressible jet, self-preservation of the mean momentum equation is possible [3] and can possibly be extended to higher moments. For the analysis, the radial-gradient terms had to be neglected for a unique set of scaling relations. Under the classification suggested by George [7] define this flow as partially self-preserving as not all the terms scale proportionally. In Fig. 3, the measured velocity moments are normalized using the scaling relations in Table 1, where all but $\overline{u^2}$ obeys the similarity relations. Whereas the current experimental data scales with b', which agrees with a previous study [4], but disagrees with the similarity scaling where $\overline{u^2} \propto U_m^2$, which also has been observed [1, 5]. For comparison, $\overline{u^2}$ is plotted with and without scaling with b'. Now, $\overline{u^3}$ is plotted scaled with U_m^3 as per the similarity scaling. When scaled this way, the curves do not perfectly overlap, which is expected as $\overline{u^2}$ appears to not scale only with U_m^2. Although not shown, the skewness, $\overline{u^3}/(\overline{u^2})^{3/2}$, is confirmed to be self-similar as the jet develops. The remaining normal stress, $\overline{v^2}$, scale with b'^2 as expected from the similarity analysis.

In classical Reynolds stress models, the anisotropy tensor, β_{ij} is a main parameter used to model the pressure-strain correlation [10], which is also strongly affected by compressibility [1]. Through the similarity scaling laws, the shear stress anisotropy

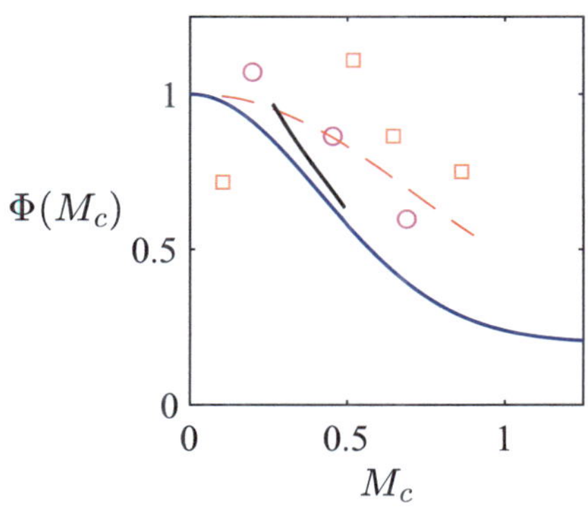

Fig. 2 Compressibility function $\Phi(M_c)$. Current study (—), (—) [8], (□) [5], (○) [4]

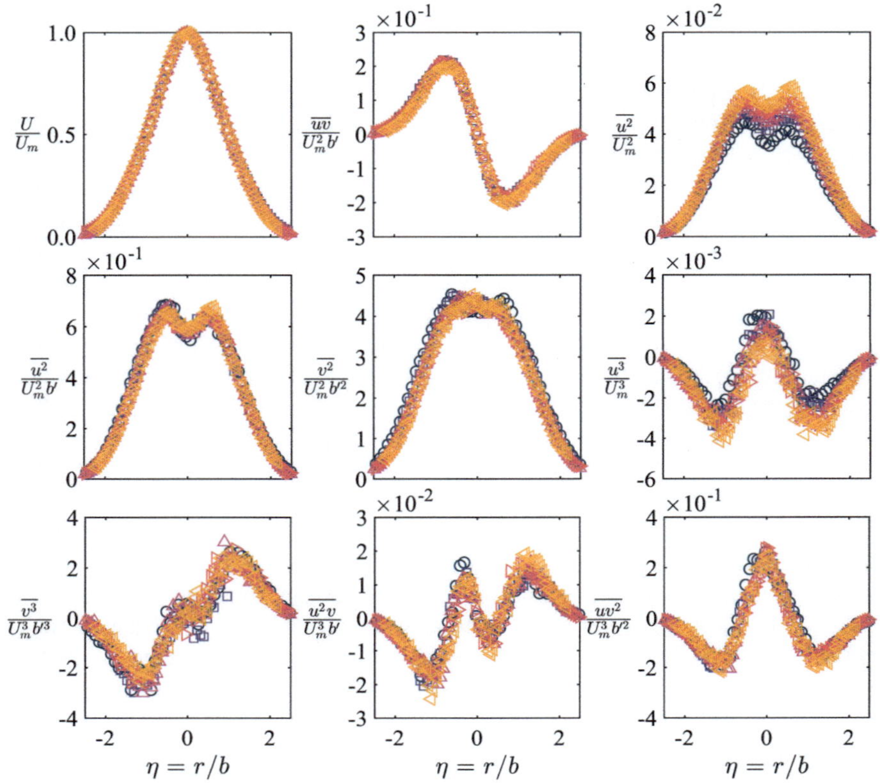

Fig. 3 Turbulence moments scaled according to similarity of the RST equations. Axial positions x/d = 12.5 (○); 15 (□); 17.5 (△); 20 (◁); 22.5 (▷). Every fifth point is shown for clarity

is suggested to scale in the jet in the form,

$$\beta_{12} = \frac{\Phi(M_c)}{C_1 + 2C_2\Phi(M_c)^2}, \tag{14}$$

where C_1 and C_2 are constants that are to be determined.

5 Conclusion

With compressibility characterized by $\Phi(M_c) \propto b'$, through similarity analysis, the second and third velocity moments containing v and w components are found to scale with powers of b'. The scaling provides an explanation of the behavior of

Reynolds stress anisotropy due to compressibility observed empirically, allowing further insight on the modelling of the pressure-strain in the attenuation of turbulence.

Acknowledgements Funding for this work was provided by the Natural Sciences Engineering Research Council (NSERC) of Canada and Atlantis Research Labs Inc. through a Collaborative Research and Development (CRD) grant.

References

1. J.B. Freund, S.K. Lele, P. Moin, J. Fluid Mech. **421**, 229–267 (2000)
2. A.J. Smits, J.-P. Dussauge, *Turbulent Shear Layers in Supersonic Flow* (Springer, 2006)
3. K.Y.-N. Hinh, R.J. Martinuzzi, C.T. Johansen, *Proceedings of TSFP12* (2022)
4. G.S. Elliot, M. Samimy, Phys. Fluids A **2**, 1231–1240 (1990)
5. S.G. Goebel, D.J Dutton, AIAA J. **29**, 538–546 (1991)
6. P. Bradshaw, Annu. Rev. Fluid Mech. **9**, 33–52 (1977)
7. W.K. George, *Advances in Turbulence*, pp. 39–73 (1988)
8. P.E. Dimotakis, *Progress in Astronautics and Aeronautics*, vol. 135 (1991), pp. 265–340
9. W.K. George, *Turbulence, Heat and Mass Transfer 1* (Begell House, 1994), pp. 13–24
10. C.G. Speziale, S. Sarkar, T.B. Gatski, J Fluid Mech. **227**, 245–272 (1991)

Experiments and Simulations on the Accelerating/Decelerating Flow on a Square Cylinder

G. Lunghi, S. Brusco, A. Mariotti, G. Piccardo, and M. V. Salvetti

Abstract We investigate the high-Reynolds accelerating and decelerating flows around a square cross-sectional cylinder through experiments and simulations. This flow has been well characterized for steady inflow conditions, but no conclusive studies are available when the inflow velocity varies. Wind-tunnel experiments are performed together with high-fidelity Large-Eddy Simulations (LES) to investigate the effect of Gaussian-type inflow accelerations and decelerations. The Reynolds number ranges between $Re = 1.720 \times 10^4$ and 6.536×10^4. The numerical results are compared with experiments, and an excellent agreement is found. During the investigated accelerations and decelerations, constant-frequency time cells are found in the wake vortex shedding. The values of the frequency between two subsequent time cells increase during the accelerations, and approaching the acceleration peak, the duration of the time cells shortens. The vortex-shedding Strouhal number decreases within the time cells since the velocity increases. The opposite occurs for the decelerating flow.

G. Lunghi (✉) · A. Mariotti · M. V. Salvetti
Dipartimento di Ingegneria Civile e Industriale, University of Pisa, Largo Lucio Lazzarino 2, 56122 Pisa, Italy
e-mail: gianmarco.lunghi@phd.unipi.it

A. Mariotti
e-mail: alessandro.mariotti@unipi.it

M. V. Salvetti
e-mail: maria.vittoria.salvetti@unipi.it

S. Brusco
Faculty of Engineering, Western University, London, Canada
e-mail: sbrusco@uwo.ca

G. Piccardo
Dipartimento di Ingegneria Civile, Chimica e Ambientale, Polytechnic School, University of Genoa, Via Montallegro 1, 16145 Genoa, Italy
e-mail: giuseppe.piccardo@unige.it

R. Örlü et al. (eds.), *Progress in Turbulence X*, Springer Proceedings in Physics 404,
https://doi.org/10.1007/978-3-031-55924-2_23

1 Introduction

The flow around elongated buildings, skyscrapers, or towers having a constant cross-section, may be analyzed with the aid of simplified geometries, such as square or rectangular prisms or cylinders. The flow around them is not as simple as their geometry, since a significant complexity may arise from the unsteadiness and three-dimensional phenomena typical of these massive-separated flows. Considering the square cross-section in a constant-wind velocity, the shear layers, once detached from the upstream edges, form the classical von Karman vortex street, which is found in the wake. The flow-features characterization around square cylinders assuming a constant inflow velocity has been provided in a wide Reynolds number range extending from $Re = 10^3$ to $Re = 10^6$, through experiments and simulations, e.g., [1–7]. On the other hand, the effects of unsteady inflow conditions, peculiar of, e.g., thunderstorm outflows, deserve further investigations since they may cause significant damage to civil buildings [8]. Indeed, acceleration significantly affects loads and the use of force coefficients evaluated under the steady-wind hypothesis may be not conservative, as shown, for instance, in the experiments in [9, 10]. In recent years, some investigations concerning flows around square cylinders have been carried out on both an experimental [11] and a numerical viewpoint [12], considering a quiescent starting condition from the accelerating flow.

We experimentally and numerically investigate herein the effects of Gaussian-type inflow accelerations/decelerations for a square cylinder in the range $Re = 1.72 \times 10^4$–6.34×10^4. The experiments [13], carried out in the framework of the ERC THUNDERR project [14], show the presence of constant-frequency time cells in the wake vortex shedding. We complement the analysis with high-fidelity Large-Eddy Simulations (LES) for comparison and cross-validation of the results. This work may be a first step to characterize wind loads on civil buildings in variable wind conditions and to better appraise the limits of the wind-load predictions obtained under the assumption of steady-wind conditions.

2 Experimental and Numerical Set-Ups

The accelerating/decelerating flows around a square cylinder are investigated through wind tunnel experiments and LES. Experimental investigations are performed in the multiple-fan wind tunnel at the Tamkang University, Taipei [13]. The wind tunnel tests have an inflow turbulence intensity equal to 2.5%. The square-cylinder model has a 60 mm long side and it is equipped with a total amount of 94 pressure taps, 46 of which are located in the midspan section with a 5 mm distance between each other in the streamwise direction. Two circular end plates of 400 mm diameter are located at the spanwise ends to mitigate three-dimensional effects. Differential pressures are evaluated through two ZOC 33 pressure transducers with 300 Hz sampling frequency.

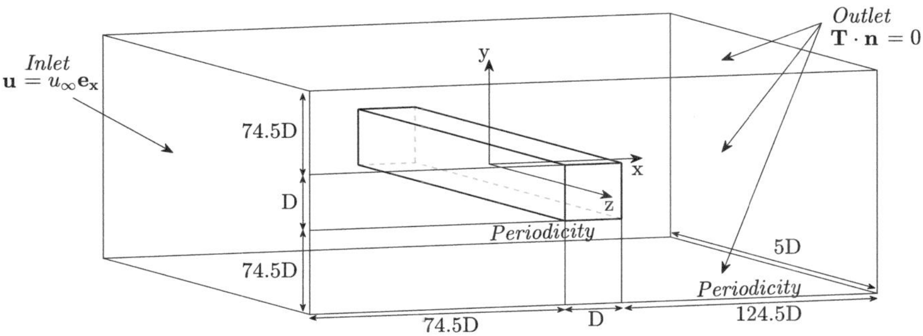

Fig. 1 Computational domain (not in scale) and boundary conditions

LES simulations are performed by employing Nek5000, a code that solves the incompressible Navier-Stokes equations through a high-order spectral-element method. The basis function inside each rectangular spectral element is the Legendre polynomials, and in particular, the order is $N - 2$ for pressure and N for velocity, being $N = 6$ as in [15, 16]. Time advancing uses a third-order backward finite-difference scheme, which is based on a high-order splitting method. A quadratic damping filter is applied to the three highest spectral modes, which introduces a subgrid-scale dissipation. The domain size is $-75 \leq x/D \leq 125$ and $-75 \leq y/D \leq 75$ in the streamwise and crossflow directions, respectively, while, the body and the domain size are extruded in the spanwise direction for a total length of $5D$, being D the cylinder lateral side (see Fig. 1). The minimum streamwise and crossflow element size is $\Delta x/D = \Delta y/D = 0.125$, and the uniform spanwise element size is equal to $\Delta z/D = 0.558$. No-slip Boundary Conditions (BC) are applied at the body surface. Periodic BCs are imposed in the spanwise direction, whereas in the remaining boundaries of the domain traction-free BCs are applied.

Simulations are performed for zero turbulence intensity of the incoming flow and considering the same time histories for the accelerating and the decelerating flows as in the experiments in [13] from $Re = 1.720 \times 10^4$ to $Re = 6.536 \times 10^4$ (see Fig. 2a, b, respectively), where the Reynolds number is defined as $Re = u_\infty D/\nu$, being u_∞ the time-varying velocity and ν the kinematic viscosity of air.

3 Results and Discussion

We compare the time behaviors of the lift coefficient, C_L, between simulations and experiments in Fig. 3a, b for the accelerating and the decelerating flows, respectively. LES and experimental results are in satisfactory agreement in terms of both the mean value and the standard deviation of the signals. For the accelerating flow, the periods of the oscillation cycles reduce in time with the increasing velocity, and the opposite occurs when the velocity decreases, i.e., the frequency decreases. The

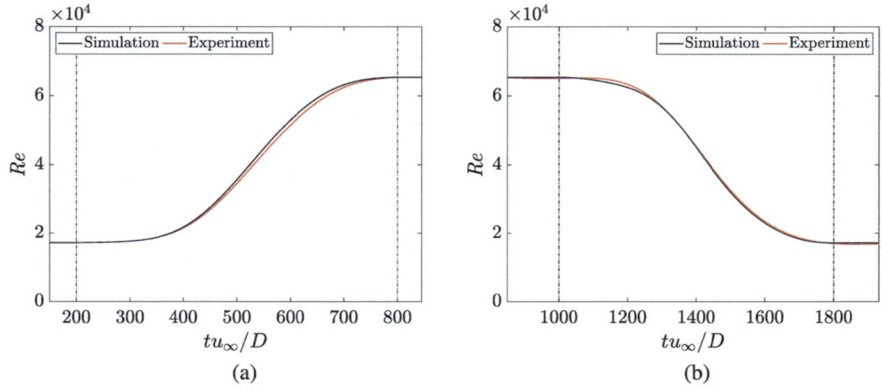

Fig. 2 Time behavior of the inflow Reynolds number for **a** the accelerating flow and **b** the decelerating flow. Experiments from [13]

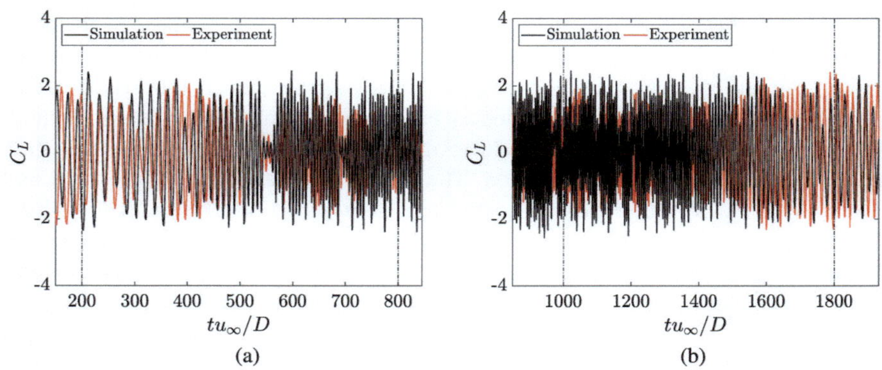

Fig. 3 Time history of the lift coefficient, C_L, for **a** the accelerating flow and **b** the decelerating flow. Experiments from [13]

increase/decrease of the frequency is not constant in time but it occurs between a constant-frequency time cell and the subsequent one.

To quantitatively describe the time behavior of these frequencies, we carried out a time-frequency analysis based on the adoption of the continuous wavelet transform, as described in [13, 17]. Wavelet transforms are preferred to other procedures, such as Fourier and Welch spectra (see, e.g., [18]) because they have a mathematically rigorous definition for the frequency peaks in time [19]. In particular, we adopt a complex Morlet wavelet with a central frequency of 6π, to enhance the frequency resolution, and we extract the dominant frequencies of the vortex-shedding, n^*, from the energy maps. The ridges of the energy maps are shown in Fig. 4 as a function of the non-dimensional time; whereas the time behavior of the vortex-shedding Strouhal number, $St = n^* D/u_\infty$, is shown in Fig. 5. The vortex-shedding frequency is constant within the time cells for both the accelerating and the decelerating flows and the

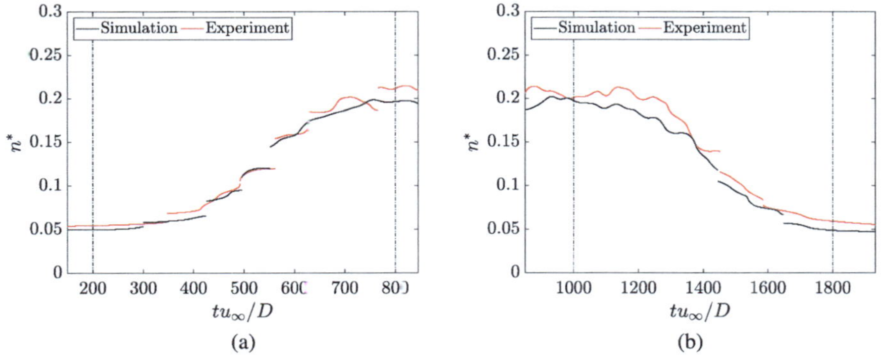

Fig. 4 Time history of the vortex-shedding frequency, n^*, for **a** the accelerating flow and **b** the decelerating flow. Experiments from [13]

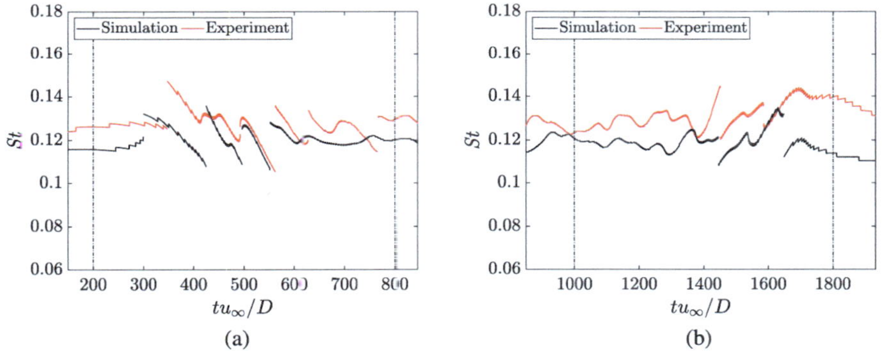

Fig. 5 Time history of the Strouhal number, St, of the vortex shedding for **a** the accelerating flow and **b** the decelerating flow. Experiments from [13]

velocity variation in time causes the vortex-shedding Strouhal number to decrease or increase inside each cell depending on whether the flow is accelerating or decelerating, followed by a sudden increase or decrease between the cells. The Strouhal number is always $0.10 \leq St \leq 0.14$. Moreover, approaching the acceleration peak, the time cells shorten.

4 Conclusion

The effects of Gaussian-type accelerations and decelerations have been addressed for a square cylinder through experiments and Large-Eddy Simulations. A good agreement is found between them. Experiments and LESs confirm the presence of vortex-shedding time cells that can be computed, for instance, from the time behavior of the lift coefficient. Inside each cell, the vortex-shedding frequency is constant. In turn, the vortex-shedding Strouhal number decreases or increases as the flow is accelerating or decelerating. The present numerical set-up will be possibly used to characterize the flow features and frequencies for different levels of acceleration.

References

1. G. Bosh, W. Rodi, Simulation of vortex shedding past a square cylinder with different turbulence models. Int. J. Num. Meth. Fluids **28**, 601–616 (1998)
2. L. Bruno, D. Oberto, Effects of cell quality in grid boundary layer on the simulated flow around a square cylinder. Comput. Fluids **238**, 105351 (2022)
3. J. Ke, RANS and hybrid LES/RANS simulations of flow over a square cylinder. Adv. Aerod. **1**, 10 (2019)
4. C. Norberg, Flow around rectangular cylinders: pressure forces and wake frequencies. J. Wind Eng. Ind. Aerod. **49**, 187–196 (1993)
5. S. Oka, T. Ishihara, Numerical study of aerodynamic characteristics of a square prism in a uniform flow. J. Wind Eng. Ind. Aerod. **97**, 548–559 (2009)
6. W. Rodi, Comparison of LES and RANS calculations of the flow around bluff bodies. J. Wind Eng. Ind. Aerod. **69–71**, 55–75 (1997)
7. F.X. Trias, A. Gorobets, A. Oliva, Turbulent flow around a square cylinder at Reynolds number 22,000: a DNS study. Comput. Fluids **123**, 87–98 (2015)
8. S. Brusco, G. Solari, Transient aeroelasticity of structures subjected to thunderstorm outflows. Eng. Struct. **245**, 112801 (2021)
9. T. Sarpkaya, Lift, drag, and mass coefficients for a circular cylinder immersed in time dependent flow. J. Appl. Mech. **30**, 13–15 (1963)
10. T. Takeuchi, J. Maeda, Unsteady wind force on an elliptic cylinder subjected to a short-rise-time gust from steady flow. J. Wind Eng. Ind. Aerod. **122**, 138–145 (2013)
11. T. Yang, M. Mason, Aerodynamic characteristics of rectangular cylinders in steady and accelerating wind flow. J. Fluids Struct. **90**, 246–262 (2019)
12. F. Guo, G. Wu, X. Du, Numerical investigation of flow around a square cylinder in accelerated flow. Phys. Fluids **33**, 104105 (2021)
13. S. Brusco, G. Buresti, Y.-L. Lo, G. Piccardo, Constant-frequency time cells in the vortex shedding from a square cylinder in accelerating flows. J. Wind Eng. Ind. Aerod. **230**, 105182 (2022)
14. G. Solari, Thunderstorm downbursts and wind loading of structures: progress and prospect. Front. Built Environ. **6**, 63 (2020)
15. A. Mariotti, L. Siconolfi, M.V. Salvetti, Stochastic sensitivity analysis of large-eddy simulation predictions of the flow around a 5:1 rectangular cylinder. Eur. J. Mech. B/Fluids **62**, 149–165 (2017)
16. G. Lunghi, E. Pasqualetto, B. Rocchio, A. Mariotti, M.V. Salvetti, Impact of the lateral mean recirculation characteristics on the near-wake and bulk quantities of the BARC configuration. Wind Struct. **34**(1), 115–125 (2022)
17. A. Mariotti, Axisymmetric bodies with fixed and free separation: base-pressure and near-wake fluctuations. J. Wind Eng. Ind. Aerod. **176**, 21–31 (2018)
18. E. Pasqualetto, G. Lunghi, B. Rocchio, A. Mariotti, M.V. Salvetti, Experimental characterization of the lateral and near-wake flow for the BARC configuration. Wind Struct. **34**(1), 101–113 (2022)
19. R. Perna, M. Abela, M. Mameli, A. Mariotti, L. Pietrasanta, M. Marengo, S. Filippeschi, Flow characterization of pulsating heat pipe through the wavelet analysis of pressure signals. Appl. Therm. Eng. **171**, 115128 (2020)

Experiments

Streamwise Scalings of the Turbulent Wake Generated by a Scaled Wind Turbine Under Laminar and Turbulent Background Flows

I. Neunaber, M. Hölling, and M. Obligado

Abstract We report an experimental study on the turbulent wake generated by a scaled wind turbine in a wind tunnel. Hot-wire anemometry is used to characterise the far wake under laminar and turbulent background flows up to 32 rotor diameters downstream the generator. We focus on the streamwise scalings of the centreline mean velocity deficit and the wake width. Following the Townsend-George theory, we verify that both parameters are properly modelled by power laws with the streamwise distance. As the exponents of the power laws are known to depend on the turbulent kinetic energy dissipation scalings, we systematically study the streamwise scalings of the three parameters (centreline mean velocity deficit, wake width, and dissipation), finding that our results are consistent with the standard Richardson-Kolmogorov energy cascade.

1 Introduction

Within a worldwide effort to increase alternative energy generation, the study of the flow downstream of one or several wind turbines has received widespread attention. Despite this growing interest, many advances in the modelling of turbulent flows have not yet been adapted to such studies [2]. They concern the understanding of the inner structure of turbulence: the energy cascade [2]. This parameter sets how energy is transferred from large to small scales and how the energy of the flow is dissipated.

I. Neunaber
Norwegian University of Science and Technology, Høgskoleringen 1, 7034 Trondheim, Norway
e-mail: ingrid.neunaber@ntnu.no

M. Hölling
Institute of Physics and For Wind, University of Oldenburg, 26129 Oldenburg, Germany
e-mail: michael.hoelling@uni-oldenburg.de

M. Obligado (✉)
University of Lille, CNRS, ONERA, Arts et Métiers ParisTech, Centrale Lille, UMR 9014 - LMFL - Laboratoire de Mécanique des fluides de Lille - Kampé de Feriet, 59000 Lille, France
e-mail: Martin.Obligado@centralelille.fr

R. Örlü et al. (eds.), *Progress in Turbulence X*, Springer Proceedings in Physics 404,
https://doi.org/10.1007/978-3-031-55924-2_24

This complicated phenomenon has been found to ultimately set important properties of turbulent wakes, such as their mean velocity deficit and how they spread in the streamwise direction [1, 5, 10].

The theory that allows to predict the streamwise far-wake scalings of the centreline mean velocity deficit u_0 and the wake width δ has been developed by Townsend and George [2, 15]. It relies on a set of hypotheses that includes the axisymmetry and the self-preservation of different averaged quantities of the flow and a scaling for the turbulent kinetic energy dissipation rate ε. The latter is usually modelled according to the Richardson-Kolmogorov cascade. In this way, the well-known predictions for the streamwise evolution (along x) of u_0 and δ are

$$u_0(x) = AU_\infty \left((x - x_0)/\theta\right)^{-2/3}, \tag{1}$$

$$\delta(x) = B\theta \left((x - x_0)/\theta\right)^{1/3}, \tag{2}$$

where A and B are dimensionless constants, θ the momentum thickness and x_0 a common virtual origin. The momentum thickness θ is defined by $\theta^2 = \frac{1}{U_\infty^2}\int_0^\infty U_\infty (U_\infty - U)\, r dr$ which is constant with x, and the wake's width is here characterised by the integral wake's width defined by $\delta^2(x) = \frac{1}{u_0}\int_0^\infty (U_\infty - U)\, r dr$. In the following, Eqs. (1) and (2) will be referred to as 'standard scalings'.

On the other hand, other scalings can be found for other types of energy cascade. For instance, the non-equilibrium cascade (see [16] for a longer discussion about this topic) predictions are

$$u_0(x) = AU_\infty \left((x - x_0)/\theta\right)^{-1}, \tag{3}$$

$$\delta(x) = B\theta \left((x - x_0)/\theta\right)^{1/2}. \tag{4}$$

The main difference in the theoretical modelling between standard and non equilibrium scalings is in the scaling of the centreline value of ε. In both cases, ε is considered to be self-similar, and to follow the law $\varepsilon_0 = C_\varepsilon K_0^{3/2}/\delta$, with ε_0 being the centreline value of ε. K_0 is the centreline turbulent kinetic energy, usually approximated, assuming axisymmetry, as $u_0'^2$ (where u_0' is the root-mean-square value of the fluctuating streamwise velocity at the centreline). The standard scalings are obtained assuming the dimensionless dissipation coefficient C_ε to be a constant for a given wake generator. On the other hand, within the non-equilibrium turbulence dissipation law, C_ε is proportional to the ratio of two Reynolds numbers, Re_G^m/Re_L^n. $Re_G \equiv U_\infty L_b/\nu$ is a global Reynolds number determined by the inlet conditions, and $Re_L \equiv u_0'\delta/\nu$ is a local Reynolds number dependent on x (ν stands for the kinematic viscosity of the flow and L_b is a characteristic length scale related to the generator). The exponents m and n are found to be equal to unity at high Reynolds number, but for lower Reynolds numbers, $0 < m < 1$ and $0 < n < 1$ are expected (that would imply slightly different exponents in the scaling laws above). Experimental measurements and DNS of bluff bodies at $Re_G \sim 40000$ have found evidence of $m = n = 1$ for $10 < x/\sqrt{\mathcal{A}} < 50$ [1, 5, 10]. $\mathcal{A}$ is the characteristic area of the plate that in previous the experiments could have either regular or irregular peripheries. Given that the differences in the exponents in both sets of scalings are small, these experimental studies have shown that quantifying the scaling laws of C_ε is essential to appropriately detect the nature of the energy cascade within the flow (and, therefore, which scaling laws of u_0 and δ hold). Furthermore, even if the exponents are close and the functional forms identical, differences can be significant for actual wind turbines and wind farms, where wake effects can be significant several rotor diameters downstream. Moreover, the properties of the energy cascade within the turbulent wake are important in terms of theoretical and numerical modelling [16].

While the theory has been developed and tested in bluff [16] and slender [11] bodies, it is also expected to apply to the far wake of wind turbines. Indeed, previous works have found that scalings laws for u_0 and δ in the form of power laws properly model the wake of a rotor in laboratory [6, 7, 9] and field [8] conditions. Nevertheless, the only work that systematically studied the connection between these scalings with the behaviour of the dissipation C_ε [9] was performed within a reduced streamwise range, covering up to 12 turbine diameters D. In the present work, we therefore propose to study the wake of a scaled wind turbine in a wind tunnel for an extended range, reporting results up to $x/D = 32$. Hot-wire anemometry was used to measure spanwise profiles for different streamwise positions, that allowed to quantify all relevant quantities. Furthermore, we tested a laminar and a grid-generated turbulent background flow, exploring also two different Reynolds numbers based on the rotor diameter Re_D (defined as $Re_D = DU_\infty/\nu$) and two tip-speed ratios (TSR). In consequence, six different operating conditions for a single wind turbine are reported in the present work.

2 Experimental Setup

Experiments were conducted at the wind tunnel located at the University of Oldenburg. The closed-loop wind tunnel has a test section with a length of 30 m and a square cross section of 3×3 m^2 [3]. For an empty test section, the background turbulence intensity is around 0.3%. The ceiling of the wind tunnel was adjusted to avoid pressure gradients within the test section.

A single three-bladed horizontal axis model wind turbine of type MoWiTO 0.6 [13] was mounted in the test section (Fig. 1). The model wind turbine has a hub height (when placed on a platform) of 93 cm and a rotor diameter of 58 cm. The turbine was controlled to optimise the power for a given blade pitch angle and incoming velocity with a control system described in a previous work [12]. The blockage of the turbine and its support remained below 5%. The latter consisted of a six-degrees of freedom Stewart platform [4], that has been used static and with the turbine rotor axis aligned to the streamwise coordinate x.

An array consisting of 16 single wire probes was used to scan the wake at hub-height (i.e., $z = 0$, see Fig. 1) between $x/D = 1$ and $x/D = 32$ in steps of $1D$. In the spanwise direction, the probes cover a range from $y/D = -1.91$ to $y/D = 1.04$.

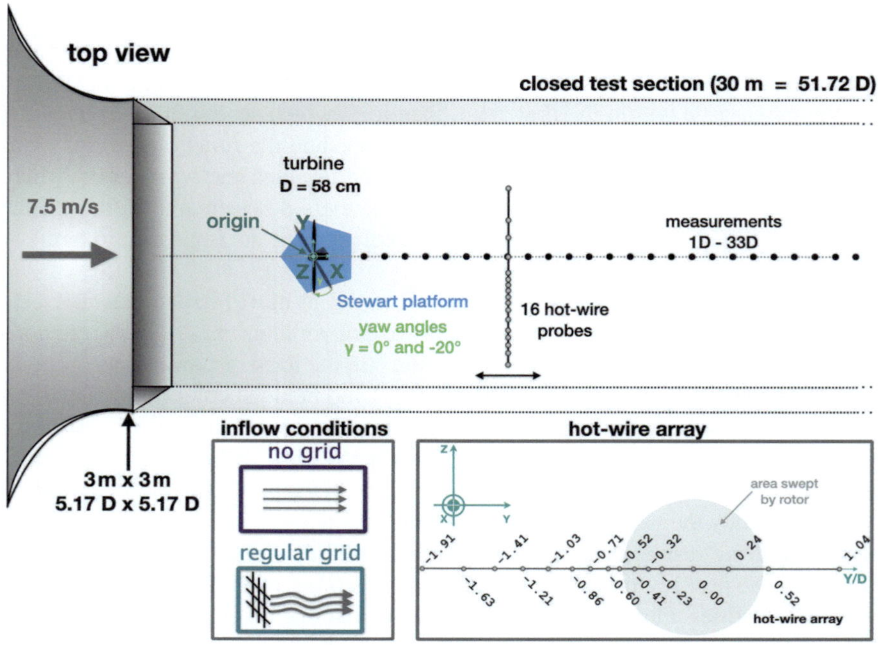

Fig. 1 Sketch of the experimental setup. The coordinate system and placing of the turbine are shown in the panel above, while the hot-wires layout is shown below

The probes were not spaced equidistantly but, as presented in Fig. 1, focusing on one side of the wake, with a high probe density around the radius of the turbine ($y/D \approx -0.5 \pm 0.25$).

The hot-wire placed at the centerline ($y = z = 0$) was operated using a StreamLine 9091N0102 frame with a 91C10 CTA (Constant Temperature Anemometry) Module. The hot-wire was sampled at $f_s = 20$ kHz with a hardware low-pass filter set to a cut-off frequency of $f = 10$ kHz. The sampling time was 240s for all cases. It has been found that these parameters allowed to resolve ε_0 in the centreline, within the range $x/D \geq 5$ (for an extended discussion on how this parameter was calculated, the reader is referred to [1]). All other hot-wires were operated by several Dantec Dynamics *multichannel CTA* 54N80. The sampling frequency was $f_s = 6$ kHz and a hardware low-pass filter with a cut-off frequency of $f = 3$ kHz was used. Data was collected synchronized using two NI USB 6211 A/D converters, and the sampling time was 240 s. The hot wires were calibrated at least twice a day.

Six operating configurations were studied, three for each different background flow, cf., Table 1. Laminar (or low-turbulent) conditions were produced for an empty test section (i.e., the wind turbine, its support and the hot-wire system were the only elements placed in the wind tunnel). The turbulent background flow has been generated by a passive grid, that generated turbulence intensities (defined as the ratio of the standard deviation and the mean values of the streamwise velocity) between

Table 1 Cases tested for each one of the two inflows considered: freestream velocity (U_∞), rotor-based Reynolds number, blade pitch angle γ with respect to the optimal pitch angle γ_o, thrust coefficient c_T and TSR

Case	U_∞ (m/s)	Re_D	γ	c_T	TSR
1	7.5	$Re_D = 2.9 \times 10^5$	γ_o	0.70	5.31
2	7.5	$Re_D = 2.9 \times 10^5$	$\gamma_o + 6°$	0.34	4.53
3	5.0	$Re_D = 1.9 \times 10^5$	γ_o	0.73	5.25

2 and 3%, depending on the streamwise position in the tunnel. When the grid is present, the turbine had to be located further downstream, so the grid-generated flow could develop before reaching the turbine. In consequence, the grid was located at the inlet of the test section and the turbine 442 cm downstream.

For the first of the three conditions tested for a given background flow, the turbine was operated with an incoming velocity of $U_\infty = 7.5$ m/s (and therefore $Re_D = 2.9 \times 10^5$) and the optimal tip-speed ratio ($TSR_o = 5.3$) [14]. In all cases, the freestream velocity was estimated with a Pitot tube located at the entrance of the test section. The second configuration consisted of the same freestream velocity but using a different, non-optimal blade pitch angle of $\gamma_{no} = \gamma_o + 6°$, and, thus, a non-optimal TSR of 4.5. For the third configuration, a lower velocity, $U_\infty = 5.0\ \mathrm{ms}^{-1}$ ($Re_D = 1.9 \times 10^5$), is investigated at optimal TSR and γ_o. Table 1 summarises the properties of the three cases studied (more information about this setup can be found in [7]).

3 Results

Figure 2 shows the streamwise scalings of u_0/U_∞ and δ/θ for all cases studied. For a given case, small differences can be observed between laminar and turbulent inflows. Indeed, for the reduced turbulence intensity generated by the passive grid, the main differences in terms of the averaged streamwise scalings are caused by difference in operating conditions from Table 1.

The solid lines in Fig. 2 correspond to fits that were made according to a general form of (1), (2), (3) and (4):

$$\frac{u_0}{U_\infty} = A\left((x - x_0)/\theta\right)^{-\alpha}, \tag{5}$$

$$\frac{\delta(x)}{\theta} = B\left((x - x_0)/\theta\right)^{\beta}. \tag{6}$$

Fits were performed in the range $9 < x/D < 28$, adjusting (5) and (6) simultaneously while keeping a fixed virtual origin for both equations. The parameters obtained

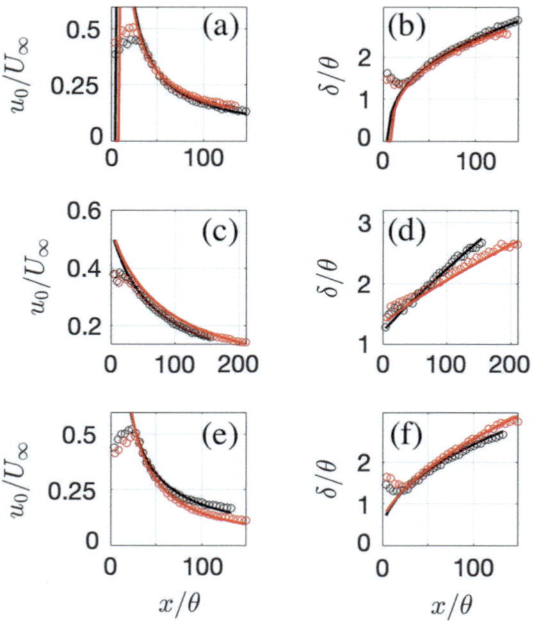

Fig. 2 Streamwise scalings of the normalised centreline velocity deficit u_0/U_∞ and the normalised wake width δ/θ for all cases studied. The normalised centreline velocity deficits for cases 1, 2 and 3 are shown in figures **a**, **c** and **e**, respectively. On the other hand, the normalized wake width for cases 1, 2 and 3 are shown in figures **b**, **d** and **f**, respectively. In all cases, circle markers correspond to experimental results and solid lines to the fitting curves, produced using the parameters from Table 2

for all the cases studied are shown in Table 2. Momentum conservation in a self-similar turbulent wake implies that $U_\infty\theta^2 = u_0\delta^2$, and therefore that $AB^2 = 1$ and $2\beta - \alpha = 0$. This relation was not imposed to the fits, but it can be observed that holds for almost all cases, with the exception of case 2 (particularly under a laminar inflow). This result may indicate that the wake has not fully developed into an axisymmetric and/or self-similar shape in this case.

Table 2 indicates no clear trend in the exponents, showing some cases close to standard scalings and other to non equilibrium ones. A broader streamwise range could help to better disentangle the exponents. Nevertheless, such an approach is challenging in a wind tunnel while keeping a realistically scaled rotor, as the latter cannot be easily miniaturised. Nevertheless, a similar scenario has been reported in studies with bluff bodies, as square plates have been found to follow standard and non-equilibrium scalings within the same range [1, 5, 10]. As discussed in Sect. 1, the only way to disentangle the nature of the streamwise scalings is via the measurement of the streamwise evolution C_ε.

In consequence, Fig. 3 shows how the centreline value of C_ε evolves in all cases studied. In the figure, we also show the values obtained with the grid alone (i.e. with

Table 2 Values of the fitting constants from (5) and (6) obtained for all 6 cases. Fits were performed within the range $6 < x/D < 30$. The values of the momentum thickness θ obtained for each case are also reported

Inflow	Case	A	B	α	β	x_0/θ	θ (cm)
Laminar	1	7.3	0.39	0.83	0.40	5.1	13.2
	2	85.7	0.043	1.16	0.76	−80.1	12.3
	3	9.10	0.30	0.84	0.45	−3.0	14.5
Turbulent	1	3.9	0.47	0.68	0.36	8.7	14.4
	2	142.0	0.077	1.21	0.62	−101.0	9.1
	3	29.1	0.21	1.13	0.53	−8.0	13.0

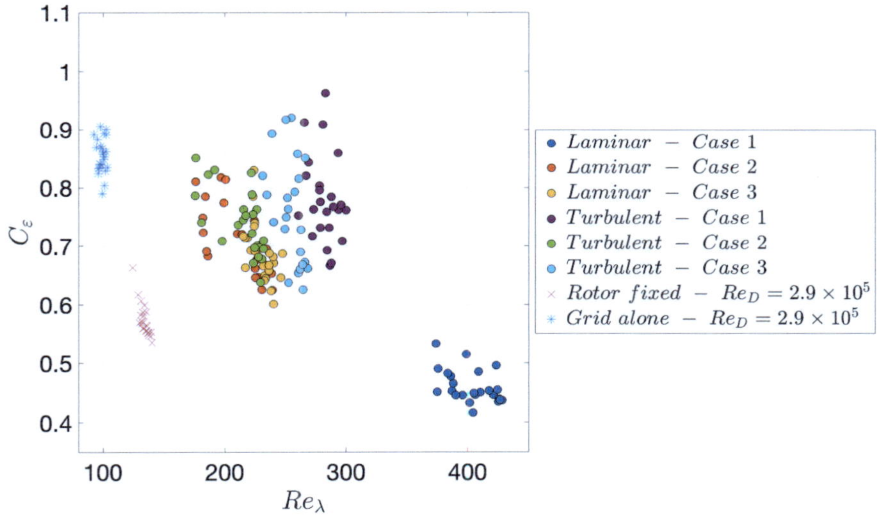

Fig. 3 Evolution of the dissipation coefficient C_ε with Re_λ for all cases studied. For reference, the results for the passive grid alone and the rotor fixed (under a laminar inlow) are also shown.We only report results that unambiguously correspond to the far wake, in the range $x/D \geq 10$

no rotor nor support) and with the rotor in a static position. Following other works, we show the evolution of C_ε with Re_λ, as it is a relation that allows to clearly see any changes in this parameter [16]. The turbulent Reynolds number is defined as $Re_\lambda = \lambda_0 u'_0/\nu$, with λ_0 the Taylor microscale of the turbulent flow at the centreline, defined as $\lambda_0 = \sqrt{15\nu u_0'^2/\varepsilon_0}$. The error in the estimation of C_ε, that comes from the uncertainty of the pitot tube reference velocity used for the calibration is of the order of 20%. In consequence, all cases studied have an approximately constant value of C_ε, of the order of 0.75. Surprisingly, the only exception is case 1 in laminar conditions, that presents both larger values of Re_λ but smaller C_ε. We remark that the wind turbine has been optimised for this case only, and the optimal blade pitch angle (and the evolution of the pressure coefficient C_p with control parameters) can be different in the turbulent inflow scenario. In consequence, and remarkably, we observe that operating the turbine in optimal conditions has a strong effect in terms of C_ε and the turbulent flow within the wake.

4 Conclusions

In this work, we have studied the behaviour of the far wake of a scaled wind turbine in a wind tunnel under different inflows, tip-speed ratios and incoming Reynolds numbers. On the one hand, we show that the streamwise scalings of the centreline velocity deficit and the wake width can be fitted simultaneously using power laws derived from the Townsend-George theory. On the other hand, the exponents, that can be related to the turbulent kinetic energy dissipation scalings, do not present a clear trend when compared to the predictions from different energy cascades. As shown in previous works, this may be related to the reduced streamwise range studied, and the evolution of the dissipation coefficient is the parameter that allows to actually disentangle the nature of the turbulent wake in terms of the energy cascades. We find that this parameter evolves, for all cases studied, according to standard scalings, consistent with a Richardson-Kolmogorov cascade. Moreover, both the dissipation coefficient and the turbulent flow in the wake seem to be strongly dependent on the operating conditions, as the optimal case (defined in terms of optimal tip-speed ratio and blade pitch angle) presents strong differences with all the other, non-optimal, operating conditions.

Acknowledgements Part of this work has been performed during the stay associated to fellowship from the Hanse-Wissenschaftskolleg (HWK Institute for Advanced Study, Delmenhorst, Germany) assigned to MO and IN. The authors also thank Joachim Peinke for fruitful discussions. Further, the authors would like to thank Julian Jüchter for providing additional measurements of the model wind turbine operation parameters and Jannis Maus and Abdulkarim Abdulrazek for assistance during the experiments.

References

1. T. Dairay, M. Obligado, J.C. Vassilicos, Non-equilibrium scaling laws in axisymmetric turbulent wakes. J. Fluid Mech. **781**, 166–195 (2015)
2. W.K. George, The self-preservation of turbulent flows and its relation to initial conditions and coherent structures, in *Advances in Turbulence*, ed. by W.K. George, R. Arndt (Springer, 1989)
3. L. Kröger, J. Frederik, J.-W. van Wingerden, J. Peinke, M. Hölling, Generation of user defined turbulent inflow conditions by an active grid for validation experiments. J. Phys. Conf. Ser. **1037**(5) (2018)
4. T. Messmer, C. Brigden, J. Peinke, M. Hölling, A six degree-of-freedom set-up for wind tunnel testing of floating wind turbines. J. Phys. Conf. Ser. **2265**, 42015 (IOP Publishing, 2022)

5. J. Nedic, J.C. Vassilicos, B. Ganapathisubramani, Axisymmetric turbulent wakes with new nonequilibrium similarity scalings. Phys. Rev. Lett. **111**, 144503 (2013)
6. I. Neunaber, M. Hölling, M. Obligado, Leading effect for wind turbine wake models. arXiv preprint arXiv:2209.04935 (2022)
7. I. Neunaber, M. Hölling, M. Obligado, Wind tunnel study on the tip speed ratio's impact on a wind turbine wake development. Energies **15**(22), 1–15 (2022)
8. I. Neunaber, M. Obligado, J. Peinke, S. Aubrun, Application of the Townsend-George wake theory to field measurements of wind turbine wakes. J. Phys. Conf. Ser. (2021)
9. I. Neunaber, J. Peinke, M. Obligado, Application of the Townsend-George theory for free shear flows to single and double wind turbine wakes-a wind tunnel study. Wind Energy Sci. **7**(1), 201–219 (2022)
10. M. Obligado, T. Dairay, J.C. Vassilicos, Non-equilibrium scaling of turbulent wakes. Phys. Rev. Fluids **1**, 44409 (2016)
11. J.L. Ortiz-Tarin, S. Nidhan, S. Sarkar, High-Reynolds-number wake of a slender body. J. Fluid Mech. **918**, A30 (2021)
12. V. Petrović, J. Schottler, I. Neunaber, M. Hölling, M. Kühn, Wind tunnel validation of a closed loop active power control for wind farms. J. Phys. Conf. Ser. **1037**(3) (2018)
13. J. Schottler, A. Hölling, J. Peinke, M. Hölling, Design and implementation of a controllable model wind turbine for experimental studies. J. Phys. Conf. Ser. **753** (2016)
14. J. Schottler, J. Bartl, F. Mühle, L. Sætran, J. Peinke, M. Hölling, Wind tunnel experiments on wind turbine wakes in yaw: redefining the wake width. Wind Energy Sci. **3**(1), 257–273 (2018)
15. A.A. Townsend, *The Structure of Turbulent Shear Flow* (Cambridge University Press, 1976)
16. J.C. Vassilicos, Dissipation in turbulent flows. Annu. Rev. Fluid Mech. **47**, 95–114 (2015)

Event-Based Imaging for Visualization and Measurement of Turbulent Boundary Layers

C. E. Willert and J. Klinner

Abstract We describe the implementation of a particle tracking velocimetry (PTV) system based on event-based vision and demonstrate its application for the characterization of a turbulent boundary layer (TBL) in air. One configuration uses a single event-camera to image a narrow light sheet extending wall-normal from the wall to obtain the time-resolved velocity profile. In another configuration a thin light sheet grazes the surface of a glass window illuminating only the viscous sublayer of the TBL. The data simultaneously captured by three synchronized event-cameras is used to reconstruct the 3d particle tracks within 250 μm of the wall. Particle movement within the viscous sub-layer permit the estimation of the local, instantaneous wall shear stress under the assumption of linearity between particle velocity and wall shear stress. Thereby, 2d distributions of the instantaneous wall shear stress are obtained. PTV tracking algorithms that are orders magnitude faster than correlation-based schemes provide data quality on par with currently used, considerably more expensive, high-speed PIV hardware.

1 Introduction—Event Based Imaging

The concept of event-based vision (EBV) is an emergent technology that is receiving increased attention with regard to possible use in widely different fields of application. Contrary to established, conventional imaging, event-cameras record only changes of intensity at the pixel level and provide an asynchronous stream of intensity-change events, consisting of the event's pixel coordinate, the time of event occurrence along with the direction of the intensity change. Further details on the underlying concepts of EBV along with an overview of applications can be found in [3, 5].

When applied to particle imaging the motion of tracer particles is captured in a compressed data stream which can be converted to sequences of pseudo-images that are suited for PIV processing and particle tracking [8]. Combining event-based

C. E. Willert (✉) · J. Klinner
Institute of Propulsion Technology, German Aerospace Center (DLR), Cologne, Germany
e-mail: chris.willert@dlr.de

R. Örlü et al. (eds.), *Progress in Turbulence X*, Springer Proceedings in Physics 404,
https://doi.org/10.1007/978-3-031-55924-2_25

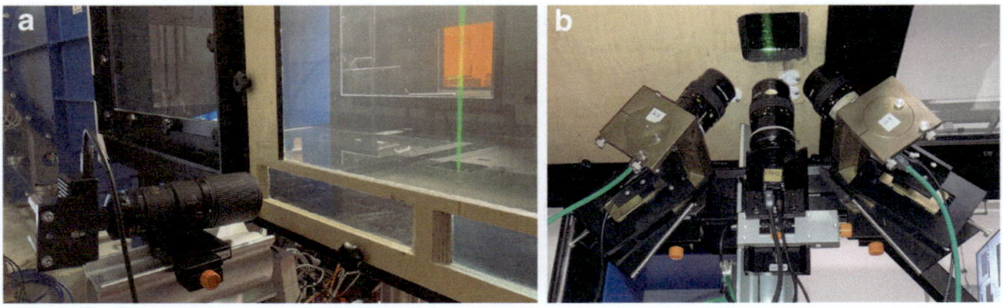

Fig. 1 **a** velocity profile measurement setup using a single event-camera at the 1 m wind tunnel of DLR in Göttingen; **b** triple event-camera setup for particle tracking in the viscous sublayer of a TBL

imaging with pulsed illumination mitigates issues associated with timing latency inherent in the sensor technology, allowing for improved flow field measurements [7]. The present study is focused on the use of event-based imaging velocimetry (EBIV) for TBL measurements in two different imaging configurations. First, a single event camera provides TBL time-resolved velocity profile statistics that are compared with conventional high-speed particle image velocimetry (PIV) measurements (Fig. 1a). The second application utilizes three event-cameras to track the 3d motion of tracer particles within the viscous sub-layer of the TBL (Fig. 1b).

2 TBL Measurement Setup

The measurements are performed in the 1 m wind tunnel of the DLR Institute of Aerodynamics and Flow Technology. The closed circuit facility features a test section of 6 m length with a cross section of 1 m width and 0.8 m height. The TBL is tripped at the end of the contraction and allowed to develop for about 2.1 m at the location of the measurement where it has a thickness $\delta_{99} \approx 40$ mm. For both the high-speed PIV and EBIV measurements a thin laser light sheet is introduced through a thin ($\approx$1.5 mm) glass window at mid-span ($z = 0$) on the bottom of the test section (Fig. 1a). With cameras aligned orthogonal to the light sheet flow field data is captured in a streamwise (x)—wall-normal (y) plane. The spanwise direction is denoted with z.

Dense seeding is provided by water/glycol based aerosol seeding from a fog generator (HazeBase Classic) that is illuminated by a diode pumped pulsed laser (NanioAir, Innolas Photonics). On the detector side, a high-speed CMOS camera (Vision Research V2640) is used for high-speed PIV, whereas two similar event-camera models (Prophesee EVK-2, EVK-4) are used for EBIV.

Near wall imaging of particles in the viscous sublayer is achieved with three synchronized event-cameras looking through the bottom of the test section and a thin light sheet grazing the surface of the glass window (Fig. 1b). Scheimpflug mounts between lenses (Nikon, Micro Nikkor 55 mm f/2.8) ensure focused particle imag-

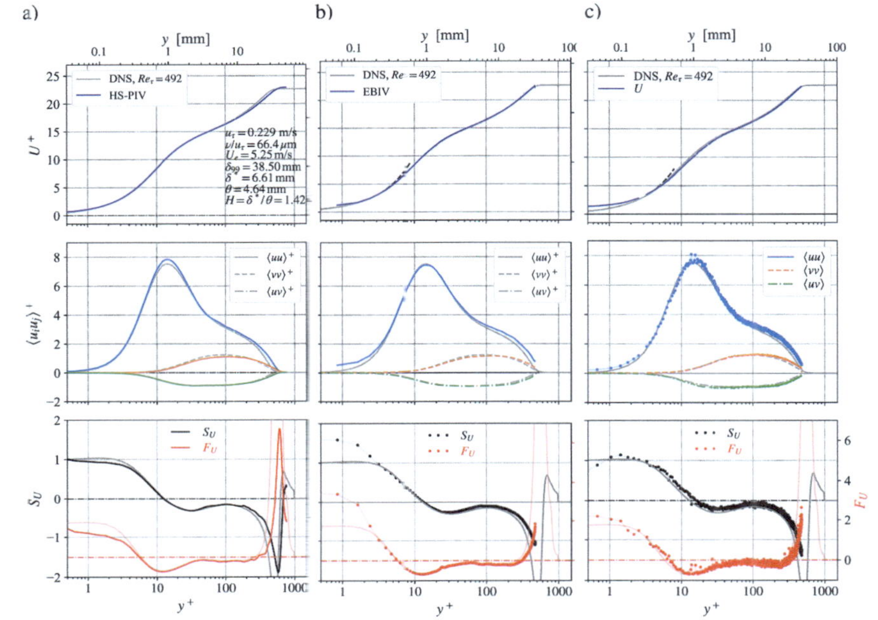

Fig. 2 Velocity profiles in viscous scaling obtained with different profile measurement techniques for $\mathrm{Re}_\tau = 585$ compared to DNS from [6]. **a** high-speed profile-PIV processed with 64×3 samples; **b** event-based imaging processed with 64×8 samples; **c** bin-averaged event-based PTV (bin-size 1 pixel, $\hat{=} 25\,\mu\mathrm{m} \hat{=} 0.4\, y^+$) Symbols used in **b**, **c** illustrate the data density

ing across the field of view of about $12 \times 8\,\mathrm{mm}^2$. The lens aperture is set at f/5.6. The illuminated volume has a thickness of about 250–300 µm as determined by the recovered 3d tracking data.

3 Results

Time resolved velocity profile

Measurements of the wall-normal profile are performed at wind tunnel free-stream velocities of $U_\infty = \mathbf{5.2}$, 7.5, 10.0 m/s with shear Reynolds numbers $\mathrm{Re}_\tau = \mathbf{585}$, 790, 1000 with data for the values in bold presented herein. For high-speed PIV the laser is operated at 10–20 kHz with the camera acquiring 8 sequences of $N = 38\,000$ images for a total record length of 30–15 s. For EBIV the laser was operated at 10 kHz with the event camera(s) streaming up to 60 s of raw event data to the host PC. Both, the PIV recordings as well as pseudo-image sequences generated from the event data are processed by a multi-frame, pyramid-based cross-correlation scheme. Mean and higher order statistics are calculated from the time-resolved velocity profile data sets to obtain the data shown in Fig. 2 and indicate very good agreement between each other and with matching DNS. The event-data is also subjected to a simple PTV algorithm that is initialized with an assumed mean displacement profile. The corresponding profiles in Fig. 2c exhibit slightly higher scatter but manage to capture the wall-normal fluctuations $\langle vv \rangle^+$ even better than PIV and EBIV which are both affected by spatial filtering.

Wall shear stress distribution

The synchronized event data from the three event cameras has to undergo several processing steps to obtain wall shear stress data. First, each view is processed with a 2d PTV scheme to obtain time-resolved particle tracks. Matching tracks from each view are then combined to reconstruct the 3d tracks using a camera calibration approach based on a two-plane camera model. Instantaneous particle velocity is estimated using a 2nd order polynomial fit on the particle track segments of length $N = 5$. Validation is based on the residuals of the 3d track reconstruction and the residuals of the least squares track fit. The wall shear stress vector $\boldsymbol{\tau}_w = [\tau_{w,x}, \tau_{w,z}]$ for each validated particle position is then obtained be dividing its estimated velocity $\mathbf{u} = [u, w]$ by its distance from the wall Δy as an approximation to the definition of wall shear stress

$$\boldsymbol{\tau}_w = \mu \left. \frac{\partial \mathbf{u}}{\partial y} \right|_{y=0} \approx \mu \frac{\mathbf{u}(\Delta y)}{\Delta y} \tag{1}$$

with μ representing the dynamic viscosity.

Figure 3a shows two realizations of recovered near-wall tracks at $U_\infty = 5.2$. The particle positions are color-coded with the mean wall stress magnitude $\|\boldsymbol{\tau}_w\|$ with the mean value gray-colored. While the tracks in Fig. 3a indicate a low-shear condition and even some flow reversal, the flow topology is completely different only 8 ms later (Fig. 3b) when it is dominated by high shear rate aligned with the mean flow direction. The shear rate partially exceeds the mean value by a factor of two.

Probability distributions of both components of the wall shear stress vector are given in Fig. 4b and closely match those obtained from DNS (see e.g. [2]). The data are obtained from 60 s of raw data and are sampled in a wall distance of 40–140 µm ($\approx 1.5\, y^+$). The sequence captures a total of 8000 eddy turn-over times of $\delta_{99}/U_e = 7.5$ ms.

The skewness and flatness of the wall shear stress components are in good agreement with data obtained by DNS at similar Reynolds numbers (e.g. Table I in [2]). The correlations for $\boldsymbol{\tau}_w$ as proposed by [4]

$$\tau^+_{x_i,rms} = \frac{\tau_{x_i,rms}}{\tau_w} = C_{0,i} + 0.018 \ln \mathrm{Re}_\tau \tag{2}$$

with $C_{0,x} = 0.298$ and $C_{0,z} = 0.164$ respectively predict $\tau^+_{x,rms} = 0.413$ and $\tau^+_{z,rms} = 0.279$. Whereas the streamwise wall shear fluctuations are overestimated by about

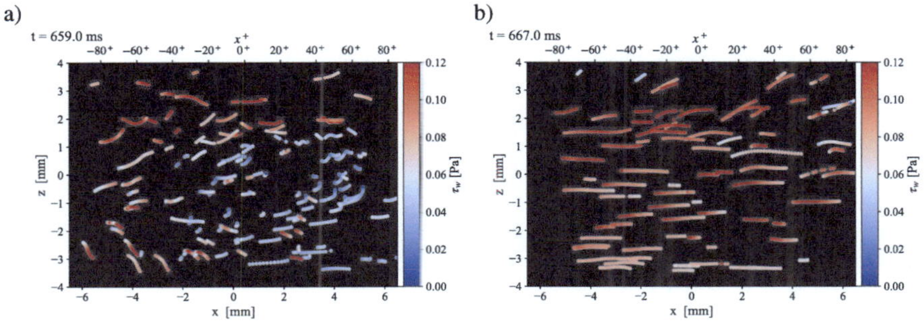

Fig. 3 Processed particle tracks color coded with wall shear stress magnitude at $Re_\tau = 585$. Each frame represents 5 ms of event-data (25 light pulses). The mean flow direction is from left to right

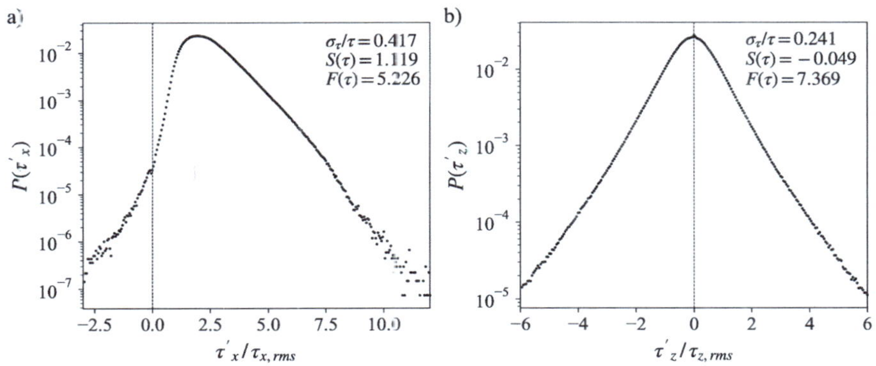

Fig. 4 PDFs of streamwise (**a**) and spanwise (**b**) wallshear stress components normalized by the rms of the respective values compiled from $1.2 \cdot 10^7$ correlated samples from 60 s of raw event data obtained at $U_\infty = 5.2$ m/s ($Re_\tau = 585$)

1% the spanwise shear stress fluctuations are underestimated by more than 10%. While not conclusively resolved, this underestimation is believed to be caused by short-comings of the implemented particle tracking scheme.

A 2d probability distribution of the wall-shear stress is plotted in Fig. 5. The contours indicate a small bulge near $\tau_w = 0$ which is caused by stationary particles that are stuck to the surface. More sophisticated tracking schemes should be able to handle this deficiency as well as recover more tracks from the raw event-data.

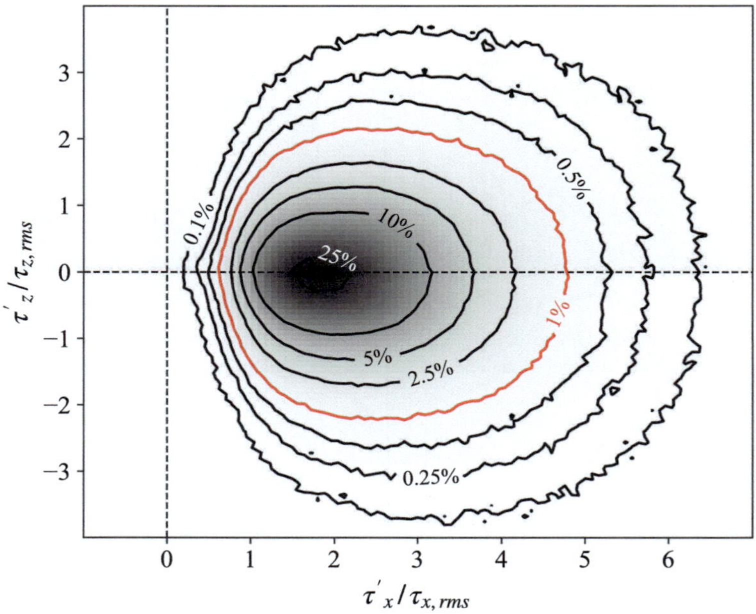

Fig. 5 2d PDFs of wallshear stress vector normalized by the rms of the respective components compiled from $1.2 \cdot 10^7$ correlated samples from 60 s of raw event data obtained at $U_\infty = 5.2$ m/s ($\mathrm{Re}_\tau = 585$). Contour levels represent probabilities of 0.1%, 0.25%, 0.5%, 1% (red), 2.5%, 5%, 10%, 25%

4 Conclusion

The material presented herein demonstrates the viability of event-based imaging velocimetry for accurate measurement of TBL properties, such as velocity profiles or wall shear stress distributions. The reduced data stream of EBV permits continuous recording on the order of minutes or longer. Uncertainties arising from the limited (1-bit) signal depth of the image data can be accounted for by making use of the available temporal resolution of the raw data which is on the order of 10 kHz (e.g. with Wiener/Kalman filtering [1]).

The time-resolved data was acquired using hardware that is considerably cheaper in comparison to conventional high-speed PIV components. Beyond this, the higher sensitivity of the EBV detectors reduce the power requirements of the laser used to illuminate the tracer particles.

References

1. D. Borer, T. Delbruck, T. Rösgen, Three-dimensional particle tracking velocimetry using dynamic vision sensors. Exp. Fluids **58**(165) (2017). https://doi.org/10.1007/s00348-017-2452-5
2. C. Diaz-Daniel, S. Laizet, J.C. Vassilicos, Wall shear stress fluctuations: mixed scaling and their effects on velocity fluctuations in a turbulent boundary layer. Phys. Fluids **29**(5), 055102 (2017). https://doi.org/10.1063/1.4984002
3. G. Gallego, T. Delbrück, G. Orchard, C. Bartolozzi, B. Taba, A. Censi, S. Leutenegger, A.J. Davison, J. Conradt, K. Daniilidis, D. Scaramuzza, Event-based vision: a survey. IEEE Trans. Pattern Anal. Mach. Intell. **44**(1), 154–180 (2022). https://doi.org/10.1109/TPAMI.2020.3008413
4. R. Örlü, P. Schlatter, On the fluctuating wall-shear stress in zero pressure-gradient turbulent boundary layer flows. Phys. Fluids **23**(2), 021704 (2011). https://doi.org/10.1063/1.3555191
5. Robotics and Perception Group: Event-based vision resources. https://github.com/uzh-rpg/event-based_vision_resources (2023)
6. P. Schlatter, R. Örlü, Assessment of direct numerical simulation data of turbulent boundary layers. J. Fluid Mech. **659**, 116–126 (2010). https://doi.org/10.1017/S0022112010003113
7. C. Willert, Event-based imaging velocimetry using pulsed illumination. Exp. Fluids **64**, 98 (2023). https://doi.org/10.1007/s00348-023-03641-8
8. C. Willert, J. Klinner, Event-based imaging velocimetry: an assessment of event-based cameras for the measurement of fluid flows. Exp. Fluids **63**, 101 (2022). https://doi.org/10.1007/s00348-022-03441-6

Augmenting PIV Temporal Resolution via Semi-Lagrangian Estimation of Velocity Fluctuations

M. Vocke, R. Kapulla, C. Morton, and R. J. Martinuzzi

Abstract This study introduces a convective velocity estimation method based on space-time correlations to enhance the temporal resolution of turbulent flow measurements using semi-Lagrangian techniques. The approach is evaluated using DNS data of a turbulent plane jet at $Re = 10{,}000$. Comparison between correlation-based convective velocity U_c and mean axial velocity U_m reveals significant differences, with U_c generally lower near the jet centreline and higher towards lateral edges. Spectral analysis demonstrates that U_c based advection yields improved predictions of the turbulence spectrum compared to U_m. Future work will explore scale-dependent convective velocities and their impact on reconstructed frequency spectra.

1 Introduction

Turbulent flows are characterized by interactions across a wide spectrum of spatiotemporal scales. Accordingly, the properties of turbulence are best described using high-frequency, high resolution data [1]. Particle Image Velocimetry (PIV) is an experimental technique that provides instantaneous flow measurements. While stan-

M. Vocke
Department of Mechanical and Manufacturing Engineering, University of Calgary, Calgary, Canada
e-mail: maegan.vocke@ucalgary.ca

R. Kapulla
Paul Scherrer Institute, Villigen, Switzerland
e-mail: ralf.kapulla@psi.ch

C. Morton
Department of Mechanical Engineering, McMaster University, Hamilton, Canada
e-mail: mortoc5@mcmaster.ca

R. J. Martinuzzi (✉)
Department of Mechanical and Manufacturing Engineering, University of Calgary, Calgary, Canada
e-mail: rmartinu@ucalgary.ca

R. Örlü et al. (eds.), *Progress in Turbulence X*, Springer Proceedings in Physics 404,
https://doi.org/10.1007/978-3-031-55924-2_26

dard systems are well suited towards statistical descriptions of turbulence, resolving the spatiotemporal scales of complex flows remains challenging due to the scarcity of high-speed PIV systems. This work introduces a strategy for increasing the temporal resolution of PIV measurements beyond the Nyquist limit.

Several methods have been employed to increase the temporal resolution of turbulent flow data. Techniques based on the amalgamation of single-point measurements and PIV datasets have been proposed by authors such as Discetti et al. [2], who utilized POD-based stochastic estimation. Machine learning techniques has become increasingly popular, despite issues relating to the selection and curation of the training data [3]. In contrast, some models rely solely on the available PIV measurements. Schienders et al. [4] utilized a vortex-in-cell method to perform temporal up-sampling, yet, divergence-free constraints limit the method's applicability towards 3D flow data. Recently, Vocke et al. [5] has demonstrated the ability of an advection-based flow reconstruction technique. Employing Rapid Distortion Theory (RDT), a quasilinear model was developed using semi-Lagrangian methods to recover spectral information up to two orders of magnitude beyond the Nyquist limit.

The present investigation extends upon [5] to improve the prediction of small-scale fluctuations. Specifically, the assumption that spatial and temporal scales are simply related by the mean flow is challenged. The quasilinear approximation within the framework of RDT reduces energy paths leading towards dissipation, as the fluctuations are limited to interactions with the convective velocity [6]. Thus, it becomes important to accurately specify the convective velocities pertaining to the turbulent fluctuations. Taylor's Hypothesis, which states that the convective velocity of the turbulent fluctuations U_c is equal to the local mean velocity U_m [7] is valid for homogeneous flows, however, it becomes less accurate when examining flows characterized by substantial mean shears, high turbulence intensity, and significant viscous forces. Shear-dominated flows such as turbulent boundary layers and jets are expected to influence the development of the eddy structure, thus, the flow patterns typically vary as they travel downstream [8]. Utilizing space-time correlations, this work describes a method to estimate broadband convective velocities for use in the semi-Lagrangian flow reconstruction technique.

2 Methodology

The DNS dataset of a turbulent plane jet with $Re = 10{,}000$ and a non-dimensional sampling rate $f_s = 24.86$ [9] is used to investigate a semi-Lagrangian flow reconstruction technique with a novel specification of convective velocity based on space-time correlations. The technique is assessed by comparing the reconstructed velocity fields and the ground truth data for a range of sub-sampling factor $S^* = f_s/f_n$, where the DNS data is re-sampled at $f_n < f_s$. Several test cases are generated based on the sub-sampling factor, where the successive flow measurements at t_f and t_b act as the initial conditions for the quasilinear model. The development the semi-Lagrangian technique, as outlined by [5], is summarized briefly. Applying a Reynolds decomposition to the Navier-Stokes equation, a relationship governing the transport of turbulent fluctuations can be established,

$$\frac{D\mathbf{u}'}{Dt} = -\nabla \cdot \left(\mathbf{u}'\mathbf{u}' - \overline{\mathbf{u}'\mathbf{u}'}\right) - (\mathbf{u}' \cdot \nabla)\, \mathbf{U_c} = \mathbf{N} + \mathbf{L}. \tag{1}$$

The nonlinear term, **N**, is neglected under Rapid Distortion Theory [10]. The problem is solved numerically using the semi-Lagrangian method, which can be described by the following system of ordinary differential equations,

$$\frac{D\mathbf{u}'}{Dt} = \mathbf{L} = -(\mathbf{u}' \cdot \nabla)\, \mathbf{U_c}, \tag{2a}$$

$$\frac{d\mathbf{x}}{dt} = \mathbf{U_c}. \tag{2b}$$

Here, $\mathbf{u}'$ represents the fluctuating velocity, while $\mathbf{x}$ describes its evolution along the Lagrangian trajectory due to the convective velocity, $\mathbf{U_c}$. Assuming the known data is separated by $\Delta t = t_b - t_f$, forward and backward time increments are established using a basic linear weighting technique,

$$\Delta t_f = \left(\frac{n}{S^*}\right) \Delta t, \tag{3a}$$

$$\Delta t_b = \left(1 - \frac{n}{S^*}\right) \Delta t, \tag{3b}$$

for $n = 1, 2, ..., S^* - 1$. Consequently, (2a) and (2b) can be formulated in relation to both the forward and backward time steps to estimate the flow at the intermediate time t_n. This estimation is accomplished through a linear temporal weighting,

$$\mathbf{u}'(\mathbf{x}, t_n) = \left(1 - \frac{n}{S^*}\right) \mathbf{u}'_f(\mathbf{x}, t_n) + \left(\frac{n}{S^*}\right) \mathbf{u}'_b(\mathbf{x}, t_n), \tag{4}$$

where $\mathbf{u}'_f$ and $\mathbf{u}'_b$ are the forward and backward estimates respectively. To estimate the convective velocity, the cross correlation function ρ_{ij} is considered. Defining the correlation of two signals separated by spatial distance $\Delta\mathbf{x}$, where the upstream signal lags by time delay τ, results in

$$\rho_{ij}(\mathbf{x} + \Delta\mathbf{x}, t + \tau) = \frac{\overline{u_i(\mathbf{x}, t) u_j(\mathbf{x} + \Delta\mathbf{x}, \tau)}}{\sqrt{u_i(\mathbf{x}, t)^2}\sqrt{u_j(\mathbf{x} + \Delta\mathbf{x}, t)^2}}. \tag{5}$$

The convective velocity relates the temporal scales to the various spatial scales within a flow,

$$U_c = \frac{\Delta x}{\tau}, \tag{6}$$

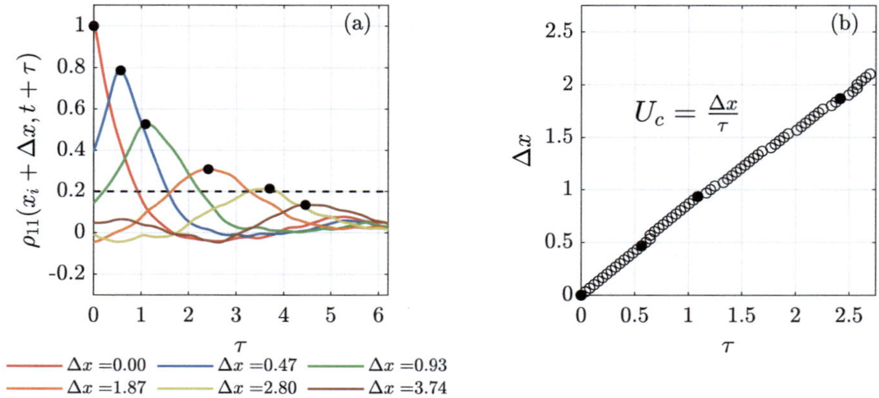

Fig. 1 (**a**) Series of cross correlation measurements for the reference point $x_i = 6.0$. Each curve represents a different downstream separation. The black dash lined indicates the limit for the significance of correlation. (**b**) Example estimation of the convective velocity for the reference point $x_i = 6.0$. Here, the time delay required to reach the maximum correlation versus the spatial separation between the measurement points is displayed. The filled black circles indicate the example spatial separations up to $\Delta x = 1.87$, as shown in Fig. 1a

where U_c is the local convective velocity. The convective velocity considers the time delay which maximizes the cross correlation coefficient for a given spatial separation. In regions subject to high turbulent intensity, the turbulent structures undergo modifications due to the combined effects of the mean and fluctuating shear stresses [8]. Consequently, the maximum amplitude of the correlation can vary as the spatial separation between the measurement points increases. A broadband U_c is obtained by considering the envelope of significant cross correlation functions. The statistical significance threshold σ_N of a correlation computed from N samples can be estimated using correlations derived from white noise. The variability among repeated cross correlations for limited sample size is utilized to gauge the lower limit of correlation significance. An illustrative example is provided in Fig. 1. The convective velocity, $U_c = \Delta x/\tau$, is estimated utilizing cross correlations that satisfy $\rho_{ij}(0, 0) < \sigma_N$, where $\sigma_N = 0.2$ for the present case.

3 Results and Discussion

A comparison between the correlation-based convective velocity and axial component of the mean velocity is displayed in Fig. 2. Significant lateral variations are noted between U_c and U_m, where $U_c < U_m$ towards the jet centreline ($-0.5 < z/D < 0.5$), while $U_c > U_m$ towards the lateral edges of the domain ($\pm 0.5 < z/D < 3$). These differences persist for axial distances up to $x/D \approx 8$. Traveling towards the fully

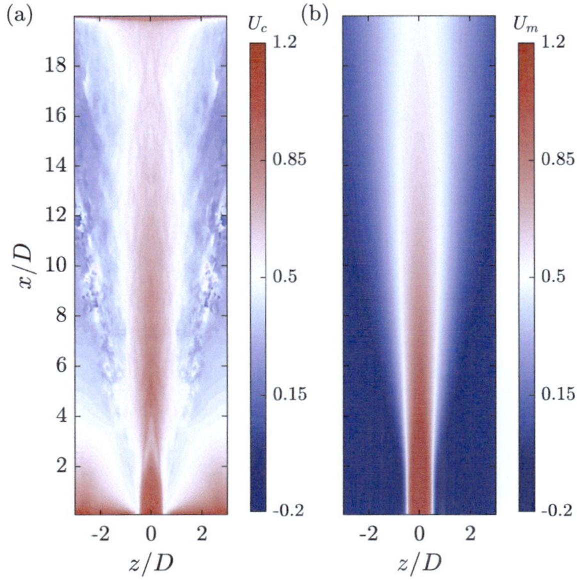

Fig. 2 Comparison between (**a**) correlation-based convective velocity U_c and (**b**) mean axial velocity U_m

developed region, $x/D > 10$, $U_c \approx U_m$ towards the jet centreline, however, U_c is still greater than U_m.

To asses the ability of the flow reconstruction technique to recover spectral content, the present work utilizes a pre-multiplied energy difference,

$$E_{ii}(\mathbf{x}, f) = 100 \, \frac{f U_0}{D} \, \frac{\widetilde{\Phi}_{ii}(\mathbf{x}, f) - \Phi_{ii}(\mathbf{x}, f)}{\langle u_i'(\mathbf{x}, t) \, u_i'(\mathbf{x}, t) \rangle}, \tag{7}$$

where $\widetilde{\Phi}_{ii}(\mathbf{x}, f)$ and $\Phi_{ii}(\mathbf{x}, f)$ represent the up-sampled and reference spectra, respectively. The spectral reconstruction errors for along the jet centreline with $S^* = 50$ and 150 are displayed for the u' components of velocity in Fig. 3 for estimations based on U_c and U_m. The difference of the errors (see Fig. 3e, f) highlight the regions where U_c estimations outperform those based on U_m. The spectral errors in the fully developed region, $x/D > 10$, (not shown in Fig. 3) are generally the same for estimations based on U_c and U_m. Both models are equally influenced by the boundary conditions, which have a notable effect up to $x/D \approx 2$. A slight improvement is noted for the reconstruction of the fundamental jet frequency when using U_c, as well as an improved prediction towards the high frequency scales.

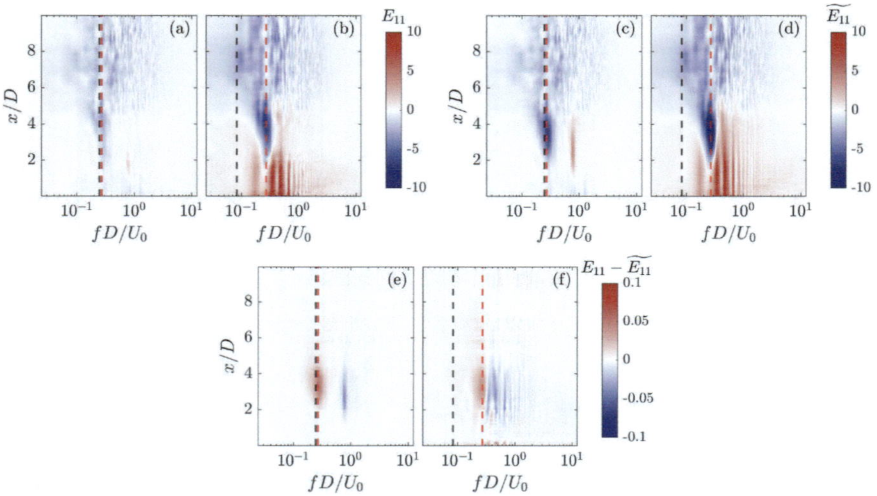

Fig. 3 Spectral reconstruction error (refer to (7)) for the u-component of velocity along the jet centreline, for the U_c based estimates E_{11}, (**a**) $S^* = 50$, (**b**) $S^* = 150$, U_m based estimates $\widetilde{E_{11}}$, (**c**) $S^* = 50$, (**d**) $S^* = 150$, and the difference between the estimates, $E_{11} - \widetilde{E_{11}}$, (**e**) $S^* = 50$, and (**f**) $S^* = 150$ The red and black lines represent the fundamental jet frequency f_v and the Nyquist frequency of the under-sampled data set, f_N, respectively

4 Conclusion

This work utilizes a convective velocity based on space-time correlations to increase the temporal resolution of turbulent flow measurements with semi-Lagrangian techniques. Preliminary findings show that models using U_c for advection outperform those using U_m, with spectral reconstruction errors less than $\pm 10\%$ for the worst case of $S^* = 150$. Future work aims to explore scale-dependent convective velocities for enhanced frequency spectrum predictions.

Acknowledgements The authors would like to thank the Natural Sciences and Engineering Research Council of Canada (NSERC) and the Nuclear Education, Skills and Technology (NEST) Framework for funding this research.

References

1. F.O. Thomas, V.W. Goldschmidt, Structural characteristics of a developing turbulent planar jet. J. Fluid Mech. **163**, 227–256 (1986)
2. S. Discetti, M. Raiola, A. Ianiro, Estimation of time-resolved turbulent fields through correlation of non-time-resolved field measurements and time-resolved point measurements. Exp. Therm. Fluid Sci. **93**, 119–130 (2017)

3. K. Fukami, K. Fukagata, K. Taira, Machine-learning-based spatio-temporal super resolution reconstruction of turbulent flows. J. Fluid Mech. **909**, A9 (2021)
4. J.F.G. Schneiders, R.P. Dwight, F. Scarano, Time-supersampling of 3D-PIV measurements with vortex-in-cell simulation. Exp. Fluids **55**, 1692 (2014)
5. M. Vocke, R. Kapulla, C. Morton, M. Klein, R.J. Martinuzzi, Advection-based temporal reconstruction technique for turbulent velocity fields. Phys. Fluids **35**, 015123 (2023)
6. J. Marston, S. Tobias, Recent developments in theories of inhomogeneous and anisotropic turbulence. Annu. Rev. Fluid Mech. **55**, 351–375 (2023)
7. V.W. Goldschmidt, M.F. Young, E.S. Ott, Turbulent convective velocities (broadband and wavenumber dependent) in a plane jet. J. Fluid Mech. **105**, 327–345 (1981). Publisher: Cambridge University Press
8. M.J. Fisher, P.O.A.L Davies, Correlation measurements in a non-frozen pattern of turbulence. J. Fluid Mech. **18**, 97–116 (1964)
9. M. Klein, A. Sadiki, J. Janicka, Investigation of the influence of the Reynolds number on a plane jet using direct numerical simulation. Int. J. Heat Fluid Flow **24**(6), 785–794 (2003)
10. J.C.R. Hunt, D.J. Carruthers, Rapid distortion theory and the 'problems' of turbulence. J. Fluid Mech. **212**, 497 (1990)

The Beauty of Active Grids and Their Infinite Possibilities of Turbulence Generation

L. Neuhaus, A. Hölling, M. Hölling, M. Wächter, and J. Peinke

Abstract Wind tunnel experiments are of major importance for fundamental turbulence research as well as investigation of objects in realistic reproducible conditions. Active grids are used to increase the possibilities of generating turbulent flows. There are several approaches to generate turbulent flows, as different needs require different excitation approaches and different grid designs. In this study, we focus on the needs of future wind turbine studies, which require a tailored turbulence intensity gradient. Experiments are conducted in the closed test section of the large wind tunnel in Oldenburg. A new half active grid is built, which consist of a classic active grid at the lower half and a completely empty upper half. In this way the new half active grid enables to generate inhomogeneous flows with strongly varying turbulence parameters. A reduction of the turbulence intensity with height, as present in the atmospheric boundary layer, can be generated. There are already many different methods to generate turbulent flows with active grids. However, there is still a wide range of possibilities to excite different turbulent flow properties and to further adapt the flows, either by adapting the shaft motion or by adapting the active grid design.

L. Neuhaus (✉) · A. Hölling · M. Hölling · M. Wächter · J. Peinke
ForWind and Institute of Physics, University of Oldenburg, 26129 Oldenburg, Germany
e-mail: lars.neuhaus@uol.de

A. Hölling
e-mail: agnieszka.hoelling@uol.de

M. Hölling
e-mail: michael.hoelling@uol.de

M. Wächter
e-mail: matthias.waechter@uol.de

J. Peinke
e-mail: joachim.peinke@uol.de

R. Örlü et al. (eds.), *Progress in Turbulence X*, Springer Proceedings in Physics 404,
https://doi.org/10.1007/978-3-031-55924-2_27

1 Introduction

Wind tunnel investigations are a key factor in fundamental turbulence research as well as an important tool when it comes to validating new concepts of e.g. wind turbines. Their importance in the future should therefore not be underestimated. To tackle future research questions and fulfill requirements of all kinds of special applications, a tremendous amount of different flow situations needs to be generated.

Starting with simple introduction of turbulence in the wind tunnel with passive grids [1], a variation of the passive grid design following fractal scaling was introduced as discussed in [2]. Another approach was introduced by [3], who changed static bars to active driven ones, allowing for a variation of the generated flow field as well as an increase of the Reynolds number corresponding to a longer range of the turbulence spectra.

In the context of wind energy research, the necessity of reproducing atmospheric wind conditions in the wind tunnel becomes apparent. Here active grids can be used to tailor shear flows with relative uniform turbulence intensity (see [4, 5]). Further investigations with active grids were undertaken to induce non uniform turbulence intensity distributions over the wind tunnel cross section (see [6, 7]). Reviews on active grids are given by [8, 9].

The objective of this work is to achieve strong turbulence intensity (TI) gradients. This is accomplished by a new, specially designed active grid. Some preliminary results are shown which indicate the adaptability of the active grid to form the desired flows.

2 Experimental Setup

The experiments are conducted in the large wind tunnel in Oldenburg [10]. The wind tunnel was specially designed for the operation with an active grid. Initially, an active grid covering the full ($3\,\text{m} \times 3\,\text{m}$) cross section was designed and used to tailor flows (see [11]). In this study, a further development of this "common" active grid is made to allow imposing strong TI gradients (Fig. 1a). The idea is to generate a flow, which is fully laminar at the upper part of the cross section and can exhibit different, turbulent properties at the lower part. These properties can be tailored based on the previous work with the "common" active grid [11].

The design and control of the half active grid (HAG) is based on the "common" active grid [10]. The outer frame is built similar and can be mounted to the wind tunnel nozzle of the large wind tunnel in Oldenburg. The upper part is kept empty, to ensure an undisturbed laminar flow. The lower part is equipped with shafts following the design of the "common" active grid with an identical mesh width M of 0.143 m. The upper and lower parts are separated at 1.22 m (0.41 wind tunnel height H) by a beam, which supports the vertical shafts. In consequence, the vertical shafts are shorter and only covering 8.5 mesh width. The support beam is equipped with

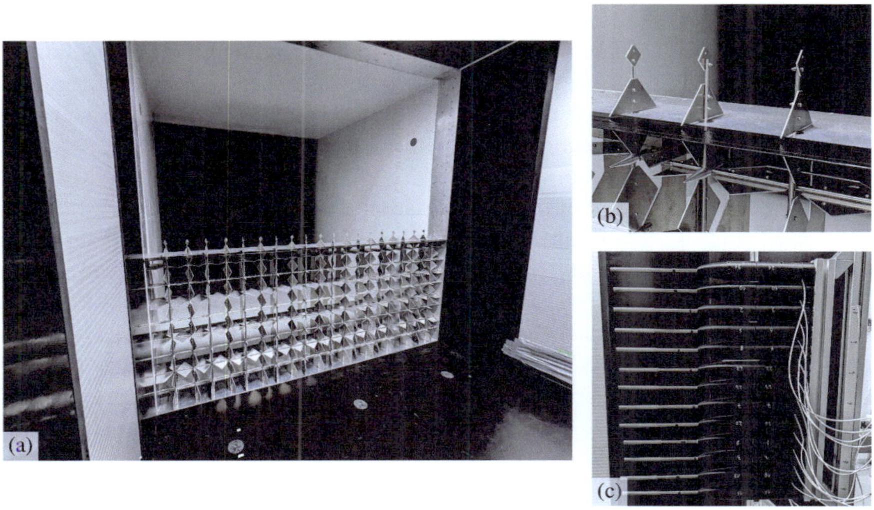

Fig. 1 Half active grid installed at the nozzle in the closed test section of the large wind tunnel in Oldenburg (**a**), detail view of *fence* above the support beam and aerodynamically optimized trailing edge of the support beam (**b**), and detail view of hot wire array mounted to support (**c**)

an aerodynamically optimized trailing edge to reduced its influence on the flow. Furthermore, a *fence* is installed to enable a smoother transition between the two different flow states (Fig. 1b). The horizontal shafts are split in the middle ($10.5M$) and both sides can be controlled individually. For the control of the servo motors (Kollmorgen AKM32H-ANMNC-00) the same hardware as for the "common" active grid is used.

Different active grid motion protocols are tested to manipulate the mixing between the different flow states (laminar at the top and turbulent at the bottom). A static motion protocol with different shaft angles and two active motion protocols, each with different standard deviations of the shaft angle's motion around a mean position, are considered. For all motion protocols, neighboring shafts point in opposite directions. By our definition, a shaft angle of 0° corresponds to an open position where the flaps are aligned with the main flow direction.

For the *static* motion protocols, horizontal shafts have an angle of 35° at the bottom, which gradually decreases to 7° for the topmost shaft. Cases with constant static angle for all vertical shafts from 0° to 30° are investigated.

In the first active motion protocol (*active I*), vertical shafts are static at 0°. The horizontal shafts are randomly flapping around a mean shaft angle of 30° with a standard deviation of 5° to 15°, depending on the case studied. However, the upper two horizontal shafts are static at 0° to allow for a smoother transition between the flow states.

In the second active motion protocol (*active II*), horizontal shafts have an angle of 35° at the bottom, which gradually decreases to 7° for the topmost shaft (corresponding to *static*). The vertical shafts are randomly flapping around a mean shaft angle of 30° with a standard deviation of 5° to 15°, depending on the case studied.

As with the HAG a strongly inhomogenous flow is expected a simple characterization by a single hot wire is not possible. The turbulent properties are assumed to vary strongly in vertical direction, whereas similar turbulent properties are assumed in the horizontal transversal direction.

To get a detailed picture of the flow variation along the vertical axis, a hot wire array consisting of 13 hot wire with a spacing of 3 cm ($0.01H$) is built (Fig. 1c). The hot wire array can be shifted in vertical direction and heights between 135 cm ($0.45H$) and 220 cm ($0.73H$) are covered. However, only a range of 36 cm ($0.12H$) of the flow is measured simultaneously. Hot wires are operated with Dantec Dynamics multichannel CTA 54N80. Signals are recorded with a sampling frequency of 6 kHz and a hardware low-pass filter with a frequency of 3 kHz. The hot wires are calibrated at least twice a day.

3 Results

Here the distribution of the turbulence intensity (TI) over height will be investigated at a position 4.75 m ($33M$) downstream of the HAG. A focus is set on the transition region between laminar flow at the top and turbulent flow at the bottom. Figure 2 shows the effect of the different motion protocols on the TI distribution and power spectral density (PSD) over height.

For the *static* motion protocols, low TIs ($<2\%$) are found in the observed height range from $0.53H$ to $0.65H$ (Fig. 2a). The flow exhibits very low TI ($<0.5\%$) for heights above $0.6H$. With increasing shaft angle α the TI at lower heights ($0.53H$) increases from 0.32% ($\alpha = 0°$) to 2% ($\alpha = 30°$). However, it is quite constant at the upper heights ($0.65H$) with TIs from 0.25% ($\alpha = 0°$) to 0.32% ($\alpha = 30°$).

The *active I* motion protocols allow for an adaption of the TI at all heights (Fig. 2b). With increasing standard deviation σ of the shaft motion the TI increases by roughly the same absolute value at all heights (0.6% at $0.48H$ and 0.4% at $0.63H$ for and increase of $\sigma(\alpha)$ by 5°). Hence, this motion protocol allows for increasing the TI at all heights and especially for setting the TI of the low turbulent flow.

The *active II* motion protocols allow for an adaption of the TI at lower heights (Fig. 2c). TIs over 9% can be reached at the lowest observed height ($0.48H$) whereas low TIs ($<1\%$) can still be found at the higher positions ($z \geq 0.65H$). Further, also the height where a significant increase of the TI (around a TI of 2%) can be observed is shifted upwards ($0.53H$ for $\sigma(\alpha) = 5°$, $0.55H$ for $\sigma(\alpha) = 10°$, and $0.58H$ for $\sigma(\alpha) = 15°$) by increasing the standard deviation of the shaft motion.

Examining the power spectral density (PSD), the differences between the motion protocols become obvious (Fig. 2d–f). For the *static* motion protocol energy is only contained in the high frequencies respectively small scales (Fig. 2d). When the shafts are flapping (*active I* and *active II*), more energy is induced in the lower frequencies respectively larger scales (Fig. 2e, f). For all cases a significant decrease of fluctua-

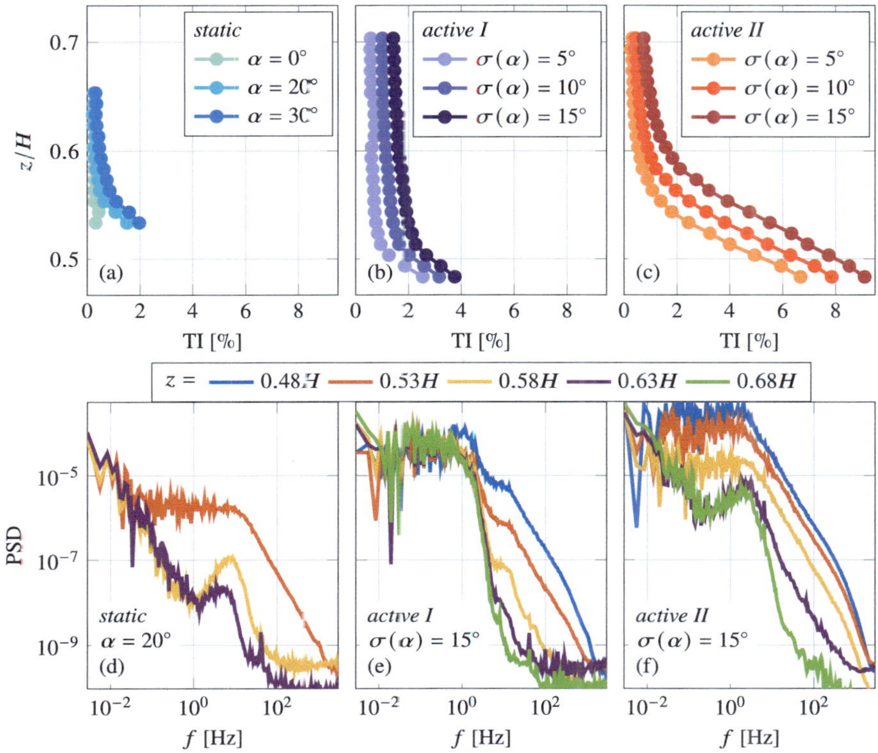

Fig. 2 Turbulence intensity (TI) (**a–c**) and power spectral density (PSD) (**d–f**) dependent on the height z for *static* (**a**, **d**), *active I* (**b**, **e**), and *active II* (**c**, **f**) motion protocols

tions/energy in the small scales with increasing height can be observed. The major difference between the two active motion protocols can be found in the region where energy is induced. For *active I*, the energy at low frequencies is quite similar for all heights, whereas high frequencies are hardly affected and exhibit similar energies as observed for the *static* motion protocol (Fig. 2d, e). In contrast, the *active II* motion protocol allows to influence both, small and large scales (Fig. 2f).

It is well known, that turbulence undergoes a continuous development downstream of a grid (see [1, 2, 12]). The downstream development of the TI by the *static* motion protocol generated flow is shown in Fig. 3. The TI is increasing downstream for almost all observed heights, but seems to be a simple shift upwards. Desired flow properties can thus also be achieved by the appropriate choice of the downstream position.

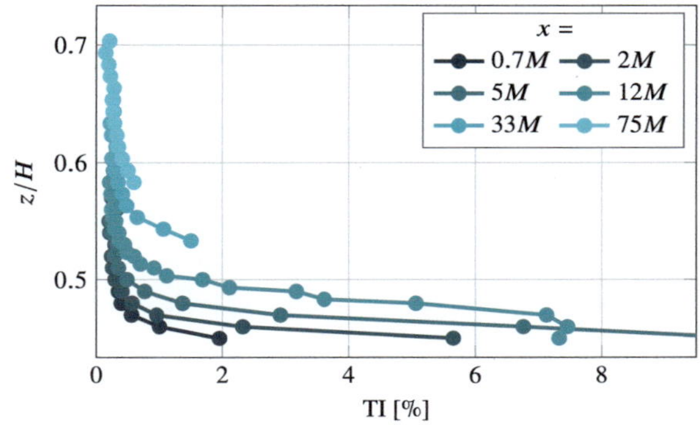

Fig. 3 Turbulence intensity (TI) dependent on the height z for different downstream positions x for the *static* $\alpha = 20°$ motion protocol

4 Conclusion

The new designed half active grid allows for generating strong TI gradients in the wind tunnel. Very low ($<1\%$) and higher ($>9\%$) TIs were imprinted in a range of 50 cm (1/6 of the wind tunnel height). The position and the strength of this gradient is adaptable by the active grid motion protocol. Thus, it is possible to adjust the TI independently at different heights.

Future studies will systematically investigate the potential for independently manipulating shear and TI. To do so, an adaption of the blockage by a fine mesh in the laminar part will be considered (following the approach of [6]). This is just another example of the infinite possibilities of using active grids to influence flow characteristics or rather tailor them to specific desires.

Acknowledgements The project has been funded by the Deutsche Forschungsgemeinsschaft (DFG, German Research Foundation)—SFB1463—434502799.

References

1. L. Simmons, C. Salter, Proc. R. Soc. Lond. Ser. A Containing Pap. Math. Phys. Charact. **145**(854), 212 (1934)
2. D. Hurst, J. Vassilicos, Phys. Fluids **19**(3), 035103 (2007)
3. H. Makita, Fluid Dyn. Res. **8**(1–4), 53 (1991)
4. R.J. Hearst, B. Ganapathisubramani, Wind Energy **20**(12), 2021 (2017)
5. L. Li, R.J. Hearst, M.A. Ferreira, B. Ganapathisubramani, Renew. Energy **149**, 735 (2020)
6. H.S. Kang, C. Meneveau, Phys. Fluids **20**(12) (2008)
7. S. Gerashchenko, G. Good, Z. Warhaft, J. Fluid Mech. **668**, 293 (2011)
8. L. Mydlarski, Fluid Dyn. Res. **49**(6), 061401 (2017)
9. R.J. Hearst, in *Progress in Turbulence VIII: Proceedings of the iTi Conference in Turbulence 2018 8* (Springer, 2019), pp. 173–178

10 L. Kröger, J. Frederik, J.W. van Wingerden, J. Peinke, M. Hölling, J. Phys. Conf. Ser. **1037**(5), 052002 (2018)
11 L. Neuhaus, F. Berger, J. Peinke, M. Hölling, Exp. Fluids **62**(6), 1 (2021)
12 L. Neuhaus, M. Hölling, W.J.T. Bos, J. Peinke, Phys. Rev. Lett. **125**(15), 154503 (2020)

Turbulent Wall-Pressure Fluctuations in the CICLoPE Facility

G. Dacome, L. Lazzarini, A. Talamelli, G. Bellani, and W. J. Baars

Abstract Wall-pressure spectra and coherence between wall-pressure and streamwise velocity in a turbulent pipe flow are presented. An experimental investigation was conducted in the CICLoPE long-pipe facility at friction Reynolds numbers in the range of $4\,700 \lesssim Re_\tau \lesssim 46\,000$. Wall-pressure energy spectra reveal an alignment of the inner-spectral peak location in terms of $\lambda_x^+ \approx 250$, as well as an increase in overall energy content with increasing Re_τ. Linear coherence spectra between wall-pressure and streamwise velocity in the logarithmic region follow a Reynolds-number-independent wall-scaling. Identification of such a universal scaling contributes to compelling evidence that wall-pressure sensing, as an input for real-time flow control, is a feasible approach for implementation in practical engineering systems.

1 Introduction

Real-time control of large-scale structures in wall-bounded turbulent flows, for skin-friction drag reduction, requires a robust sensing architecture and scalability to application-level conditions. For Reynolds numbers of practical relevance, the friction Reynolds number $Re_\tau \equiv \delta/l^* = \delta U_\tau/\nu$ (the ratio of inertial-to-viscous length scales) is in the order of $\mathcal{O}\left(10^3\right)$ to $\mathcal{O}\left(10^6\right)$. Here, δ is the boundary layer thickness (in our work equal to the pipe radius R), $U_\tau \equiv \sqrt{\tau_w/\rho}$ is the friction velocity (with τ_w being the wall-shear stress and ρ the fluid density) and ν is the kinematic viscosity. At these conditions, large-scale motions in the logarithmic region become (1) energetically dominant over the near-wall structures, and (2) the primary contributor to the bulk turbulence kinetic energy production. And so, large-scale turbulence forms

G. Dacome (✉) · W. J. Baars
Faculty of Aerospace Engineering, Delft University of Technology, 2629HS Delft, The Netherlands
e-mail: g.dacome@tudelft.nl

L. Lazzarini · A. Talamelli · G. Bellani
Dipartimento di Ingegneria Industriale, Università di Bologna, 47121 Forlì, Italy

R. Örlü et al. (eds.), *Progress in Turbulence X*, Springer Proceedings in Physics 404,
https://doi.org/10.1007/978-3-031-55924-2_28

an attractive target for streamwise-persistent control [1]. Given the requirement to minimize form drag, the question arises as to whether wall-pressure (which is a relatively easy quantity to measure) can serve as an input for detecting large-scale velocity fluctuations.

Studies of the wall-pressure in turbulent wall-bounded flows have been carried out by means of both numerical and wind tunnel experiments. Literature (e.g. Farabee and Casarella [2] and Klewicki et al. [3] for experimental work, and Panton et al. [4] for a numerical study) reports the Reynolds-number scaling of wall-pressure energy spectra and its intensity. An inner-spectral peak was found to reside around a streamwise wavelength of $\lambda_x^+ \equiv \lambda_x U_\tau/\nu \approx 250$. Its peak-magnitude increases as a function of Re_τ and so does the large-scale energy content. Relations between wall-pressure events and turbulent velocities have also been investigated. For instance, Thomas and Bull [5] revealed characteristic wall-pressure signatures associated with near-wall burst-sweep events. Spatio-temporal correlation between streamwise velocity and wall-pressure have also been studied [6]. These have revealed a low but significant coherence between the low-frequency content of the wall-pressure signal and the velocity in the logarithmic region of a Turbulent Boundary Layer (TBL) caused by the passage of large-scale motions.

With the goal of analysing the potential of wall-pressure as an input to real-time control for skin-friction drag reduction in wall-bounded turbulence, we present an experimental investigation of the Reynolds-number scaling of the wall-pressure spectrum in high-Reynolds-number pipe flow. Additionally, we investigate the scaling of the scale-dependent correlation between wall-pressure and streamwise velocity in the logarithmic region. This will provide insight into whether wall-pressure can be employed as a wall-based input for real-time control when considering both laboratory- and application-level Reynolds numbers.

2 Experimental Methodology

Experiments were carried out in the Centre for International Cooperation in Long-Pipe Experiments (CICLoPE, see Fig. 1a, b). The facility comprises a 111.15 m-long circular pipe with a radius of $R = 0.4505$ m; for the reported experiments the friction Reynolds number was in the range $4\,700 \lesssim Re_\tau \lesssim 46\,000$. All testing conditions considered are reported in Table 1. The flow in the test section (at the downstream end of the pipe) can be considered fully developed. Additional information on the design of the facility can be found in the work of Bellani and Talamelli [7].

Pressure sensors were integrated in the wall of the pipe within a cavity communicating with the flow by means of a pinhole orifice, having a diameter of $d_p = 0.3$ mm. The resonance frequency of the cavity was found via an acoustic characterization experiment, whereby inlet and cavity pressures were measured simultaneously in quiescent flow and correlated in the frequency domain. A second-order model was fit to the gain of the linear transfer function, $|H_p|$, relating inlet to cavity pressure [8]. Figure 1c presents the gain curve and shows resonance at $f_r \approx 4$ kHz.

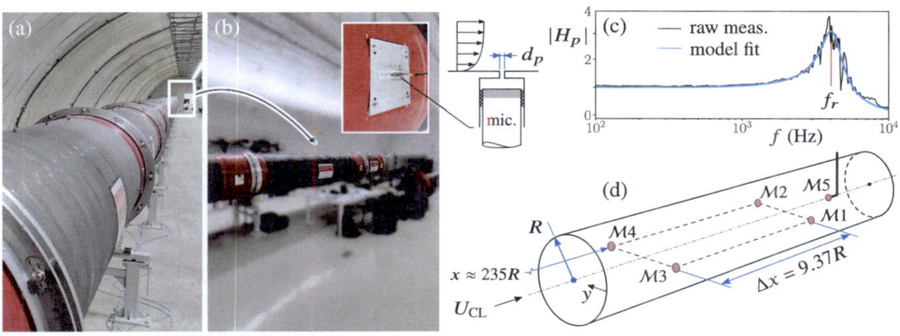

Fig. 1 **a**, **b** Photographs of the CICLoPE facility and test section; **b** one of the four wall plugs with the microphone sensor; **c** gain of the transfer function between inlet and cavity pressures, measured in quiescent flow; **d** microphone lay-out with $x = 0$ corresponding to the pipe inlet

Time-resolved pressure sensing was performed using GRAS 46BE 1/4-in. CCP free-field microphones. Their dynamic range is 35–160 dB and their accuracy is ±1 dB in the range of 10 Hz to 40 kHz. Each measurement was performed with an acquisition frequency of $f_s = 51.2$ kHz and with a 24-bit A/D conversion for an uninterrupted duration of $T_a = 240$ s. Five microphones were employed in our study (labelled $\mathcal{M}1$ to $\mathcal{M}5$ in Fig. 1d): four were integrated in the wall of the pipe ($\mathcal{M}1$ to $\mathcal{M}4$) for wall-pressure measurements and one was mounted along the pipe's centreline ($\mathcal{M}5$) for monitoring the acoustic noise of the facility. The wall-mounted microphones were arranged in two streamwise pairs, separated by a distance of 4.22 m ($\Delta x = 9.37R$). Microphones in one pair were located in azimuthally-opposite positions on the pipe wall to facilitate noise-filtering in post-processing.

Time series of streamwise velocity at a fixed wall-normal location in the logarithmic region ($y_{\text{HW}} = 0.011$ m) were acquired using Hot-Wire Anemometry (HWA), simultaneously with all pressure signals. A Dantec Streamline 90C10 CTA module was used, with a standard Dantec 55P15 boundary layer probe.

3 Wall-Pressure Filtering and Scaling

Before analysing the hydrodynamic wall-pressure signal and, in particular, its energy spectrum ϕ_{pp}, the effects of cavity resonance as well as facility acoustic noise must be removed. The correction for cavity resonance is applied directly in the frequency domain following $\phi_{pp} = \widetilde{\phi}_{pp}/|H_p|^2$ (with $\widetilde{\phi}_{pp}$ being the spectrum of the raw pressure time series). The result of this correction is reported in Fig. 2a, where the raw wall-pressure spectra are displayed as a function of streamwise wavelength, $\lambda_x^+ \equiv U_c^+/f^+$ ($U_c^+ = 10$ is the convection velocity and f^+ the frequency, both inner-scaled). For the three highest Reynolds number data sets 5 to 7, a wiggle in the spectrum appears around the location of the inner-spectral peak at $\lambda_x^+ \approx 250$. This wiggle is ascribed

Table 1 Test conditions in the CICLoPE long-pipe facility. Superscript '+' indicates inner-scaling

	Pipe flow conditions						Sensor parameters				
Data set	Colour	Re_τ	τ_w (Pa)	U_τ (m/s)	l^* (μm)	U_{CL} (m/s)	d_p^+	l_{HW}^+	y_{HW}^+	f_s^+	$T_a U_{CL}/R$
1	—	4797	0.031	0.161	93.8	3.837	3.17	13.2	116	4.84	2036
2	—	7079	0.068	0.238	63.6	5.833	4.68	19.5	172	3.28	3129
3	—	13647	0.265	0.470	32.3	12.11	9.23	38.4	338	1.67	6487
4	—	22789	0.710	0.769	19.7	20.71	15.1	62.9	553	1.02	11079
5	—	31439	1.358	1.065	14.3	29.50	20.9	87.0	766	0.74	15792
6	—	37966	1.983	1.287	11.9	36.13	25.3	105	927	0.61	19325
7	—	45810	2.902	1.559	98.2	44.60	31.0	129	1138	0.50	24081

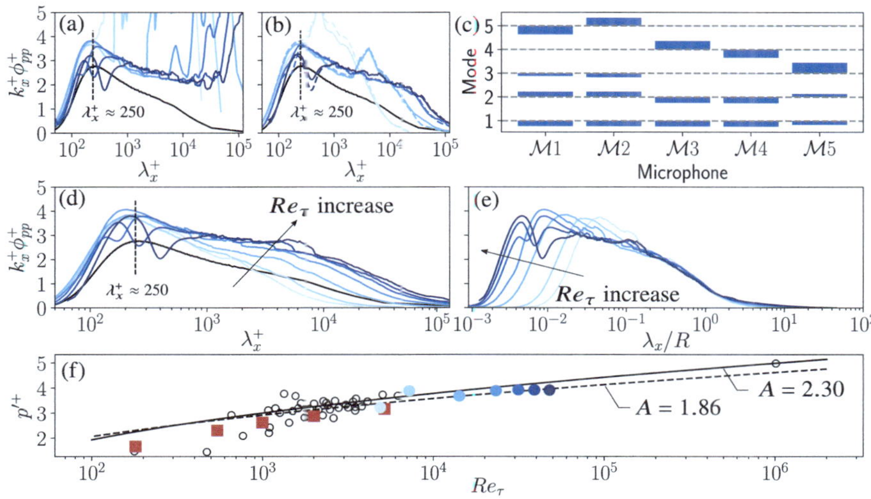

Fig. 2 **a** Wall-pressure spectra prior to application of noise removal procedures but with Helmholtz resonance correction applied; **b** noise-corrected spectra with LSE- (solid) and Wiener filtering-procedures (dashed); **c** spatial POD mode shapes of the pressure signals; **d**, **e** noise-corrected spectra with POD technique as a function of inner- and outer-scaled wavelength; **f** wall-pressure intensity taken as the integral of the spectra (filled blue markers), compared to literature data (hollow markers) including the DNS data by Panton et al. [4] (filled red markers) and empirical relations reported by Klewicki et al. [3] (solid and dashed curves)

to an imperfect correction for cavity resonance: the presence of the flow slightly alters the resonance frequency of the pinhole cavity in comparison to the acoustic calibration experiment. Note that at low Reynolds numbers the resonance frequency is far beyond the energetic frequency range of the wall-pressure spectrum.

Facility noise removal is a crucial pre-processing step [9] and three different strategies were attempted. The first two procedures rely on the estimation of a correction time series $\widetilde{p}(t)$ from a single microphone pair ($\mathcal{M}1$ and $\mathcal{M}2$), which is subtracted from the raw signal (here $\mathcal{M}1$) to obtain the noise-corrected wall-pressure signal, $p_w(t) = p_1(t) - \widetilde{p}(t)$, with $p_1(t)$ being the raw pressure time series. Both Linear Stochastic Estimation (LSE) and an adaptive Wiener filtering approach were applied to generate $\widetilde{p}(t)$ and yielded similar results, displayed in Fig. 2b. For the low Reynolds-number data sets an unexpected hump is still present in the range $300 < \lambda_x^+ < 10^4$.

A third and most effective noise removal approach implemented is based on Proper Orthogonal Decomposition (POD). This method is superior in performance with respect to the other methods both in virtue of its lower overhead and the fact that it employs signals from all five microphones. First, all time series are temporally aligned to compensate for excessive lags in the cross-correlation. Following a conventional POD implementation [10], the five spatial modes are shown in Fig. 2c. After conducting an analysis on the coupling of POD modes, through scrutinizing the correlation of the (temporal) expansion coefficients, it was concluded that only the 4th and 5th modes capture the pressure fluctuations that are incoherent between sensors. As such, noise-corrected wall-pressure signals are reconstructed by considering these two modes only and the resulting spectra are reported in Fig. 2d. Alignment of the inner-spectral peak at $\lambda_x^+ \approx 250$ is observed, as well as a progressive increase in large-scale energy content with increasing Re_τ, in accordance with literature. Additionally, the large-scale tails of the spectra collapse when the streamwise wavelength is scaled in outer units (Fig. 2e). Finally, validation of the noise-removal approach is provided by comparing the wall-pressure intensity to empirical relations [3]: Fig. 2f shows good agreement between the intensity inferred from integrating the spectra after POD-filtering and the findings of the literature.

4 Wall-Pressure–Velocity Coupling

The linear coherence spectrum $\gamma^2_{up_w}$ is employed to describe the correlation between streamwise velocity in the logarithmic region and wall-pressure. Baars et al. [11] report an analysis of this coherence for a relatively low Reynolds number range up to $Re_\tau \approx 5\,200$. A comparison between $\gamma^2_{up_w}$ extracted from Direct Numerical Simulation (DNS) data [12] and experiments [11] is displayed in Fig. 3a for $Re_\tau \approx 2\,000$. Here, the spectral coherence is plotted versus the distance-from-the-wall–scaled streamwise wavelength. Curves are shown for several wall-normal locations within the logarithmic region. The broadband hump centered at $\lambda_x/y \approx 14$ reasonably collapses, with a maximum at $\gamma^2_{up_w} \approx 0.1$.

An identical analysis applied to the high-Reynolds-number CICLoPE data results in the spectra shown in Fig. 3b. Coherence spectra reveal a trend similar to the lower-Re_τ data in Fig. 3a, and also peak near $\lambda_x/y \approx 14$. Despite the relatively low value in coherence, results suggest Reynolds-number-independent wall-scaling. These scaling results, together with an in-depth analysis of wall-pressure-based

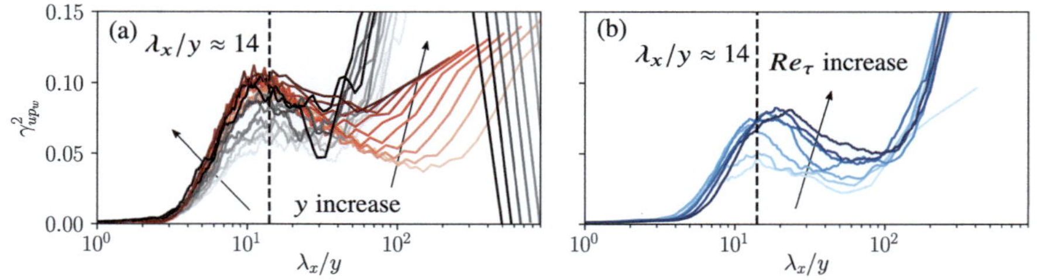

Fig. 3 **a** Coherence spectra of wall-pressure and streamwise velocity in the logarithmic region for DNS channel flow data (red, [12]) and experimental data (black, [11]) at $Re_\tau \approx 2\,000$ and for 10 logarithmically spaced locations in the range $y^+ \in (80, 200)$. **b** Coherence spectra of wall-pressure and velocity for data acquired in the CICLoPE facility

velocity-estimation [11], illustrate how wall-pressure sensing for real-time flow control is viable at high-Reynolds-number conditions.

5 Concluding Remarks

Upon removing facility noise from the wall-pressure time series, by means of a multi-sensor POD-based approach, energy spectra of wall-pressure fluctuations reveal good alignment of the small-scale energy peak at $\lambda_x^+ \approx 250$, as well as an expected increase in large-scale energy content with increasing Reynolds number. The intensity of the noise-corrected wall-pressure signals also agrees with empirical relations derived in the literature, further validating the noise-removal approach. Coherence spectra of wall-pressure and velocity further proved a Reynolds-number-independent wall-scaling, which is encouraging when considering wall-pressure for real-time control in practical engineering systems.

Acknowledgements The work of the UniBo team was carried out within the MOST—Sustainable Mobility National Research Center and received funding from the European Union Next-Generation EU (PIANO NAZIONALE DI RIPRESA E RESILIENZA (PNRR)—MISSIONE 4 COMPONENTE 2, INVESTIMENTO 1.4—D.D. 1033 17/06/2022, CN00000023).

References

1. N. Renard, S. Deck, A theoretical decomposition of mean skin friction generation into physical phenomena across the boundary layer. J. Fluid Mech. **790**, 339–367 (2016)
2. T.M. Farabee, M.J. Casarella, Spectral features of wall pressure fluctuations beneath turbulent boundary layers. Phys. Fluids A **3**(10), 2410–2420 (1991)
3. J.C. Klewicki, P.J.A. Priyadarshana, M.M. Metzger, Statistical structure of the fluctuating wall pressure and its in-plane gradients at high Reynolds number. J. Fluid Mech. **609**, 195–220 (2008)
4. R.L. Panton, M. Lee, R.D. Moser, Correlation of pressure fluctuations in turbulent wall layers. Phys. Rev. Fluids **2**(9), 094604 (2017)
5. A.S.W. Thomas, M.K. Bull, On the role of wall-pressure fluctuations in deterministic motions in the turbulent boundary layer. J. Fluid Mech. **128**, 283–322 (1983)
6. B. Gibeau, S. Ghaemi, Low- and mid-frequency wall-pressure sources in a turbulent boundary layer. J. Fluid Mech. **918**, A18 (2021)
7. G. Bellani, A. Talamelli, The final design of the long pipe in CICLoPE, in *Progress in Turbulence VI* (2016), pp. 205–209
8. Y. Tsuji, J.H.M. Fransson, P.H. Alfredsson, A.V. Johansson, Pressure statistics and their scaling in high-Reynolds-number turbulent boundary layers. J. Fluid Mech. **585**, 1–40 (2007)
9. A.M. Naguib, S.P. Gravante, C.E. Wark, Extraction of turbulent wall-pressure time-series using an optimal filtering scheme. Exp. Fluids **22**, 14–22 (1996)
10. C.E. Tinney, P. Jordan, The near pressure field of co-axial subsonic jets. J. Fluid Mech. **611**, 175–204 (2008)
11. W.J. Baars, G. Dacome, M. Lee, Reynolds-number scaling of wall-pressure–velocity correlations in wall-bounded turbulence. J. Fluid Mech. **981**, A15 (2024)
12. M. Lee, R.D. Moser, Direct numerical simulation of turbulent channel flow up to $Re_\tau = 5200$. J. Fluid Mech. **774**, 395–415 (2015)

Comparing Hot-Wire Measurements and Particle Image Velocimetry of Turbulence Fields Generated by an Active Grid

I. Neunaber, M. Asadi, L. Li, and R. J. Hearst

Abstract The potential for flow manipulation through the use of an active grid is vast. In this study, we investigate the effects of six distinct flapping excitation protocols on turbulent flows, employing both hot-wire anemometry and particle image velocimetry (PIV) with a large field of view, measuring close to 1 m in the streamwise direction. Our primary focus is on comparing the results, particularly examining spectra and the integral length scale L. While the spectra derived from hot-wire measurements, transformed into wave number space through Taylor's hypothesis, exhibit strong agreement with spatial spectra from PIV data, we observe a significant dependency of L on the length of the data set.

1 Introduction

Active grids have fascinated the turbulence community for the past three decades, as the possibilities of flow manipulation with a single device are manifold (e.g., [1–4]), and formerly unachievable turbulent Reynolds numbers of $\mathcal{O}(10^3 - 10^4)$ can be reached [5, 6]. This makes active grids an interesting tool both for turbulence research and for applications that require customized inflows such as gusts (e.g., [7]), or atmospheric-like turbulence (e.g., [8]).

Different philosophies of actuation have been established for active grids. Studying actuation by pure rotation, [3] and [5] have investigated the impact of different rotation frequencies on the turbulence, inter alia on the turbulence intensity TI and the integral length scale L. In addition, [3] also investigated the impact of different flap geometries on TI and L. In [6], the flapping-based excitation protocol is coupled with the wind tunnel motor, so that unprecedentedly high Reynolds numbers were achieved.

I. Neunaber (✉) · M. Asadi · L. Li · R. J. Hearst (✉)
Norwegian University of Science and Technology, Høgskoleringen 1, 7034 Trondheim, Norway
e-mail: ingrid.neunaber@ntnu.no

R. J. Hearst
e-mail: jason.hearst@ntnu.no

R. Örlü et al. (eds.), *Progress in Turbulence X*, Springer Proceedings in Physics 404,
https://doi.org/10.1007/978-3-031-55924-2_29

In this work, we explore different excitation protocols based on flapping motion to see the impact on TI and L, and we also compare results obtained using particle image velocimetry with a 1 m field of view with hot-wire measurements.

2 Experimental Set-Up and Methodology

The experiments were conducted in the Large-Scale Wind Tunnel at the Norwegian University of Science and Technology. The wind tunnel features an inlet measuring 2.7 m $\times$ 1.8 m (width $\times$ height) and a test section with a length of 11 m. At the entrance of the test section, an active grid, fitting the dimensions of the wind tunnel inlet, was installed (see Fig. 1b and additional information on the grid given in [9]). The mesh length of the grid is $M = 100$ mm, and the shaft angle α is set to 0° when the wings align with the flow.

The grid is subjected to a randomized flapping actuation, where the axes are flapped consecutively in a randomized order from $-\alpha$ to $+\alpha$. After a randomized waiting period $T \pm \Delta T$, which follows a top-hat distribution, the axes are flapped back, and the cycle repeats. Six distinct actuation protocols were investigated in this study. These protocols involve three wing angles, namely $\alpha_1 = 15°$, $\alpha_2 = 30°$, and $\alpha_3 = 45°$, combined with two waiting periods, $T_A = (0.005 \pm 0.005)$s and $T_B = (0.5 \pm 0.5)$s. For clarity, we refer to these cases as 1-3 A or B, based on the wing angle and waiting period combination. The mean velocity is $U \approx 10\,\mathrm{ms}^{-1}$, and Fig. 1a provides the TIs for all cases.

Hot-wire measurements were performed 61M downstream of the grid at the center of the wind tunnel, utilizing a Dantec Dynamics StreamLine Pro for wire operation. The sampling frequency was $f_s = 75$ kHz, and a hardware low-pass filter with a cut-off frequency of $f_{LP} = 30$ kHz was employed. The measurement duration was $\Delta t = 600$ s, and the hot-wire was frequently calibrated. Additionally, temperature correction, following the procedure described in [10], was applied.

As depicted in Fig. 1c, four cameras, consisting of two 25-megapixel (MP) cameras and two 16MP cameras, were precisely aligned to create a field of view (FOV) measuring 964 mm in length and 208 mm in width. This FOV extended from 65M to 74.6M downstream of the grid and from $Z = -0.55M$ to $Z = -2.55M$. Illumination of the FOV was achieved using two Nd:YAG lasers with nominally 200 mJ per pulse, and 2000 double-frame images were captured at a sampling frequency of $f_{PIV} = 1.5$ Hz, resulting in an acquisition time of $t_{PIV} = 1333$ s. The vector fields were processed using DaVis 10.1, employing multiple passes, with the final pass employing a 48px $\times$ 48px interrogation window (IW) and 75% overlap. This processing resulted in a field of 1330 $\times$ 288 vectors, with a vector spacing of 0.73 mm. To merge the images captured by the four cameras, adjustments were made for camera alignment and differences in the mean values of overlapping regions. A weighted average of velocities in the overlapping region was then computed, with the weightings adjusted according to the X coordinate.

a) Measured TI (%)

	1A	1B	2A	2B	3A	3B
α	15°		30°		45°	
TI_{HW}	3.1	3.4	5.1	5.2	7.2	7.9
TI_{PIV}	2.3	2.5	4.5	4.7	6.3	6.8

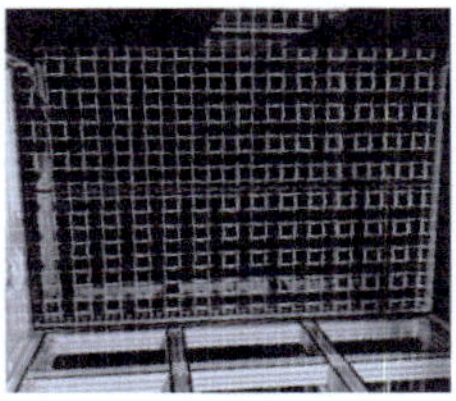

b) photograph of the active grid

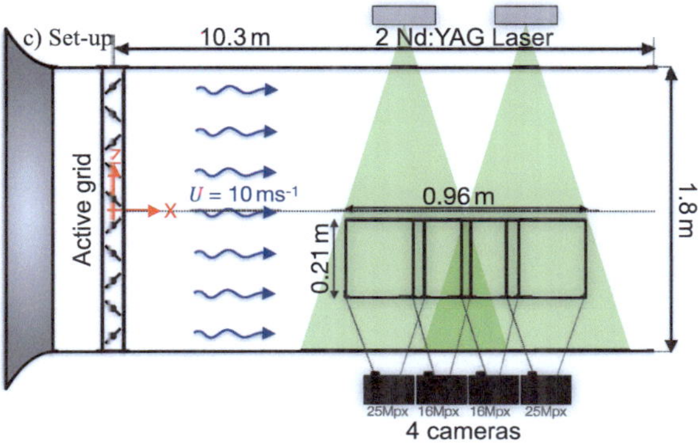

Fig. 1 **a** Measured TI; **b** Photograph of the active grid; **c** PIV and wind tunnel set-up

To facilitate a meaningful comparison between the high-resolution temporal hot-wire data and the spatial PIV data, a two-step procedure was implemented. First, the PIV data was downsampled by selecting every fourth vector to mitigate the influence of the 75% IW overlap, resulting in a new vector spacing of $\Delta r = 2.9$ mm. Subsequently, the hot-wire data was binned, and Taylor's hypothesis was locally applied, averaging over multiple samples to convert the data into spatial data with a 2.9 mm resolution. To ensure consistency with the spatial extent of the PIV field, 2000 independent sections of 960 mm length were extracted for analysis.

3 Results and Discussion

To provide a general overview, Fig. 2 displays instantaneous streamwise and spanwise velocity fields for cases 1A, 2A, and 3A. It is evident that the increase in TI from case 1 to 3 is visually apparent in the instantaneous fields, as indicated by the amplified velocity fluctuations and the presence of larger structures.

Figure 3 presents a comparison of the wave number spectra. The full-resolution hot-wire spectra, transformed into wave number space via Taylor's hypothesis, are collapsed with the down-sampled hot-wire spectra and the PIV spectra. For the spatial hot-wire data, spectra across 2000 snippets were averaged, while for the PIV data, a line of vectors at $Y = -0.82M$ was extracted from each of the 2000 images, with subsequent spectrum calculation and averaging.

The full-resolution hot-wire spectra in Fig. 3 exhibit the typical evolution of turbulence spectra. As noted previously in the discussion of TI, the spectra shift towards higher power spectral densities (PSDs) for higher case numbers, while configurations A and B converge. Comparing these spectra with the down-sampled hot-wire spectra, it is evident that the latter match well across most of the inertial sub-range but begin to deviate slightly around $k \approx 700\ \text{m}^{-1}$. Similarly, the spatial spectra derived

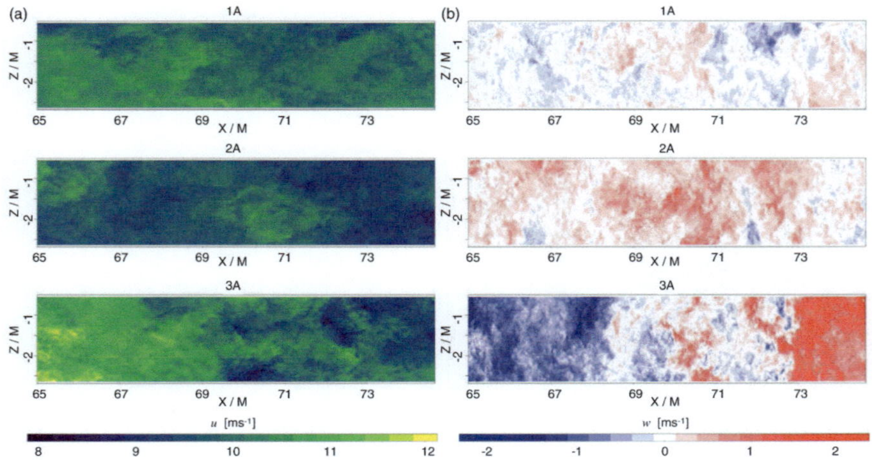

Fig. 2 Fields of instantaneous velocity for cases 1A, 2A, and 3A **a** instantaneous streamwise velocity u; **b** instantaneous wall-normal velocity w

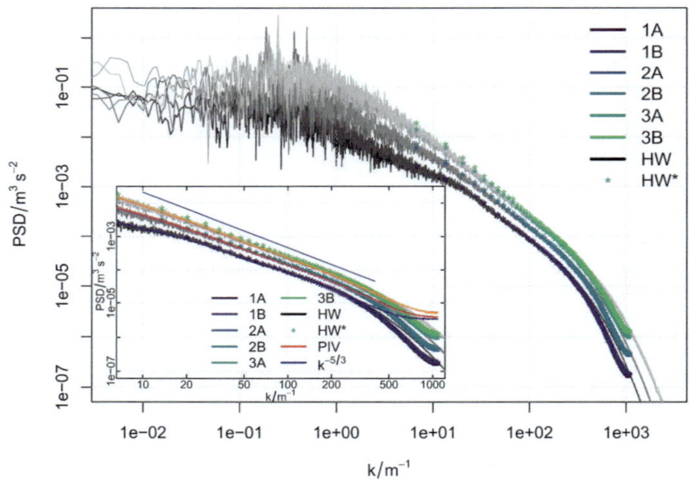

Fig. 3 Spectra in space: In grey-scale, the full-resolution hot-wire spectra, converted to wave number spectra, are plotted. In green color scale, the down-sampled spatial hot-wire spectra are plotted (marked as HW^* in the legend). In orange-red, the PIV spectra are included. A darker color indicates a lower case number

from the PIV fields align with the spectra obtained from hot-wire data within the inertial sub-range. However, a deviation becomes noticeable at $k \approx 300\ \mathrm{m}^{-1}$, where noise becomes significant. These findings lead to several conclusions. Firstly, the global application of Taylor's hypothesis to convert full-resolution hot-wire spectra into spatial data is a valid assumption for the tested flows. Secondly, despite the

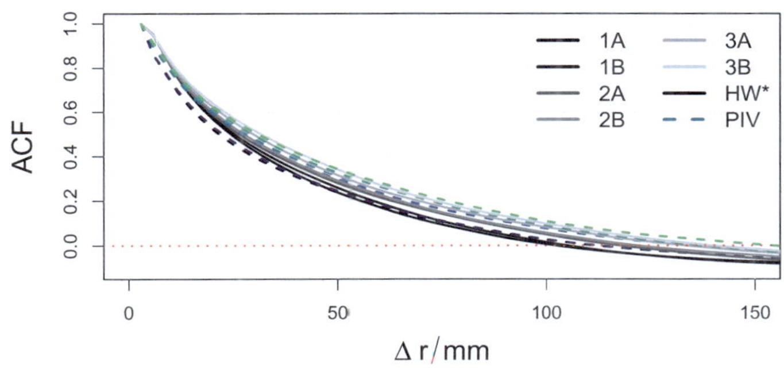

Fig. 4 Comparison of the autocorrelation function (ACF) of the down-sampled hot-wire data (grey) with that of the PIV data (green, dashed) for all cases

substantial 1-m FOV, the PIV data spectra do not encompass the entire inertial subrange. Therefore, the spectral method proposed by [11] for calculating the integral length scale cannot be applied.

The next step involves the investigation of Autocorrelation Functions (ACFs). Due to the limited 0.96-m extent of our PIV field, a modified ACF approach is employed. For each image, a 450 mm segment (approximately 50% of the field length) of the same line of vectors used for the PSD is correlated with the original line, shifted by Δr, and this process is iterated. The average is then calculated. Figure 4 compares the ACFs for the six cases using a 450 mm snippet size. ACFs derived from PIV and spatial hot-wire data exhibit reasonably good agreement, revealing a subtle trend towards longer correlations with increased TI in the flow.

Finally, Fig. 5 compares the ACFs obtained from PIV data and spatial hot-wire data with those from full-resolution hot-wire data. Additionally, ACFs obtained from fragments of the high-resolution hot-wire data with durations of 0.1, 1.0 and 10.0 s, corresponding to structures of 1, 10 and 100 m, are compared. Cases 1–3*A* are presented as examples. Notably, the ACF of the full-resolution hot-wire data decays significantly more slowly than that of the PIV data and the spatial hot-wire data, crossing 0 at $\Delta r \approx 6$ m (compared to $\Delta r \approx 0.15$ m). Examining the ACF of the full-resolution 0.1 s fragments, we observe that it closely resembles that of the PIV data and the spatial hot-wire data, while the ACF of the 10 s fragment aligns closely with the ACF of the complete hot-wire time series. The ACF of the 1.0 s fragment crosses 0 between these two extremes. This illustrates the significant impact of available data length on certain analyses.

Table 1 summarizes the integral length scales calculated by integrating the various ACFs from Fig. 5 up to the first zero-crossing. It also provides a comparison with values obtained from the spectra. Depending on the length of the data set, the calculated integral length scale exhibits variations by a factor of 20. Integral length scales determined through the spectral method indicate structure sizes of approximately 40 cm, which should be within the PIV FOV's capability. However, values obtained from the PIV data are an order of magnitude smaller.

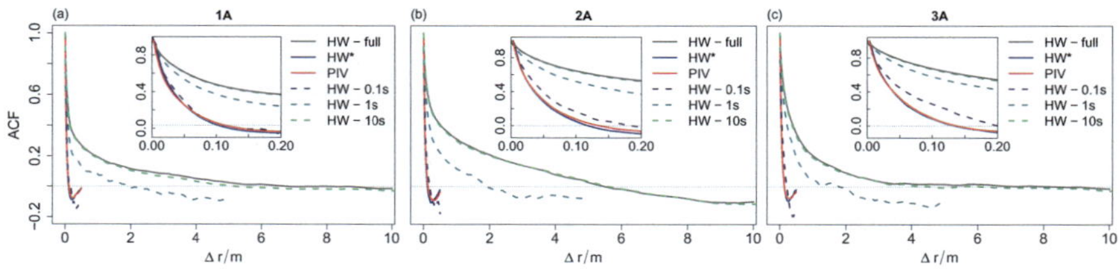

Fig. 5 Comparison of autocorrelation functions of high-resolution hot-wire data with 0.1, 1.0 and 10.0 s fragments of the high-resolution hot-wire data, the down-sampled spatial hot-wire data and the PIV data for **a** case $1A$, **b** case $2A$, and **c** case $3A$

Table 1 Integral length scales calculates in centimeters using different methods and for the different data sets: L obtained from integrating the ACF of 0.1, 1.0 and 10.0 s fragments of the high-resolution hot-wire data are in columns 1–3, L from the down-sampled spatial hot-wire data HW^* in column 4, L from the PIV data in column 5, L from the full hot-wire time series in column 6, and L obtained by the spectral method from the full-resolution hot-wire data in column 7

	0.1s	1s	10s	HW*	PIV	HW full	Spectrum
1A	3.59	21.71	58.98	2.99	2.90	67.92	36.5
1B	5.07	25.47	59.20	3.13	2.96	69.03	34.4
2A	5.00	33.97	103.45	3.38	3.55	104.56	48.7
2B	5.38	38.98	66.72	3.48	3.68	69.03	41.6
3A	6.18	27.90	59.53	3.76	3.80	67.42	41.1
3B	6.18	27.90	59.53	3.99	4.07	67.42	41.9

4 Conclusions

In this study, we conducted a comparative analysis of large FOV PIV measurements and hot-wire measurements to explore the strengths and limitations of both methods. The examination of instantaneous PIV fields revealed the presence of spatial structures, which became more pronounced as TI increased. A scrutiny of spatial spectra indicated strong agreement between those derived from PIV data and those obtained from full-resolution hot-wire data, following their transformation into wave number space through Taylor's hypothesis. Additionally, we observed that the structures present in the 1-m FOV PIV field largely encompassed the high-frequency portion of the inertial sub-range of the high-resolution hot-wire spectra. However, due to the inability of PIV data to capture a significant portion of large-scale structures identified from the hot-wire time series, integral length estimates based on ACFs exhibited significant bias towards smaller values.

References

1. H. Makita, Realization of a large-scale turbulence field in a small wind tunnel. Fluid Dyn. Res. **8**, 53–64 (1991)
2. A. Thormann, C. Meneveau, Decay of homogeneous, nearly isotropic turbulence behind active fractal grids. Phys Fluids **26**, 025112 (2014)
3. R.J. Hearst, P. Lavoie, The effect of active grid initial conditions on high Reynolds number. Exp. Fluids **56**, 185 (2015)
4. L. Neuhaus, F. Berger, J. Peinke, M. Hölling, Exploring the capabilities of active grids. Exp. Fluids **62**, 130 (2021)
5. J.V. Larssen, W.J. Devenport, On the generation of large-scale homogeneous turbulence. Exp. Fluids **50**, 1207–1223 (2011)
6. L. Neuhaus et al., Generation of atmospheric turbulence with unprecedentedly large Reynolds number in a wind tunnel. Phys. Rev. Lett. **125**(15), 154503 (2020)
7. T.T.B. Wester et al., High speed PIV measurements of an adaptive camber airfoil under highly gusty inflow conditions. J. Phys.: Conf. Ser. **1037**, 072007 (2018)
8. M. Wächter et al., The turbulent nature of the atmospheric boundary layer and its impact on the wind energy conversion process. J. Turbul. **13**, N26 (2012)
9. O. Kildal et al., On the use of an active turbulence grid in wind tunnel testing of bridge decks. J. Wind Eng. Ind. **233**, 105331 (2023)
10. M. Hultmark, A.J. Smits, Temperature corrections for constant temperature and constant current hot-wire anemometers. Meas. Sci. Technol. **21**, 105404 (2010)
11. J. Hinze, *Turbulence*, McGraw-Hill Classic Textbook Reissue Series (McGraw-Hill, New York, 1975)

Study of Dust Devils in a Laboratory Experiment

R. du Puits and C. Kaestner

Abstract In our paper, we report an alternative idea, how dust devils can be generated and studied in a well-controlled laboratory experiment. We used a large-scale Rayleigh-Bénard experiment with air heated from below and cooled from above to mimic the convective atmospheric boundary layer. We measured the Lagrangian flow field by tracking the flow with helium-filled, and thus, virtually neutrally buoyant soap bubbles. Vortical structures with a vertical updraft and a significant pressure drop in their center have been identified as dust devil-like structures. Generally, they are smaller and weaker in our experiment than real dust devils in their natural environment. Nevertheless, they share many characteristic properties. This pertains e.g. to the lateral migration with the background wind, the arbitrary sense of rotation, or the relation between the size and the lifetime.

1 Introduction and Motivation

Dust devils are strong vortical wind phenomena that grow close to the surface of the earth. Due to their rare and spontaneous appearance, it is hard to study them in their natural environment. The work presented here shows up a way to study their typical behavior under well-controlled laboratory conditions.

Unlike their bigger counterparts, the tornados, dust devils appear close to the surface of the earth. One of the main ingredients of their growing is a strong solar radiation which heats the surface locally and generates a hot spot. The air above becomes warmer, expands and rises up. Cold air is sucked horizontally above the ground, and under certain circumstances, it can be set into a strong vortical and upward motion [1]. Dust devils can achieve heights as large as 1,000 m. Their

R. du Puits
Aerodynamics Group, Technische Universitaet Ilmenau, Ilmenau, Germany
e-mail: ronald.dupuits@tu-ilmenau.de

C. Kaestner (✉)
Production Technology Group, Technische Universitaet Ilmenau, Ilmenau, Germany
e-mail: christian.kaestner@tu-ilmenau.de

R. Örlü et al. (eds.), *Progress in Turbulence X*, Springer Proceedings in Physics 404,
https://doi.org/10.1007/978-3-031-55924-2_30

diameter can amount up to 100 m and tangential wind speeds of up to 25 m/s have already been measured. The typical lifetime of these phenomena ranges from a few seconds up to several minutes. Occasionally, dust devils have been observed lasting even hours [2]. One other typical feature is a pressure drop in the core of the structure which usually remains below 1.5 mPa. Dust devils in nature have been extensively studied in the past [3–6], but due to their rare and spontaneous appearance, accurate and representative measurements are hard to perform. On the other hand, it is also not trivial to mimic their growing in a laboratory experiment. Great efforts have been made to setup such experiments [7–9], but almost all of them lack in one specific issue: The vortical structures were set into rotation by artificial forcing either by e.g. a rotating plate or an appropriate arrangement of baffles. In the experiment we report in this work, dust devils or more precisely dust devil-like flow structures emerge in turbulent convection in air without such forcing.

2 Experimental Set-Up and Measurement Technique

Figure 1 shows a sketch of the experimental facility we used for our work. It represents a classical Rayleigh-Bénard (RB) experiment, where a fluid layer of a certain thick-

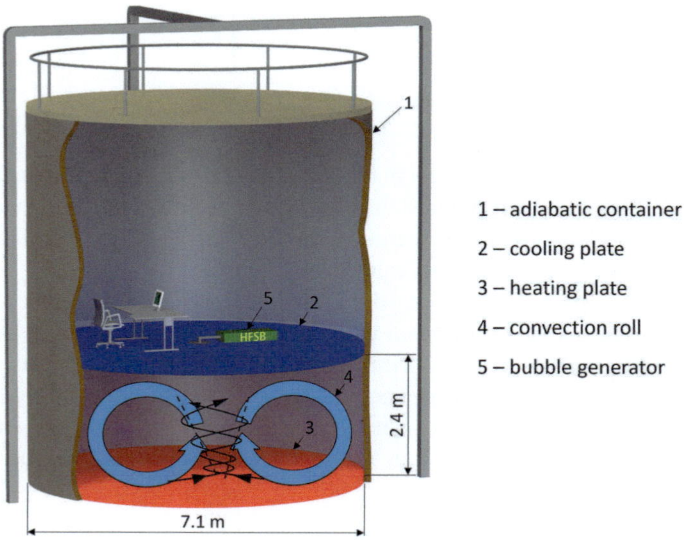

Fig. 1 Sketch of the large-scale Rayleigh-Bénard experiment "Barrel of Ilmenau". The distance H between the heating and the cooling plates is variable, but for the experiments reported here, it is fixed at $H = 2.4$ m. This setting supports the evolution of a global flow pattern with multiple convection cells. Dust devils or dust devil-like vortices preferentially form at the interconnection of these cells. The maximum Rayleigh number in this configuration amounts to $Ra = 5.6 \times 10^{10}$

ness H ($H = 2.4$ m for the work reported here) is heated from below and cooled from above. The diameter of the container amounts to 7.1 m and it is filled with plain air (for more detailed information we refer to [10]). The temperature gradient across the fluid layer triggers a highly turbulent, convective re-circulation which is similar to that in the (convective) atmospheric boundary layer. The strength of this re-circulation is described by the Rayleigh number $Ra = (\beta g \Delta T H^3)/(\nu\kappa)$. Here, β, g and ΔT represent the thermal expansion coefficient, the gravitational acceleration, and the temperature difference across the fluid layer, while ν and κ stand for kinematic viscosity and thermal diffusivity. For the atmospheric boundary layer, this definition yields to $Ra \approx 10^{18}$. The settings of our experiment, $H = 2.4$ m and $\Delta T = 20$ K, result, however, in a Rayleigh number of only $Ra = 2.5 \times 10^{10}$. We are aware of this huge difference, but Direct Numerical Simulations (DNS) of turbulent RB convection by Giersch and Raasch show that this is still sufficiently high that dust devil-like structures may appear [11].

In order to find such structures in the flow, we measured the Lagrangian flow field using Particle Tracking Velocimetry (PTV). We used Helium-filled soap bubbles of 3...5 mm diameter as virtually neutrally buoyant tracers. They have a typical lifetime of 25 s and are capable to follow even small turbulent flow structures [12]. They are illuminated using multiple high-power LED modules and tracked by four JAI GO-5000M-USB cameras (5 MPix, 12 bit, 31 fps). LEDs and cameras are mounted at the inner sidewall of the test section. In combination with either 8 mm or 25 mm lenses, almost the entire volume of the test section or a smaller measurement volume of about 4 m length, 2 m width and 1 m height could be monitored. We mainly used the second optical set-up. This provides a higher spatial resolution of the measured flow field and enables us to detect flow structures down to the decimeter scale. The Lagrangian particle tracks have been reconstructed using the DaVIS software and the Shake-the-box algorithm (LaVision) [13]. Potential dust devil-like flow structures have been searched by an in-house detection algorithm, whose details can be found in [14]. After detecting such a structure, the Lagrangian flow field has been transformed into an Eulerian one with an equal grid spacing. For this, we do not use a simple interpolation scheme, but a highly sophisticated algorithm that is based on the equation of vorticity transport [15]. This is essential to compute the pressure drop in the core of the vortex, another typical feature of a dust devils. We also use the Eulerian velocity data to determine other characteristics of the structures found, like core position, radius, sense of rotation etc.

3 Results and Discussion

We start this section showing the largest and longest-living dust devil-like structure found in totally two hours of observation. We call it Primary Dust Devil (PDD) and show it in Fig. 2. The PDD has a typical extent of about 800 mm in horizontal direction (x, y-coordinates) and about 600 mm in vertical direction (z-coordinate). It lives for almost 14 s (see movie in [16]) and moves from a Position A at

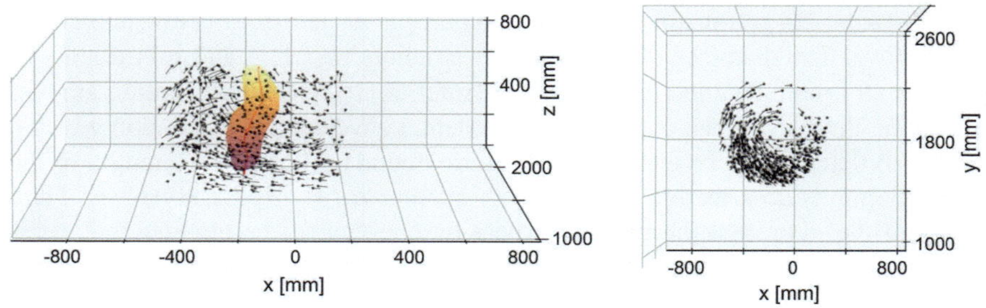

Fig. 2 Lagrangian particle tracks of a very large and long living example of a dust devil-like vortex in the Rayleigh-Bénard cell "Barrel of Ilmenau". The left part of the figure shows the structure from the side, where the right part is a view from the top. The red line marks the minimum pressure in each horizontal plane. The coloured contour is an isosurface of 50% of the maximum pressure drop in the core with respect to the ambient pressure. It is usually considered as the radius of a dust devil

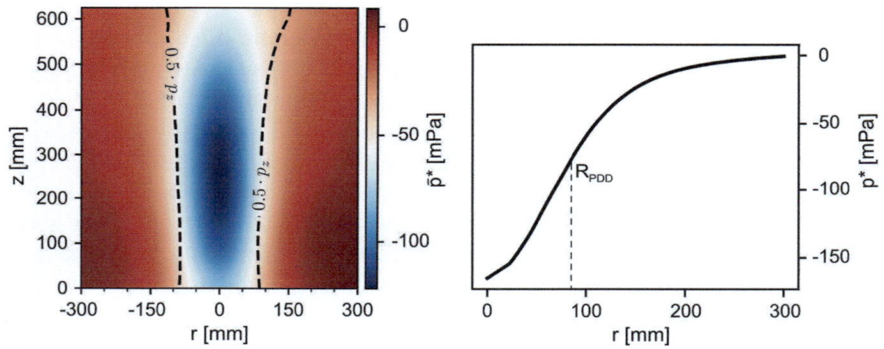

Fig. 3 Azimuthally averaged pressure drop $\bar{p}^*$ over the radius r and the height z of the PDD (left subfigure). The same quantity at a fixed height of $z = 62$ mm (right subfigure). R_{PDD} indicates the radius of the vortex

$x = -500$ mm, $y = 1400$ mm to a Position B at $x = 450$ mm, $y = 2400$ mm. This path essentially coincides with the orientation of the prevailing background wind. The central axis of the dust devil and its radius are usually defined using the pressure field around the vortex. This is shown in Fig. 3 as the azimuthal mean of the pressure drop $\bar{p}^*$ over the distance from the central axis and the height of the vortex. The radius R is then defined as the location, where $\bar{p}^* = 0.5\bar{p}^*(r = 0)$ [17]. For the PDD, $R \approx 80$ mm. It is quite obvious that the defined radius is much smaller then the lateral extent of the rotational motion of the vortex (see Fig. 2), but it fairly coincides with the location of the maximum of the azimuthal velocity [14]. Insofar, it seems to be a reasonable definition.

A lateral motion often referred to as migration is, likewise, a typical feature of atmospheric dust devils [2]. We use this feature to summarize all dust devil-like structures observed in our experiment in two plots (see Fig. 4). The left one shows the migration pathes of those structures that evolve using an additional heating element

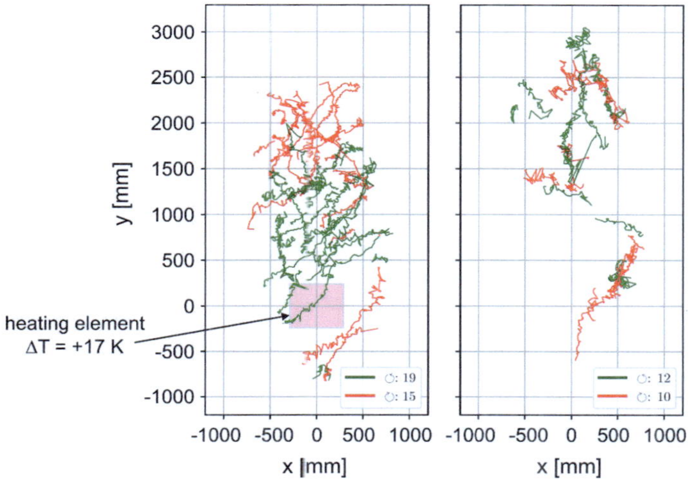

Fig. 4 Migration pathes of the dust devil-like vortices observed in the Rayleigh-Bénard cell "Barrel of Ilmenau". The total observation period was 2×1 h. The left plot shows the pathes of those vortices that appear with an additional heating element at the surface of the heating plate on, the right plot shows the pathes, if the heating element was off

(480 mm by 580 mm) fixed at the surface of the heating plate. It generates a local hot spot with an overheat of +17 K and should increase the probability of the appearance of dust devils [18]. And indeed, this seems to be the case. While we observed 34 dust devil-like structures having the heating element switched on (see the left subfigure of Fig. 4), only 22 appeared in the same period, if the heating element was off. Another interesting feature of dust devils is the sense of rotation. In the atmosphere, clockwise as well as anticlockwise rotating dust devils appear [19]. This can be attributed to the fact that they are too small to get affected by the Coriolis force. In almost all previous experiments, the sense of rotation of the artificially generated dust devils has been set by the outer forcing [7–9] and this feature could not be evaluated explicitly. In our experiment, dust devil-like structures grow without applying any rotational forcing. Insofar, our measurements confirm for the first time, that purely buoyancy-driven dust devil-like vortices do neither exhibit any preferential sense of rotation and clockwise as well as anticlockwise rotating vortices appear.

Acknowledgements We wish to acknowledge the financial support of the German Research Foundation under the Grant number 387703749. Moreover, we thank Alice Loesch, Sabine Scherge and Julien David Schneider for their technical assistance to run the experiments and to analyze the data.

References

1. M. Balme, R. Greeley, Dust devils on Earth and Mars. Rev. Geophys. **44**(3) (2006)
2. D. Reiss, R. Lorenz, M. Balme, L. Neakrase, A.P. Rossi, A. Spiga, J. Zarnecki, *Dust Devils* (Springer, 2016)
3. W.D. Flower, *Sand Devils* (HM Stationery Office, 1936)
4. F. Bell, Dust devils and aviation, report, Meteorol. Note 27, Aust. Bur. of Meteorol., Melbourne, Victoria (1967)
5. P.C. Sinclair, General characteristics of dust devils. J. Appl. Meteorol. **8**, 32–45 (1969)
6. S.M. Metzger, J.R. Carr, J.R. Johnson, T.J. Parker, M.T. Lemmon, Dust devil vortices seen by the Mars Pathfinder camera. Geophys. Res. Lett. **26**, 2781–2784 (1999)
7. D.E. Fitzjarrald, A field investigation of dust devils. J. Appl. Meteorol. **12**, 808–813 (1973)
8. T. Maxworthy, On the structure of concentrated, columnar vortices. Astronaut. Acta **17**, 363–374 (1972)
9. M. Balme, R. Greeley, B. Mickelson, J. Iversen, G. Beardmore, S. Metzger, A laboratory scale vortex generator for simulation of Martian dust devils. *AGU Fall Meeting Abstracts* (2001), pp. P31A–0542
10. R. du Puits, C. Resagk, A. Thess, Thermal boundary layers in turbulent Rayleigh-Bénard convection at aspect ratios between 1 and 9. New J. Phys. **151**, 013040 (2013)
11. S. Giersch, S. Raasch, Evolution and features of dust devil-like vortices in turbulent Rayleigh-Bénard convection–A numerical study using direct numerical simulation. J. Geophys. Res. Atmos. **126**, e2020JD034334 (2021)
12. A. Loesch, R. du Puits, The Barrel of Ilmenau: a large-scale convection experiment to study dust devil-like flow structures. Meteorol. Z. **30**(1), 89–97 (2021)
13. D. Schanz, S. Gesemann, A. Schroeder, Shake-The-Box: Lagrangian particle tracking at high particle image densities. Exp. Fluids **57**(5), 1–27 (2016)
14. C. Kaestner, J.D. Schneider, R. du Puits, Evolution and features of dust devil-like vortices in turbulent Rayleigh-Bénard convection–an experimental study. J. Geophys. Res. Atmos. **128**, e2022JD037466 (2022)
15. Y.J. Jeon, J.F.G. Schneiders, M. Mueller, D. Michaelis, B. Wieneke, 4D flow field reconstruction from particle tracks by VIC+ with additional constraints and multigrid approximation, in *Proceedings 18th International Symposium on Flow Visualization* (2016)
16. C. Kaestner, J.D. Schneider, R. du Puits, PDD migration during the whole observation time in turbulent Rayleigh-Bénard convection, https://ftp.tu-ilmenau.de/hpc-private/mb/Kaestner_et_al_2022_dust_devil-like_vortex_in_turbulent_Rayleigh_Benard_convection.avi (2022)
17. R.D. Lorenz, B.K. Jackson, Dust devil populations and statistics. Space Sci. Rev. **203**(1), 277–297 (2016)
18. S. Giersch, M. Brast, F. Hoffmann, S. Raasch, Toward large-eddy simulations of dust devils of observed intensity: effects of grid spacing, background wind, and surface heterogeneities. J. Geophys. Res. Atmos. **124**, 7697–7718 (2019)
19. J.J. Carroll, J.A. Ryan, Atmospheric vorticity and dust devil rotation. J. Geophys. Res. Atmos. **75**, 5179–5184 (1970)

Study of the Upstream Influence of the Diffuser of CICLoPE "Long Pipe" Using Oil Film Interferometry

L. Lazzarini, G. Bellani, and A. Talamelli

Abstract Wall-bounded turbulence is an extremely relevant topic for engineering and natural science application even though many aspects of this physics are not clear yet. Due to the relevance of these flows, in recent times, several facilities were developed to investigate turbulent flows at high-Reynolds number under fully developed conditions. In particular, the "Long Pipe" in CICLoPE is a 111.5 m full carbon fibre pipe with a diameter of 0.9 m, minimising spatial and temporal filtering issues, a common problem that arises as the Reynolds number increases. Several measurements documenting fully developed conditions in the test section of this facility have been previously reported. A detailed investigation of the turbulent flow in this location is important not only because this is the position where measurements are typically performed; but also to exclude that there is no upstream influence of the diffuser which is mounted downstream. While previous work reported on the invariance of the velocity profiles in the near-wall region, in this work we focus on the centerline velocity and the local wall-shear stress.

1 Introduction

At very high Reynolds numbers typical of aerospace applications, the boundary layer is turbulent and hence associated to large values of friction. For instance, it is estimated that the skin friction drag of conventional airliners, in cruise conditions, can be $\approx 50\%$ of the total drag. Due to its chaotic nature, the study of turbulent boundary layers is a very complex process. On one hand, the Reynolds number must be large enough to reproduce the actual physical mechanisms in applications. On the other hand, the measurement of the different physical quantities becomes complicated due

L. Lazzarini (✉) · G. Bellani · A. Talamelli
Department of Industrial Engineering, University of Bologna, Bologna, Italy
e-mail: lorenzo.lazzarini7@unibo.it

G. Bellani
e-mail: gabriele.bellani2@unibo.it

A. Talamelli
e-mail: alessandro.talamelli@unibo.it

R. Örlü et al. (eds.), *Progress in Turbulence X*, Springer Proceedings in Physics 404,
https://doi.org/10.1007/978-3-031-55924-2_31

to the necessity of measuring velocity gradients extremely close to the wall surface. Standard facilities are therefore forced to choose between providing high Reynolds numbers, or high accuracy and high resolution. The "long pipe" wind tunnel at the CICLoPE Laboratory [1, 2], thanks to the large dimensions of its test section (111.5 m long and 0.9 m diameter) was conceived to tackle this problem, thus permitting to reach a sufficiently high range of Reynolds numbers while simultaneously enabling high fidelity measurements with standard measurement systems [3–5]. In pipe flow wall shear stress is typically obtained through static pressure drop, allowing the calculation a global wall shear stress value through an "indirect" measurement (i.e., static pressure evolution throughout a pipe section). Oil film interferometry (OFI) [6, 7], is a technique that allows the measurement of a localised value of the wall shear stress, which can be inferred directly from the spreading rate of an oil drop at the wall. The aim of the present study is to develop and validate a technique to measure wall shear stress using OFI, and by means of this technique investigate the potential interference effect of the diffuser installed immediately downstream of the TS.

2 Methodology

2.1 Oil Film Interferometry

Wall shear stress is a key quantity that needs to be measured in wall-bounded turbulent flows. There are many techniques that use different measurement principles. The classical methodology in pipe flows consists in measuring static pressure drop through a known section of the pipe. It represents a robust technique, that anyhow is limited by the resolution of the pressure transducer, thus requiring relatively high speed. Moreover, it gives only a spatial average of the wall shear stress for the pipe section considered. Oil film interferometry, on the other hand, represents an efficient methodology to calculate wall shear stress in a localised region of the pipe using a thin oil drop.

A thin oil film that can move freely on a solid surface is deformed under the effect of the skin friction resulting from the flow over the liquid surface. Variation in film thickness is evaluated from the dynamics of the interference fringes formed on the film illuminated with a monochromatic light source, resulting in the estimate of the local wall shear stress, as described in Fig. 1.

Tanner and Blows [8] and Squire [9] are among the first to investigate the possibility of using OFI in turbulence research aiming at measuring skin friction. Although the oil moves under the effects of many parameters, the dominant contribution is caused by the tangential stress at the wall. Under this hypothesis, it is possible to link the evolution of the thickness of the oil film, h_{oil}, to the wall shear stress τ_w.

$$\frac{\partial h_{oil}}{\partial t} + \frac{\partial}{\partial x}\left(\frac{\tau_w h_{oil}^2}{2\mu_{oil}}\right) = 0, \tag{1}$$

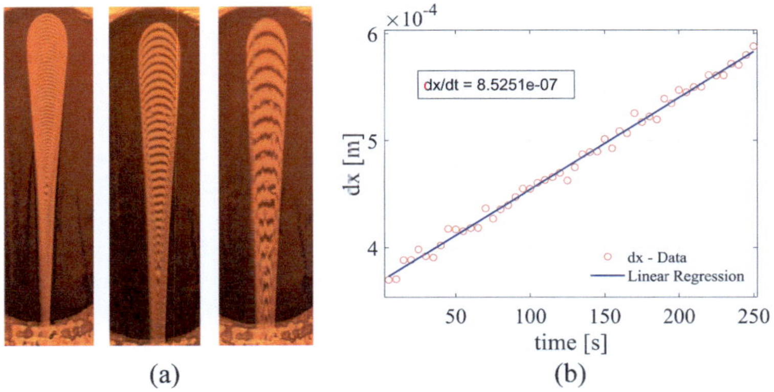

Fig. 1 **a** Evolution of fringes of the oil drop under the effect of the skin friction inside the test section of CICLoPE "long pipe". **b** Fringes distances dx as function of time; blue line highlights the linear regression while red circles are experimental data points computed over each snapshot

where μ_{oil} is the oil viscosity. Being the flow mainly developed in a single direction, we can neglect the transversal component of velocity and consider the motion of the oil mono-dimensional and aligned with the flow direction. The deformation of the oil is then a function of time and of one space coordinate. We can integrate (1) for the thickness of the oil, h_{oil} to obtain the tangential stress at the wall. Moreover, using some optic manipulation focused on interferometry theory, we are able to write the following equation:

$$\Delta h_{oil} = \frac{\lambda_L}{2 \cdot \sqrt{n_{oil}^2 - n_{air}^2 \cdot \sin^2 \Omega_1}}; \tag{2}$$

λ_L indicates the wavelength of the light source, n_{oil} and n_{air} refers to the refractive index of the oil and air respectively, Ω_1 refers to the observation angle between camera and oil film. Thus, we can compute the thickness variation between two successive fringes. Δh_{oil} is constant over time, while relative Δx which represents the distance between two successive fringes changes. This growth can be computed as follows:

$$\Delta x = \Delta h_{oil} \cdot \frac{\tau_w t}{\mu_{oil}} \Rightarrow \tau_w = \frac{\mu_{oil}}{\Delta h_{oil}} \frac{\partial x}{\partial t}. \tag{3}$$

2.2 Experimental Setup

The experimental setup requires a camera, a monochromatic sodium lamp, and oil with known viscosity. Camera and lamp are equally spaced by 90° due to the position of the access port of the pipe. A small drop of oil under the effect of the skin friction is placed at the bottom of the test section of the long pipe as described in Fig. 2.

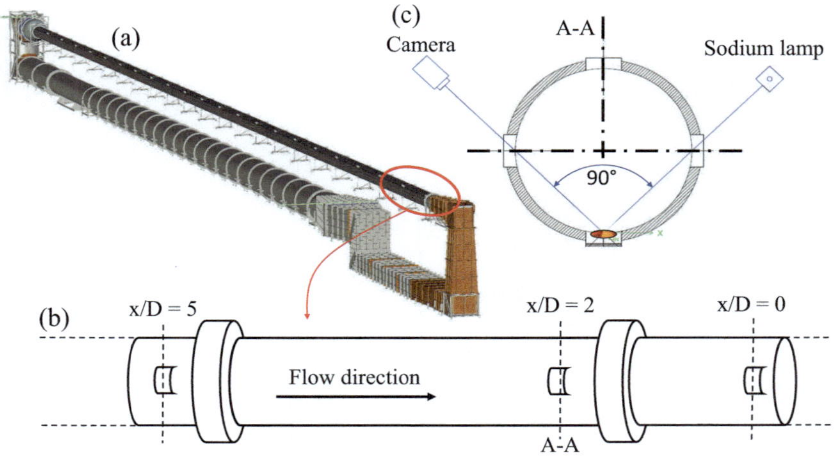

Fig. 2 **a** Schematic of the "long pipe" in CICLoPE, **b** Schematic of the measurement positions: x/D = 0 represents the location of test section, TS. **c** Schematic of the experimental setup. The oil drop is placed on top of the A3 window, covered with a black background to facilitate the acquisition

3 Results and Discussion

In Fig. 3a, "global" wall shear stress distribution, obtained from static pressure drop, is compared with "local" values coming from OFI measurements, for two different measurement positions. The data distribution presents only a small deviation from the ideal line, suggesting an apparent good agreement between the two methodologies. To better explore possible deviations, Fig. 3b shows the percentage difference for the two sets of measurements. The figure shows that apart from a few points, the two distributions seem to present a small systematic deviation from the pressure drop. The measurements performed in the TS show an average of $\approx 2\%$ underestimating the reference value, while the opposite happens at x/D = 5.

From different sets of measurements, it is possible to assess that $\approx 1\%$ (20% of the total error) is due to the choice of the OFI interrogation area, thus providing potential improvements in accuracy of such technique. Finally, local wall shear stress measurements between the TS and the station x/D = 5 were found to vary by $\approx 2\%$. These differences are of the same order of the measurement error, therefore, within this limit, we can conclude that the diffuser appears not to bias the wall-shear stress at the TS location.

Mean flow distribution inside the "long pipe" was measured by means of a Prandtl tube placed at the centreline in three different sections: x/D = 0 (Test section), x/D = 2 and x/D = 5. In order to investigate possible interference from the diffuser downstream of the TS, the centreline velocity data was compared to measurements by McKeon et al. [10]. Good agreement was observed, as exhibited in Fig. 4a. With a maximum observed deviation of $\approx 1\%$ from the friction law proposed by McKeon

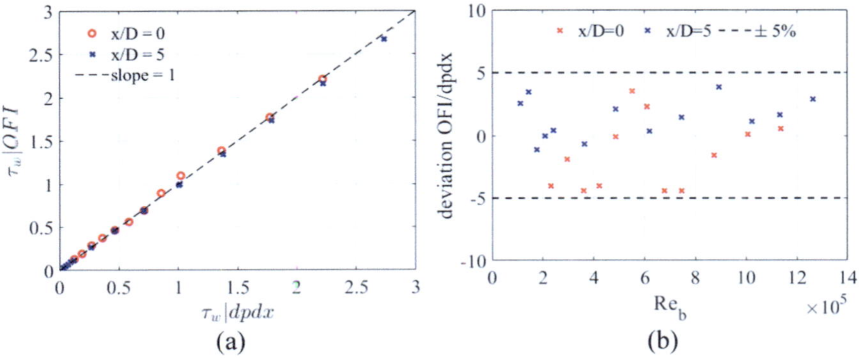

Fig. 3 **a** Comparison of "local" (OFI) and "global" (static pressure drop) wall shear stress measurements. Red circles represents OFI measurement in TS, blue cross, x/D = 5. **b** Percentage deviation between wall shear measurements from OFI and static pressure drop

et al. [10], it seems that the downstream diffuser has very little influence on the measured data. Looking in more detail, the error distribution, presented in Fig. 3b, shows again a small systematic offset. However, the maximum variation between the different sections lies well below a ±1% band which is coherent with the overall measurement uncertainty, suggesting the fact that the flow may be considered fully developed at least from the x/D = 5 station onwards.

4 Concluding Remarks

Wall shear stresses have been successfully measured in the "long pipe" in CICLoPE implementing a new technique based on oil-film interferometry, OFI. Measurements performed by OFI and static pressure drop, show good agreement between each other. Furthermore, an invariance in wall shear stress was observed between measurements performed at different streamwise positions, viz, x/D = 0 and x/D = 5. The centerline velocity invariance was assessed using a Pitot probe at different sections of CICLoPE "long pipe" finding agreement with the literature results. It was shown that less than 1% variation in centre-line velocity up to x/D = 5 downstream has been found.

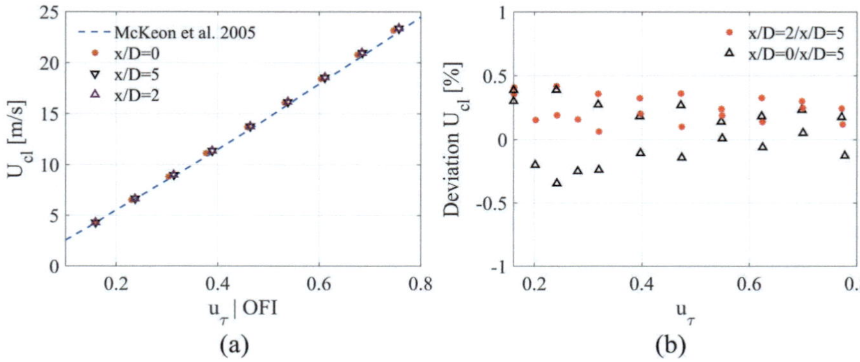

Fig. 4 **a** Centreline velocity measured in different position of the "long pipe" plotted against the friction velocity measured using OFI technique. Blue dashed lines represents behaviour proposed by McKeon et al. [10]. **b** Centreline velocity variation plotted against friction velocity. Red dots: variation between x/D = 2 and x/D = 5, black upward triangles: variation between x/D = 0 and x/D = 5

References

1. G. Bellani, A. Talamelli, *The Final Design of the Long Pipe in CICLOPE*. Prog. Turb. **VI**, 205–209 (2016)
2. E.-S. Zanoun, E. Öngüner, E. Christoph, B. Gabriele, T. Alessandro, One-dimensional flow spectra and cumulative energy from two pipe facilities. Prog. Turb. **VIII**, 209–214 (2019)
3. X. Zheng, G. Bellani, L. Mascotelli, R. Örlü, A. Ianiro, C.S. Vila, S. Discetti, J. Serpieri, M. Raiola, A. Talamelli, Y. Li, N. Jiang, Inter-scale interaction in pipe flows at high Reynolds numbers. Exp. Thermal Fluid Sci. **131**, 110529 (2022)
4. H. Nagib, P. Monkewitz, L. Mascotelli, G. Bellani, A. Talamelli, Uncertainty analysis of the Von Kármán constant for the mean centerline velocity in CICLoPE. Prog. Turbul. **VIII**, 197–202 (2019)
5. E.-S. Zanoun, C. Egbers, R. Örlü, T. Fiorini, G. Bellani, A. Talamelli, Experimental evaluation of the mean momentum and kinetic energy balance equations in turbulent pipe flows at high Reynolds numbers. J. Turbul. **20**(15), 285–299 (2019)
6. H.C.H. Ng, I. Marusic, J.P. Monty, N. Hutchins, M.S. Chong, Oil film interferometry in high Reynolds number turbulent boundary layers. In: *16th Australasian Fluid Mechanics Conference Crown Plaza, Gold Coast*, Australia, 2–7 Dec. 2007
7. R. Vinuesa, R. Örlü, *Measurement of Wall Shear Stress*. Experimental Aerodynamics, Chapter 12, pp. 393–429
8. L.H. Tanner, L.G. Blows, A study of the motion of oil films on surfaces in air flow, with application to the measurement of skin friction. J. Phys. E: Sci. Instrum. **9**(3), 194 (1976)
9. L.C. Squire, The motion of a thin oil sheet under the steady boundary layer on a body. J. Fluid Mech. **11**(2), 161–179 (1961)
10. B.J. McKeon, M. Zagarola, A. Smits, A new friction factor relationship for fully developed pipe flow. J. Fluid Mech. **538**, 429–443 (2005)

Flow Control

Effect of Streamwise Travelling Waves on Coherent Vortices and Fine Structure of Wall Turbulence

M. Umair and S. Tardu

Abstract In this work we analyse the effect of streamwise travelling waves wall forcing on the near-wall quasi-streamwise vortical structures. Six different amplitudes were considered at fixed frequency and wavelength that produced drag reduction margin ranging from 30–60%. The main effect of the large amplitude STW is to attenuate the spanwise vorticity at the wall. The near-wall quasi-streamwise vortices (QSVs) are shifted away from the wall with a drift of up to 10 wall units for the largest amplitude case considered here. This drift of QSVs results in their weakened signature at the wall, and hence reduced drag. The contribution of the quadrant events to the Reynolds shear stress show a significant increase in the sweep events by up to 130% for the large amplitude cases. Furthermore, the intensity of outward ($Q1$) and inward ($Q3$) interaction events show an increase of about 80% up to the edge of the buffer layer, implying increased incoherence of the near-wall flow field.

1 Introduction

Wall oscillations based techniques, streamwise travelling waves (STW) in particular, have received a growing interest due to their capability of producing significantly large drag reduction (DR) margins even at large Reynolds number (Re). Despite the huge effort put forward by the control community, the mechanism of drag reduction remains elusive. The reader is directed to [1] for a recent review.

Following [2], the STW are imposed at the wall using

$$W_{\text{wall}} = A \sin(\kappa x - \omega t), \tag{1}$$

where A is the amplitude, $\omega = 2\pi/T$ is the angular velocity, and $\kappa = 2\pi/\lambda$, is the wavenumber (T and λ represent the time period and wavelength of the travelling

M. Umair · S. Tardu (✉)
LEGI, Univ. Grenoble Alpes, CNRS, Grenoble-INP, 38000 Grenoble, France
e-mail: sedat.tardu@legi.grenoble-inp.fr

M. Umair
e-mail: mohammad.umair@legi.grenoble-inp.fr

R. Örlü et al. (eds.), *Progress in Turbulence X*, Springer Proceedings in Physics 404,
https://doi.org/10.1007/978-3-031-55924-2_32

wave, respectively). Here, x, y, z represent the streamwise, wall-normal, spanwise directions, respectively, and t is time. The corresponding instantaneous velocities in the streamwise, wall-normal, and spanwise directions are U, V, and W, respectively.

In the present work, we focus primarily on the effect of STW forcing on the near-wall quasi-streamwise vortical structures (QSVs). We use the canonical channel configuration with the streamwise and spanwise extent of $6\pi h$ and $3\pi h$, h being the channel half-height. The Reynolds number based on the centreline velocity (U_c) was fixed at $Re(= hU_c/\nu) = 4200$, where ν is the kinematic viscosity of the fluid. It corresponds to a friction Reynolds number $Re_\tau = hu_\tau/\nu = 180$ for the uncontrolled case (u_τ being the friction velocity). The details about the numerical schemes can be found in [3, 4]. Seven different controlled cases are considered, the first one is based on the homogeneous wall oscillation control (HWO) at its optimal configuration ($A = 12$, $T = 100$) and the rest six are for the STW forcing where we impose six different amplitudes (A/U_c) ranging from 0.15 to 1.25, keeping the forcing frequency (ω) and wavenumber (κ) fixed at 0.16 and 1.66 (outer units), respectively. This results in a DR margin of up to 60% for the largest amplitude case.

2 Response of Skin-Friction Coefficient and Spanwise Vorticity

Figure 1a shows the time traces of the normalized skin-friction coefficient (C_f/C_{f0}) immediately after the onset of actuation for different cases. Here, C_f is the skin-friction coefficient for the controlled flow and C_{f0} is the skin-friction coefficient for the uncontrolled flow. The amplitude of the STW has a remarkable effect on the time history of the skin-friction coefficient. It can be clearly seen that

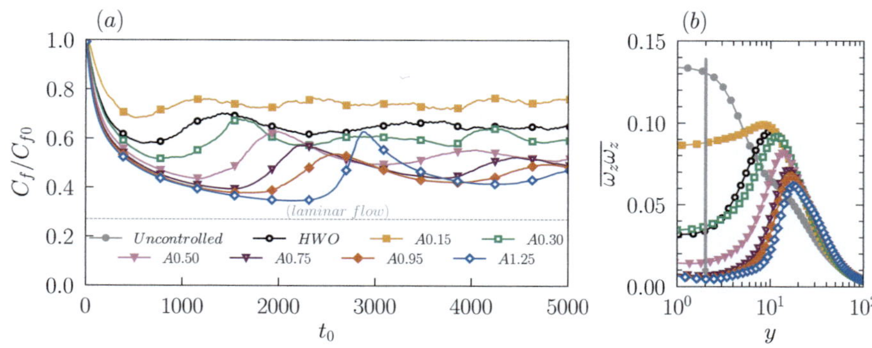

Fig. 1 **a** Initial response of the normalized skin-friction coefficient (C_f/C_{f0}) after the actuation of control (adapted from [4] with permission from American Physical Society), and **b** the mean spanwise turbulent enstrophy for all cases (adapted from [5] with permission from Cambridge University Press)

the time needed by the flow to adapt to its new quasi steady state appears to become longer for large amplitude STW. The most interesting is the response of the large amplitude STW, where the flow reaches quite close to the relaminarization limit before returning to the turbulent state. As pointed out in [4], this type of behaviour has close resemblance with the optimal wall turbulence control.

All the quantities are normalized by the inner scale of the controlled flow. The spanwise turbulent vorticity (ω_z) diffuses to the wall to generate the streamwise fluctuating wall shear stress (τ'). In the large amplitude STW cases, $\overline{\omega_z\omega_z}$ is spectacularly annihilated up to $y \approx 8$, with entirely negligible turbulent activity ($\overline{\tau'\tau'}$) at the wall (Fig. 1b). This is one of the outstanding effects of STW control compared to HWO control - the $\overline{\omega_z\omega_z}$ profile of which collapse rather with the small amplitude STW case ($A0.30$). In the case of HWO $\overline{\omega_z\omega_z}$ is damped in the viscous sublayer with a maximum at $y \approx 10$, while the peak of $\overline{\omega_z\omega_z}$ in the case of large amplitude STW is pushed towards the high buffer layer at $y \approx 20$.

3 Drift of Quasi-streamwise Vortices

Quasi-streamwise vortices (QSVs) are an essential feature of the near-wall turbulent flow field and have been recognized to play a significant role in the regeneration cycle of near-wall turbulence [6]. Earlier studies [7, 8] studied the effect of wall oscillations on the structural modifications of QSVs and report no effect on the wall-normal location of the peak population density of QSVs. Reference [4], on the other hand, point out that the QSVs reach their full maturity well above $y = 20$ for the large amplitude STW. To clarify this point, we examine the effect of STW actuation on the near-wall QSVs, focusing on detecting the drift (Δd_s) in QSVs because of the control, if any. The details of the eduction scheme are provided in [5].

Figure 2 shows the λ_2 structures for a few of the cases studied here, and also the respective wall-normal distribution of the number of vortices, normalized by the maximum number of vortices. For the canonical turbulent channel flow, the majority of the mature QSVs are located at $y \approx 24$, which is in agreement with [9]. For the STW cases, however, we observe a systematic drift of QSVs away from the wall. The drift is as large as 10 wall units for the $A1.25$ case. This is in disagreement with the earlier studies of [7, 8], but conforms to the finding of [4].

4 Quadrant Contributions to the Reynolds Shear Stress and Skewness of Fluctuating Velocity Fields

Following [10], the total Reynolds shear stress can be divided into four different categories depending on the signs of u and v. The second ($Q_2 : u < 0, v > 0$) and fourth ($Q_4 : u > 0, v < 0$) quadrant events contribute positively to the production of the Reynolds shear stress and are associated with the organized (coherent) structures in the near-wall turbulence, while the first ($Q_1 : u > 0, v > 0$) and third ($Q_3 : u <$

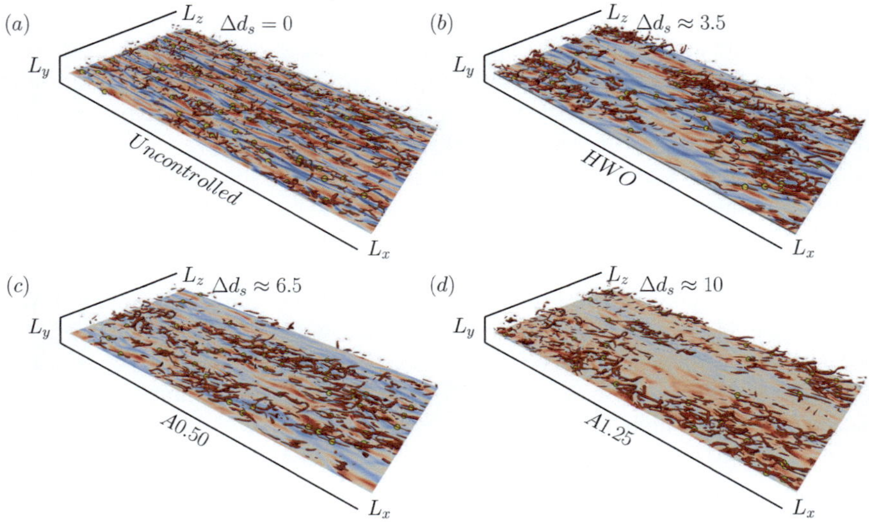

Fig. 2 The near-wall QSVs along with the instantaneous field of streamwise velocity fluctuations (u) (red: $u > 0$; blue: $u < 0$) on a wall-parallel plane at $y = 15$ for **a** Uncontrolled, **b** HWO, **c** $A0.50$, and **d** $A1.25$ case, respectively (adapted from [5] with permission from Cambridge University Press)

0, $v < 0$) quadrant events contribute negatively to the production of Reynolds shear stress. In the canonical as well as controlled cases, the main contribution to the total Reynolds shear stress clearly comes from the $Q2$ and $Q4$ events throughout the generation region (Fig. 3). Their relative importance vary with the distance from the wall. In the canonical case, below roughly $y \approx 15$ the contribution of sweep ($Q4$) events is larger while ejections ($Q2$) dominate above $y \approx 15$. In the large amplitude case, the contribution of sweeps increases up to 130% and the region where sweep dominates extends up to $y = 40$. This can also be observed by looking at the skewness profiles S_u and S_v (Fig. 4b), that show a large negative peak in S_v and positive peak in S_u at about $y \approx 18$. Quite interestingly, the contributions of Q_1 and Q_3 events, that amount to about 20% in total in the canonical case, increased to about 80% up to the edge of the buffer layer. An increase in the contribution of these events that act to destroy the production of Reynolds shear stress can be attributed to the increased incoherence of the near-wall flow field. This can also be clearly observed by looking at the correlation coefficient of Reynolds shear stress $C_{uv} = -\overline{uv}/u'v'$ (here, prime indicates the root-mean-square values), shown in Fig. 4d, that shows a significant decline in the same region for the large amplitude STW cases. The ensemble of these results are in agreement with the drift of the near-wall quasi-streamwise vortices discussed above.

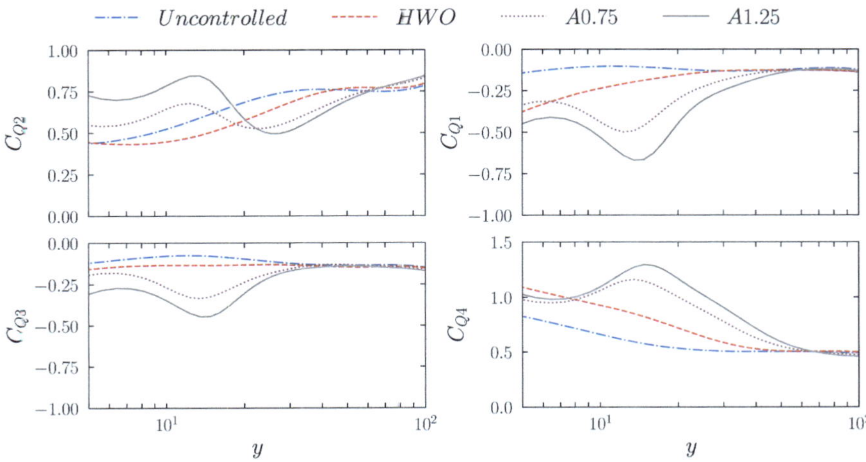

Fig. 3 Reynolds shear stress from each quadrant normalized by the local mean Reynolds shear stress

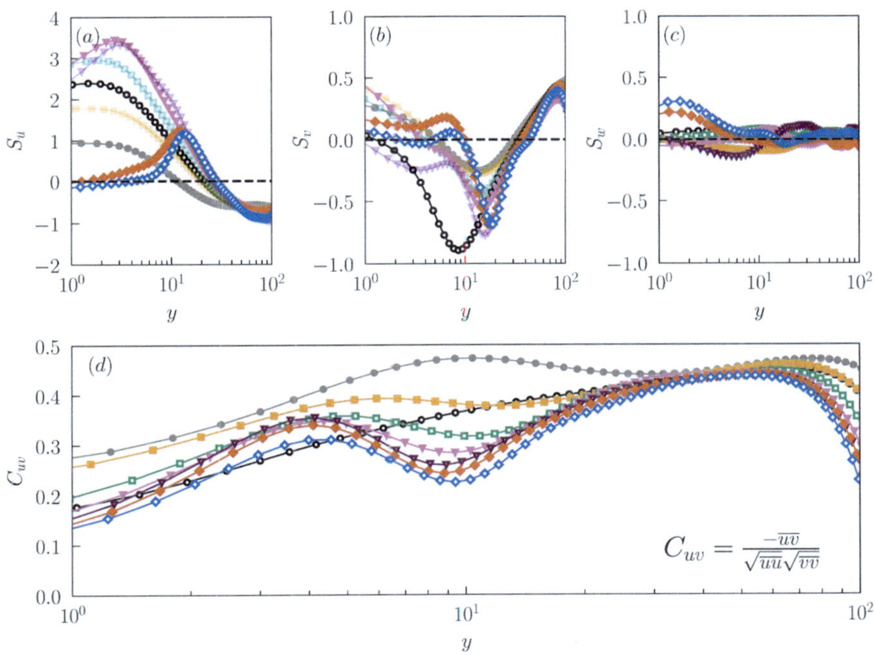

Fig. 4 **a**, **b**, **c** Skewness factor for the fluctuating velocity components in the uncontrolled and controlled flows, and **d** the correlation coefficient of uv. Refer to Fig. 1 for markers corresponding to different cases

5 Conclusions

In this study, we examined the effect of STW on the near-wall QSVs in a turbulent channel flow at Reynolds number $Re_\tau = 180$. The frequency and wavelength of the imposed travelling wave were kept fixed, and only the amplitude was varied to examine the direct effect arising because of travelling wave wall oscillations. At the largest amplitude, the flow almost reach the relaminarization limit. Such a significant level of drag reduction cannot be achieved in case of homogeneous wall oscillation. The main effect of the STW control appears in the spanwise turbulent enstrophy component that shows a significant decline in its intensity, almost vanishing at the wall for large amplitude cases. The near-wall QSVs drift away from the wall by up to 10 wall units, resulting in their weakened signature at the wall and hence reduced drag. The contribution of the quadrant events to the Reynolds shear stress show a significant increase in the sweep events by up to 130% for the large amplitude cases. Moreover, the contributions of Q_1 and Q_3 events show an increase of about 80% up to the edge of the buffer layer, implying increased incoherence of the near-wall flow field.

Acknowledgements The computational facilities for the present work were provided by the GRICAD infrastructure (https://gricad.univ-grenoble-alpes.fr).

References

1. P. Ricco, M. Skotes, M. Leschziner, A review of turbulent skin-friction drag reduction by near-wall transverse forcing. Prog. Aerosp. Sci. **123**, 100713 (2021)
2. M. Quadrio, P. Ricco, C. Viotti, Streamwise-travelling waves of spanwise wall velocity for turbulent drag reduction. J. Fluid Mech. **627**, 161–178 (2009)
3. F. Bauer, S. Tardu, O. Doche, Efficiency of high accuracy DRP schemes in direct numerical simulations of incompressible turbulent flows. Comput. Fluids **107**, 123–140 (2015)
4. M. Umair, S. Tardu, O. Doche, Reynolds stresses transport in a turbulent channel flow subjected to streamwise traveling waves. Phys. Rev. Fluids **7**(5), 054601 (2022)
5. M. Umair, S. Tardu, Vorticity transport in a turbulent channel flow subjected to streamwise travelling waves. J. Fluid Mech. **967**, A9 (2023)
6. J.M. Hamilton, J. Kim, F. Waleffe, Regeneration mechanisms of near-wall turbulence structures. J. Fluid Mech. **287**, 317–348 (1995)
7. A. Yakeno, Y. Hasegawa, N. Kasagi, Modification of quasi-streamwise vortical structure in a drag-reduced turbulent channel flow with spanwise wall oscillation. Phys. Fluids **26**, 085109 (2014)
8. E. Gallorini, M. Quadrio, D. Gatti, Coherent near-wall structures and drag reduction by spanwise forcing. Phys. Rev. Fluids **7**(11), 114602 (2022)
9. J. Jeong, F. Hussain, W. Shoppa, J. Kim, Coherent structures near the wall in a turbulent channel flow. J. Fluid Mech. **332**, 185–214 (1997)
10. J.M. Wallace, H. Eckelmann, R.S. Brodkey, The Wall Region in Turbulent Shear Flow. J. Fluid Mech. **54**(1), 39–48 (1972)

Plasma Jets in a Turbulent Channel Flow

F. M. Ricci, E. Amico, G. Cafiero, G. Iuso, and J. Serpieri

Abstract To investigate the effect of friction drag manipulation, an array of plasma actuators was designed to induce a wall jet oscillating along the opposite crosswise directions. Besides, steady mono-directional and bi-directional forcing were performed too. Pressure measurements and optical flow diagnostics were deployed to inspect the effect of the operated actuation on a fully-developed turbulent flow.

1 Introduction

In recent efforts, plasma actuators (PAs) have been considered as plausible flow actuators to induce a Stokes-like wall flow, similar to what caused by oscillating the flow-exposed surfaces along the crosswise direction, and capable of reducing the friction drag exerted by turbulent flows (e.g. [1–4]). The interest they collected is motivated by their simpler embodiment compared to mechanical, pneumatic or piezo-electric devices (e.g. [5–9]). Nevertheless, the flow control effect that they induce differs from the more studied and understood strategies based on displacing the flow exposed surfaces and deserves more investigations. This study is motivated by this need.

2 Experimental Setup

The experimental campaign took place in a channel flow facility (0.42 m $\times$ 0.035 m $\times$ 10 m, in width, height h and length), operated at the friction Reynolds numbers of $Re_\tau = \frac{u_\tau h/2}{\nu} = 190$ (where u_τ is the friction velocity and ν the kinematic viscosity). Pressure taps every 0.20 m along the streamwise direction are installed in the channel walls. These pressure taps were connected to a piezo-resistive

F. M. Ricci · E. Amico · G. Cafiero · G. Iuso · J. Serpieri (✉)
Department of Mechanical and Aerospace Engineering, Politecnico di Torino, Corso Duca degli Abruzzi, 24, 10129 Turin, Italy
e-mail: Jacopo.Serpieri@me.com

R. Örlü et al. (eds.), *Progress in Turbulence X*, Springer Proceedings in Physics 404,
https://doi.org/10.1007/978-3-031-55924-2_33

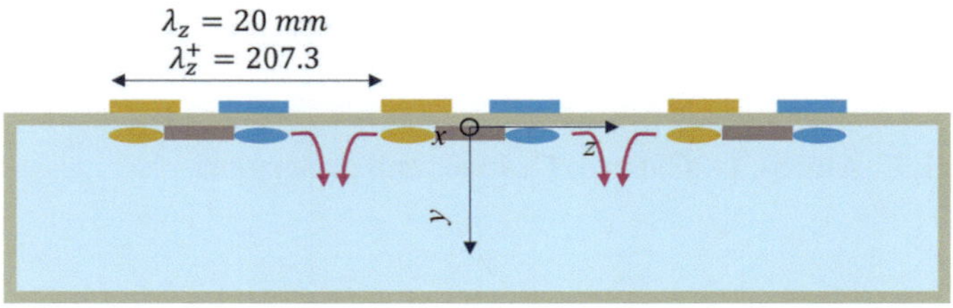

Fig. 1 Schematics of the setup as seen in a cross-stream plane (not to scale): actuators spacing (λ_z, λ_z^+), operated electrodes (yellow and blue rectangles), exposed electrodes (brown rectangles), plasma discharges (oval shapes colored as the related HV electrodes), colliding plasma jets (magenta arrows)

pressure transducer (*DSA*) operated at 20 Hz and set to acquire for 200 s for statistical significance of the temporal averages.

The flow actuator hereby considered featured an array of six dielectric barrier discharge (DBD) PAs with the electrodes aligned with and extending 600 mm along the streamwise direction. The PAs considered 0.5 mm thick polyethylene terephthalate sheets as the dielectric layer. The electrodes were made with 0.35 mm thick copper tape and the isolation of the encapsulated electrodes was guaranteed by multiple layers of *Kapton* tape. Every actuator featured one exposed electrode connected to the ground and two encapsulated electrodes, one per side of the exposed electrode, each connected to a high voltage (HV) source. The latter were four *GBS Elektronik Minipuls 4* supplied with a 7 kHz signal and set to deliver a sinusoidal signal of peak-to-peak amplitude of 8 kV. A schematic of the actuator installed in the channel is shown in Fig. 1, together with the coordinates system used in this study: x is the streamwise direction, while y and z are the wall normal and the crosswise ones, respectively. x originates at the actuator upstream edge, y at the wall and z at the channel center-line. The velocity components along these directions are denoted with u, v, w, respectively. Reynolds decomposition will be considered: $[u(t), v(t), w(t)] = [U, V, W] + [u'(t), v'(t), w'(t)]$, where t is time and upper case letters and overlines, in the remainder, imply time-averages.

According to the chosen actuation strategy, the HV signals were opportunely modulated. A schematic of the voltage-modulation effect can be extracted from Fig. 1. For the actuation strategy aiming at replicating the oscillating wall forcing, either the yellow or blue actuators were operated at the same instant and each set was operated for 50 % of the modulation cycle. Consequently, for this case, either the yellow or blue colored plasma discharges took place. The modulation frequency was set to meet the literature-deemed-optimal non-dimensional period of $T^+ = 100$ (where the + labeled quantities are non-dimensionalised with inner-layer scales) (see [6]). When instead the steady monodirectional forcing is considered, only the blue electrodes were constantly operated, therefore leading only to the blue-colored plasma discharges. This case is referred to as $w_{PAs} > 0$. Finally, the case where both the electrodes sets were constantly operated was also considered. In these conditions, the plasma-induced jets from both the electrodes sets collided halfway and generated

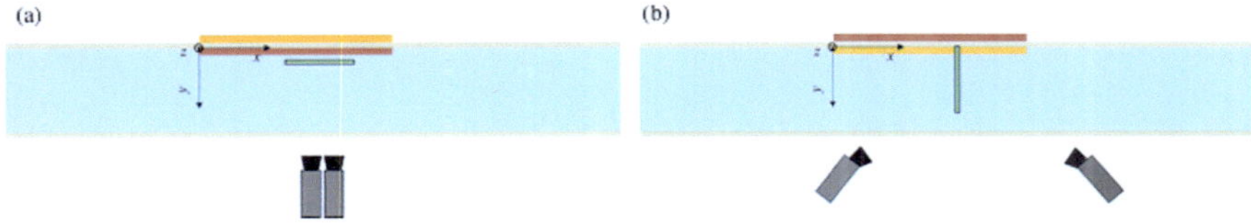

Fig. 2 Schematics side view of the PIV (**a**) and SPIV **b** setups (not to scale): channel and actuators shown as in Fig. 1, laser plane (light green) and cameras (grey shapes)

upwash (i.e. $y \rightarrow h/2$) fluid motions (see [7, 10]). This is also shown in Fig. 1 with the magenta arrows, together with the crosswise spacing of the actuators: $\lambda_z = 20$ mm ($\lambda_z^+ = 207.30$ for the reference flow; i.e. about twice the average streaks' spacing).

Two different optical flow diagnostics experiments were set up: a xz plane PIV setup (see Fig. 2a) and a yz plane stereographic PIV (SPIV) setup (see Fig. 2b). Both setups consisted of two *Andor Zyla* sCMOS cameras (5.5 Mpx, 16 bits) and a *Litron* double cavity laser (200 mJ/pulse, 15 Hz maximum repetition rate). The cameras were equipped with *Tokina* 100 mm macro lens with green filters and, for the SPIV setup, with Scheimpflug-condition adapters from *LaVision*. The system synchronization was guaranteed by a *National Instrument PCI-6602* board. The acquisition frequency was 15 Hz and a total of 1000 image pairs was acquired.

The PIV setup featured the two cameras placed one next to the other along the streamwise direction such to image a total field of view (FOV) extending 111.90 mm $\times$ 48.10 mm along the $x - z$ directions. This corresponds, in non-dimensional units, to 1550 $\times$ 497.40 along $x^+ - z^+$. The illuminated plane was set at $y = 5$ mm ($y^+ = 51.82$). The final velocimetry resolution was 3.67 vectors/mm, i.e. 0.35 vectors per viscous length scale. The thickness of the laser sheet was set to ≈ 0.5 mm using light knives.

The SPIV setup investigated a domain extending 29.50 mm $\times$ 22 mm along z and y, thus 305.70 $\times$ 227.98 along $z^+ - y^+$. The plane was set at $x = 546$ mm ($x^+ = 5.67\ 10^3$) from the actuator upstream edge. The velocimetry resolution was 9.26 vectors/mm (0.89 vectors per viscous length scale). The thickness of the laser sheet was increased to ≈ 1.5 mm to guarantee a good dynamic range.

3 Results and Discussion

The effect of the performed flow actuation measured by the differential pressure transducer shows rather similar behaviors for all the three actuation strategies compared to the reference unforced flow. First, at the upstream stations, all the actuated flows showed, for the same fan speed, a higher level of pressure. This could be explained as a flow-blockage effect induced by the plasma jets. Then, moving along the streamwise direction, as the flow undergoes actuation, the slope of the differential pressure decreased whereas, moving downstream of the actuation region, the

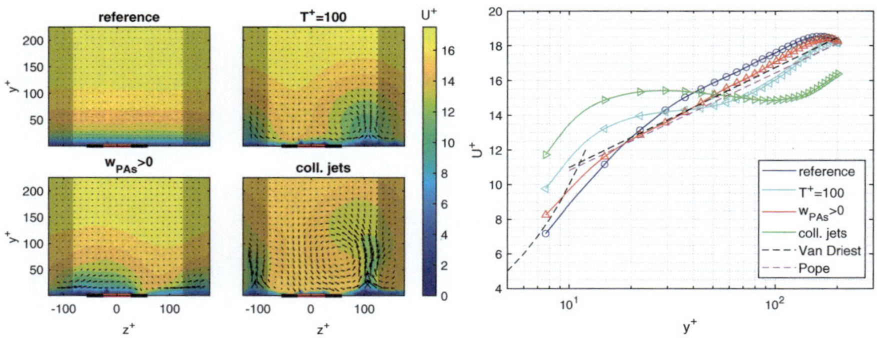

Fig. 3 SPIV time-averaged fields: **a** $V - W$ fields (vectors) on U contours; 1:10 vectors shown (arbitrary length); exposed electrodes (red line), HV electrodes (black lines); **b** λ_z-averaged U profiles (un-shaded areas of **a**); 1:7 measurements shown; black dashed line from [11]; red dashed line from [12]

slope recovered to the unforced value (see also [4]). Such behaviour of the pressure trend is thought to be initially caused by the downwash ($y \to 0$) motions of higher-momentum fluid needed to comply with mass continuity and the plasma-jets-induced crosswise momentum. On the other side, at more downstream stations, the beneficial effect of the performed actuation is likely overcoming this negative (drag enhancing) effect and a lower pressure drop was observed for the actuated flow cases.

The SPIV setup allows to inspect the effect of the different actuation strategies on cross-stream planes. The time-averaged velocity fields are shown in Fig. 3a. The fields are made non dimensional using the reference flow friction velocity u_τ. While, the reference flow features a crosswise-uniform streamwise velocity field with non-appreciable in-plane velocity components, all the forced flows present strong in-plane motions with related crosswise modulations of the streamwise velocity component. In particular, the $T^+ = 100$ case, shows time-averaged wall normal fluid motions between the actuators ($z^+ \approx \pm 103$). At the same time, a downwash sets over the exposed electrode, or at slightly more negative z positions (the reason of this mis-alignment might be due to small differences in the induced jets' strength). This flow scenario appears similar to the "colliding jets" case where both electrodes' sets are constantly operated; albeit this appears somewhat milder for the $T^+ = 100$ case. The similarity between these two fields suggests that for the considered actuators' spacing and jets strength, the induced flow does not fully resemble, for the crosswise-oscillating case, the Stokes-like flow where a wall layer of crosswise velocity was desired. The $w_{PAs} > 0$ case also shows a crosswise modulation but with weaker wall-normal fluid motions.

Wall-normal profiles, averaged along one λ_z, of streamwise velocity are shown in Fig. 3b together with the [11, 12] law of the wall profiles. While for the reference case, the measured values nicely lay on the logarithmic distribution, all the forced flow curves feature more non-homogeneous distributions. Namely, the log-layer velocities

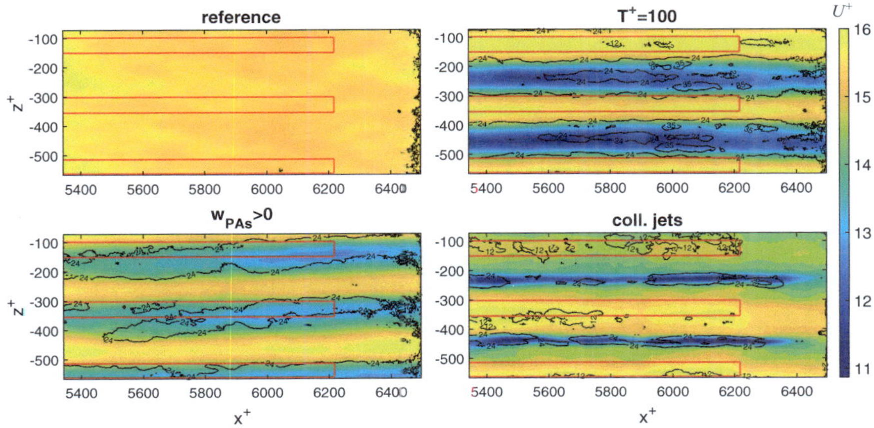

Fig. 4 U^+ (colors) and $\overline{u'^{2+}}$ (isolines) fields and PAs' exposed electrodes (red lines)

attain to lower values compared to the reference flow; whereas, the buffer-layer appears inflated by the performed flow actuation.

The velocity fields measured by the PIV setup are also useful to describe the effect of the plasma jets. The time-averaged U^+ fields are presented in Fig. 4: a strong modulation from the performed actuation is here evident too, especially for the $T^+ = 100$ and the "colliding jets" cases. The $w_{PAs} > 0$ field shows the velocity isolines tilted towards larger z as a consequence of the monodirectional actuation. The variance $\overline{u'^{2+}}$ velocity fields are shown too in Fig. 4. While for the $T^+ = 100$ case, two larger $\overline{u'^{2+}}$ regions are evident, due to the temporal modulation of the induced motions, the $w_{PAs} > 0$ flow features more compact regions of larger $\overline{u'^{2+}}$, related to the shears induced by the strong upwash motion between the actuators.

The occurrence and strength of low-speed streaks (LSS) of streamwise velocity are directly connected to larger skin friction events. As such, the LSS features are analysed extracting connected regions of $u' < 0$, extending in the streamwise direction for at least 100 viscous length scales. The velocity fields are then conditionally averaged to derive LSS statistical features. These are shown in Fig. 5. Figure 5a shows that the $T^+ = 100$ and $w_{PAs} > 0$ forcing reduce, at the measurement plane, the velocity deficit within the LSS. Figure 5b reveals that the wall-normal vorticity ω'_y within the LSS is much reduced by all the considered forcing. The flow actuation increases the probability (pdf) of larger streaks' length L_s^+ (Fig. 5c) whereas, has milder effects on their average width W_s^+ (Fig. 5d).

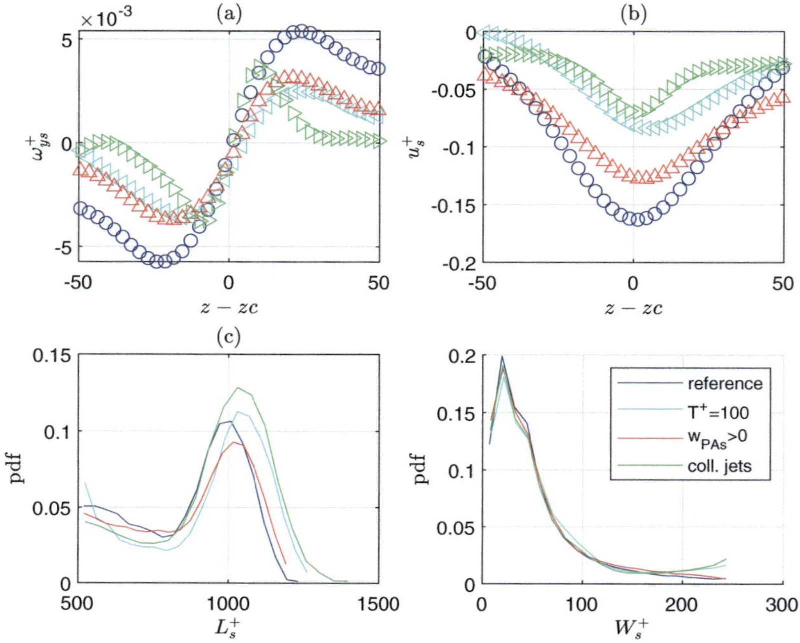

Fig. 5 LSS features for the considered flow cases: ${\omega'_{ys}}^+$ (**a**); ${u'_s}^+$ (**b**); pdf of L^+_s (**c**) and pdf of W^+_s (**d**); z_c is the LSS center along z

The preliminary results hereby presented show strong effects from the performed actuation on the operated time-averaged flow as well as on the coherent turbulent structures. These will be analysed and discussed in future studies.

Acknowledgements The authors are grateful to T. Astarita for the (S)PIV software.

References

1. K.S. Choi, T. Jukes, R. Whalley, Philosophical transactions of the royal society a: mathematical. Phys. Eng. Sci. **369**(1940), 1443 (2011). https://doi.org/10.1098/RSTA.2010.0362
2. R.D. Whalley, K.S. Choi, Exp. Fluids **55**(8) (2014). https://doi.org/10.1007/S00348-014-1796-3
3. M.T. Hehner, D. Gatti, P. Mattern, M. Kotsonis, J. Kriegseis, AIAA J. **59**(2), 763 (2021). https://doi.org/10.2514/1.J059802
4. J. Serpieri, M.T. Hehner, J. Kriegseis, Flow, Turbul. Combust. (2023). https://doi.org/10.1007/s10494-023-00463-w
5. G.M. Di Cicca, G. Iuso, P.G. Spazzini, M. Onorato, J. Fluid Mech. (2002). https://doi.org/10.1017/S002211200200157X
6. G.E. Karniadakis, K.S. Choi, Annu. Rev. Fluid Mech. **35**, 45 (2003). https://doi.org/10.1146/annurev.fluid.35.101101.161213

7. G. Iuso, G.M. Di Cicca, J. Turbul. (2009). https://doi.org/10.1080/14685240601110088
8. F. Auteri, A. Baron, M. Belan, G. Campanardi, M. Quadrio, Phys. Fluids **22**(11) (2010). https://doi.org/10.1063/1.3491203
9. I. Marusic, D. Chandran, A. Rouhi, M.K. Fu, D. Wine, B. Holloway, D. Chung, A.J. Smits, Nat. Commun. **12**(1), 1 (2021). https://doi.org/10.1038/s41467-021-26128-8
10. X.Q. Cheng, C.W. Wong, F. Hussain, W. Schröder, Y. Zhou, J. Fluid Mech. **918**, 1 (2021). https://doi.org/10.1017/jfm.2021.311
11. E. Van Driest, J. Aeronaut. Sci. (1956). https://doi.org/10.2514/8.3713
12. S.B. Pope, Cambridge University Press (2000)

On the Wave-Induced Stokes Sublayer and Drag Reduction in the Turbulent Wind

A. Cimarelli, F. Romoli, and E. Stalio

Abstract The interactions of a turbulent wind with a water surface represents a very fundamental problem for many atmospheric processes. The momentum and heat exchanges across the interface with oceans abruptly affects the atmosphere and the understanding of the driving mechanisms would certainly improve weather predictions capabilities. We performed a Direct Numerical Simulation of the wind-wave interaction problem using realistic values of the fluid properties of air and water [3]. The simulation reveals that at low Reynolds numbers, an interesting wind-wave pattern propagating at an angle in the upstream direction is generated. This pattern is recognized to be at the basis of the generation of a spanwise oscillating Stokes sublayer that is responsible for a drag reduction mechanism in the turbulent wind. Despite the simulated flow conditions are far from the intense events occurring at the ocean-atmosphere interface, this basic flow phenomenon may actually explain the large scatter of the drag coefficient data in field measurements where swell waves of arbitrary directions are often present.

1 Introduction

Surface fluxes of momentum and energy through the ocean-atmosphere interface are of primary importance for the characterization of geophysical flows and climate formation mechanisms. The presence of a somewhat compliant water surface gives rise to complex nonlinear interaction mechanisms between the atmospheric boundary layer and the wave field developed in response to the turbulent wind forcing itself [1]. These interactions determine the flux of momentum and energy trough the air-sea interface thus representing a fundamental process for the climate behaviour. However,

A. Cimarelli (✉) · F. Romoli · E. Stalio
DIEF, University of Modena and Reggio Emilia, 41125 Modena, Italy
e-mail: andrea.cimarelli@unimore.it

F. Romoli
e-mail: federica.romoli@unimore.it

E. Stalio
e-mail: enrico.stalio@unimore.it

R. Örlü et al. (eds.), *Progress in Turbulence X*, Springer Proceedings in Physics 404,
https://doi.org/10.1007/978-3-031-55924-2_34

the physical mechanisms behind the wind-wave interactions still lack a sufficiently general and complete theory [2]. The main reason is the multiplicity of complex phenomena involved and the lack of suitable experimental and numerical data. In order to address the latter issue, we performed a Direct Numerical Simulation of the wind-wave interaction problem in the flow settings of a two-phase open channel driven by a constant pressure gradient where the wind is turbulent and the water is almost quiescent [3]. To the author' knowledge, the simulation represents one of the very first attempts to get the fully-coupled solution of the wind-wave problem based on first principles and on realistic values of the fluid properties of air and water.

2 Flow Settings and Numerical Details

The flow case considered is an open channel composed of a water layer on the bottom of an air layer with gravity acting in the wall-normal direction. The evolution of the flow is governed by the continuity and momentum equations where the different density ρ and kinematic viscosity $\nu = \mu/\rho$ of the two fluids is taken into account by solving a transport equation for the volume fraction,

$$\frac{\partial \alpha}{\partial t} + \frac{\partial \alpha u_j}{\partial x_j} + \frac{\partial}{\partial x_j}\left[\alpha\left(1-\alpha\right) u_{r_j}\right] \tag{1}$$

so that

$$\rho = \rho_w \alpha + \left(1-\alpha\right)\rho_a \qquad \mu = \mu_w \alpha + \left(1-\alpha\right)\mu_a \tag{2}$$

where the subscripts w and a are used to denote quantities computed for water and air, respectively. Here, we have $\rho_a = 1$, $\rho_w = 1 \cdot 10^3$, $\mu_a = 1.48 \cdot 10^{-5}$ and $\mu_w = 1 \cdot 10^{-3}$. A compression velocity defined as

$$u_{r_i} = c_\alpha |\boldsymbol{u}| \hat{n}_i \tag{3}$$

is used for keeping the interface region sharp. Here, $\hat{\boldsymbol{n}} = \nabla\alpha/|\nabla\alpha|$ is the unit vector normal to the interface and $c_\alpha = 0.5$ is the value of the compression coefficient used in the present work. The volume fraction α is also used to compute the forcing f_{σ_i} appearing in the momentum equation related with surface tension as

$$f_{\sigma_i} = \sigma \kappa \hat{n}_i \tag{4}$$

where $\kappa = -\partial n/\partial x_i$ is the local curvature of the interface and $\sigma = 0.07$ is the surface tension coefficient.

The flow is solved in a numerical domain whose extension is $(L_x, L_y = h_a + h_w, L_z) = (25.6h_a, 1.6h_a, 25.6h_a)$, where (x, y, z) denote the streamwise, vertical and spanwise coordinates. The problem has been discretized numerically by means of a structured Cartesian grid of hexahedral cells. The number of volumes used

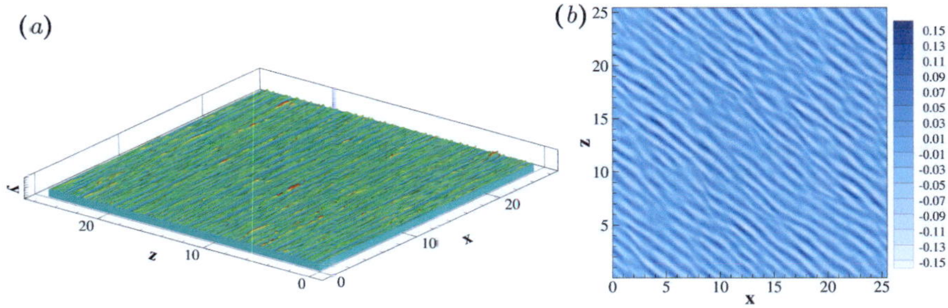

Fig. 1 **a** Iso-surface of the instantaneous streamwise velocity $u = 0.4U_{max}$ colored with the distance from the water surface. **b** Iso-contours of the instantaneous pattern taken by the wave elevation $\eta^+(x, z)$

is $(N_x, N_y, N_z) = (820, 277, 1232)$ that are distributed homogeneously in the horizontal directions, while stretching laws have been applied in the vertical direction in order to obtain a better resolution in the highly inhomogeneous region around the wind-wave interface. The resulting grid spacing in friction units is $(\Delta x^+, \Delta y^+_{min}, \Delta z^+) = (9.9, 0.1, 6.6)$, where Δy^+_{min} is achieved at the wind-wave interface. Throughout the paper, variables in friction units imply normalization of lengths with the wind friction length ν_a/u_{τ_a}, and velocities with the wind friction velocity u_{τ_a} evaluated on top of the interface region between the two fluids.

The convective and diffusive fluxes at the numerical volume faces are evaluated through a second-order central difference scheme, whereas time integration is performed with a first-order implicit Euler scheme. The time step is variable throughout the simulation to obtain a condition $CFL < 1$ in each point of the domain. Periodic boundary conditions are applied in the streamwise and spanwise directions. No-slip and free-slip boundary conditions are imposed at the water bed and top boundary, respectively. Finally, a zero gradient condition is imposed in the vertical direction for the volume fraction. The flow is driven by imposing a constant pressure gradient that leads to a friction Reynolds number in the air boundary layer of $Re_{\tau_a} = 317$ and in the water layer of $Re_{\tau_w} = 106$. Hence, a turbulent wind over an almost quiescent and laminar water layer is developed, see Fig. 1a.

3 Results

As shown in Fig. 1b, the dynamical evolution of the flow system leads to a water wave pattern that at equilibrium with the turbulent wind, propagates at an angle $(\gamma = 38.6°)$ with respect to the mean wind direction. Given the low Reynolds number considered, both the wave height and the wave steepness are small, $\delta^+_w = 0.3$ and $S = 7.3 \cdot 10^{-4}$ respectively. On the other hand, the phase speed is significant, $\mathbf{c}^+ = (10, 8)$. Despite the small size of the water wave pattern, its effect on the

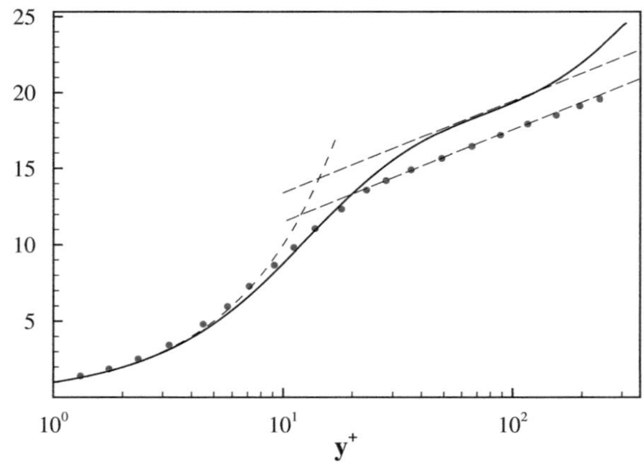

Fig. 2 Mean velocity profile U^+ of the wind-wave problem (solid line) compared with the mean velocity profile of wall turbulence in an open channel (circles) from [4]. The mean wind velocity U^+ is computed as the difference with respect to the mean water interface velocity located at $y^+ = 0$

turbulent wind is far from being negligible. Indeed, the mean velocity profile exhibits a significant reduction of drag when compared with wall turbulence, see Fig. 2. In particular, we measure that the von Kármán constant is almost unaltered with respect to wall turbulence, $k \approx 0.38$, while the additive constant is substantially increased from $B = 5.7$ to $B = 7.3$. Hence, its increment $\Delta B = 1.6$ can be interpreted as a measure of the drag reduction experienced by the wind-wave problem with respect to wall turbulence.

As shown in [3], the origin of drag reduction can be associated with the presence of a wave-induced Stokes sublayer. Given the very low steepness of the water waves, it is possible to assume that the air flow does not separate and by following the wave pattern accelerates on the windward side and decelerates on the leeward side. This behaviour can be associated with a pattern for the pressure field that is minimum above the wave crests and maximum within the trough regions. Because the water-wave pattern is skewed with respect to the mean wind direction, these pressure variations give rise to periodically distributed pressure gradients in both the streamwise and spanwise directions. Accordingly, in the near-interface region the flow oscillates both in the streamwise and spanwise direction as highlighted by the pattern of the streamwise and spanwise velocity shear reported in Fig. 3a, b, respectively. Of interest for drag reduction is the periodic flow oscillation in the spanwise direction reported in Fig. 4. Indeed, such type of modulation resembles the spanwise Stokes layer [5] that is widely recognized to give rise to a weakening of the self-sustaining processes of turbulence and, hence, to drag reduction. In the context of drag-reducing techniques based on oscillating walls, it is also known that the penetration length of the Stokes layer is the key quantity for the effectiveness of the viscous shearing action of the moving wall in weakening the near-wall turbulence interactions. Here, we measure a very small thickness of the Stokes layer $\ell_s^+ \approx 2$ and a very weak intensity of the associated spanwise motion $|w|_{max}^+ = \mathcal{O}(10^{-2})$. It is then important to highlight that in the context of drag-reducing techniques based on oscillating walls these very small values associated with the Stokes layer would

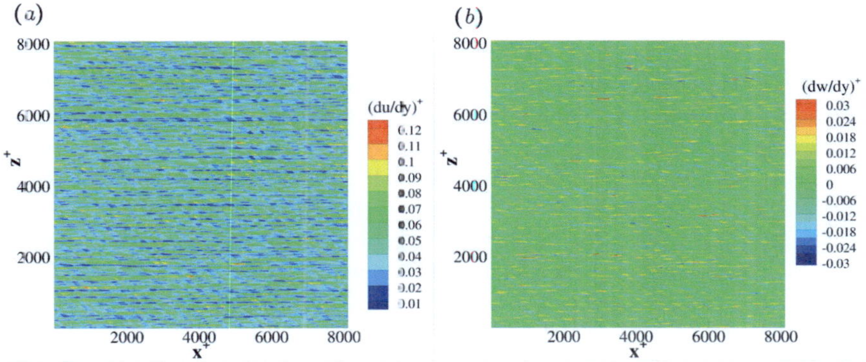

Fig. 3 Iso-contours of the instantaneous **a** streamwise velocity shear $\partial u^+/\partial y^+$ and **b** spanwise velocity shear $\partial w^+/\partial y^+$ evaluated at the water interface

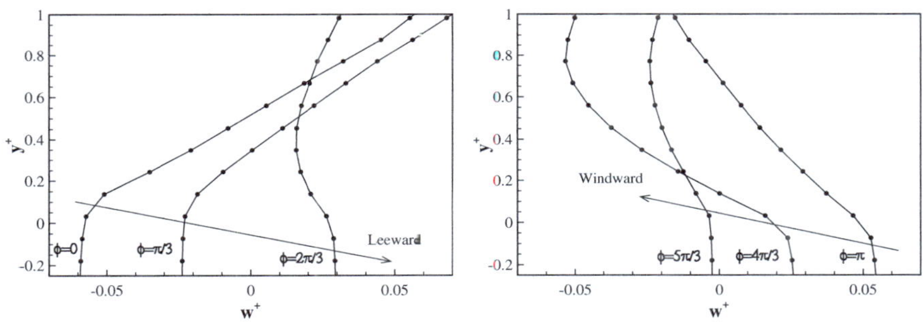

Fig. 4 Profiles of the instantaneous spanwise velocity taken at different locations within a single water-wave length

lead to an almost negligible drag reduction [5] thus unveiling the high efficiency of the wave-induced Stokes sublayer in weakening the near-interface self-sustaining processes of turbulence.

4 Conclusions

The present work reports results from a Direct Numerical Simulation of the wind-wave interaction problem. This database represent one of the very first attempts to obtain the fully coupled solution of the wind-wave problem based on first principles and on realistic values of the fluid properties of air and water [3]. Here, we focused the analysis to the spanwise oscillating Stokes sublayer induced by the water-wave pattern that is skewed with respect to the mean wind direction. Despite its weakness, this oscillating layer is identified as the responsible for the drag reduction measured in the turbulent wind. This insight can have an impact on our understanding of ocean-atmosphere interface mechanisms. Indeed, it is well known that field realizations are characterized by the presence of swell waves generated from remote wind events that move in arbitrary direction with respect to the local wind. As unequivocally shown here, this misalignment can give rise to a wave-induced Stokes sublayer and to a modulation of the momentum flux at the air-water interface thus possibly explaining the large scatter of the drag coefficient data reported in field measurements. In particular, it suggests that these data should be analysed by conditioning the statistics with the angle between local wind and waves.

Acknowledgements We wish to acknowledge the financial support given by the Department of Engineering "Enzo Ferrari" of the University of Modena and Reggio Emilia through the action "FAR dipartimentale 2020/2021". The work is also funded by the project ECOSYSTER—SPOKE 6 under the National Recovery and Resilience Plan, Mission 04 Component 2 Investment 1.5—NextGenerationEU, Call for tender no. 3277 dated 30 December 2021. We acknowledge the Italian supercomputing centre CINECA that, under the ISCRA B project WWW (turbulent Wind and Water Waves interaction: multiscale analysis and modelling) and the ISCRA C project WWIF (simulations of Wind-Wave Interfacial Flows), provided the high-performance computing resources for the present work.

References

1. P.P. Sullivan, J.C. McWilliams, Dynamics of winds and currents coupled to surface waves. Ann. Rev. Fluid Mech. **42**, 19–42 (2010)
2. A. Ayet, B. Chapron, The dynamical coupling of wind-waves and atmospheric turbulence: a review of theoretical and phenomenological models. Bound.-Lay. Meteorol. 1–33 (2021)
3. A. Cimarelli, F. Romoli, E. Stalio, On wind-wave interaction phenomena at low Reynolds numbers. J. Fluid Mech. **956**, A13 (2023)
4. R. Nagaosa, R.A. Handler, Statistical analysis of coherent vortices near a free surface in a fully developed turbulence. Phys. Fluids **15**, 375–394 (2003)
5. M. Quadrio, P. Ricco, The laminar generalized Stokes layer and turbulent drag reduction. J. Fluid Mech. **667**, 135–157 (2011)

Effects of Sinusoidal Riblets on Turbulent Boundary Layer Flow Structures

G. Cafiero, E. Amico, J. Serpieri, and G. Iuso

Abstract We investigate experimentally the effects of sinusoidal microgrooves on the flow structures developing in a zero pressure gradient turbulent boundary layer. The sinusoidal, or 3D riblets, have been found to be more effective than the widely investigated straight riblets [1, 2]. Particle image velocimetry measurements carried out in the streamwise-wall normal plane allow us to infer on the effect of sinusoidal riblets on the near-wall structures, such as hairpin vortices, along with their vorticity and wall distance. In addition to the proposed uplift mechanism induced by the straight grooves, the sinusoidal pattern is expected to introduce a further share of wall normal displacement, in agreement with the further share of drag reduction. Implementing a vortex detection method based on the second invariant of the velocity gradient tensor, we provide statistical evidence for this further uplift, in the case of the sinusoidal groove, at least in the case characterized by the smaller amplitude. Furthermore, the available data suggest that the spanwise vorticity associated with these structures is attenuated as an effect of the groove profile.

1 Introduction

The study of effective drag reduction techniques plays a key role in meeting the ambitious targets of reducing the pollution of the transportation sector. In the aeronautical sector, for instance, the skin friction drag may be as large as 50% of the overall drag count.

The interruption of the turbulence self-sustained cycle leads to the suppression of the exchange of momentum between the outer and inner regions of the boundary layer, and as such to the reduction of the turbulent fluxes near the wall, in turn promoting an attenuation of the skin friction drag. A possible approach to the problem is the use of active methodologies for flow control. Examples are large scale control [3], plasma actuators [4], spanwise wall oscillation [5]. The main drawback of these techniques

G. Cafiero (✉) · E. Amico · J. Serpieri · G. Iuso
Politecnico di Torino, Corso duca degli Abruzzi 24, 10129 Torino, Italy
e-mail: gioacchino.cafiero@polito.it

R. Örlü et al. (eds.), *Progress in Turbulence X*, Springer Proceedings in Physics 404,
https://doi.org/10.1007/978-3-031-55924-2_35

is that they need power to function, in addition to the increased risks associated with moving parts, or high voltages.

On the other hand, among the passive solutions, riblets have been often proposed as one of the most promising to achieve satisfactory levels of friction drag reduction, and as such are widely investigated in the literature [6–8].

The key parameters of the riblet geometry are represented by the depth of the micro-groove, the spacing between the grooves and its cross-section. Bechert et al. [9] showed that blade riblets aligned to the streamwise direction yield the highest drag reduction, although their implementation can be cumbersome in realistic environments owing to the complexity of their manufacturing.

A non-dimensional riblet spacing ($s^+ = su_\tau/\nu$, with u_τ being the friction velocity and ν the kinematic viscosity; with the superscript $+$ we indicate the scaling in inner units) of approximately 15–16 demonstrated the highest values of drag reduction.

The underlying mechanism leading to drag reduction has been addressed in the literature by [10, 11] among others. The authors showed that the optimal value of the spacing corresponds to the condition where the near-wall structures cannot fit within the groove; hence, the structures are displaced towards the outer layers of the boundary layer. Near the wall, a region of nearly stagnating flow is then produced, with a corresponding reduction of the wall shear stress and skin friction drag.

Sinusoidal riblets have been proposed as an alternative solution that could join the benefits of straight riblets with the spanwise-induced motion of the wall oscillation [1, 2, 12–14]. The sinusoidal riblets are obtained by considering grooves that are not aligned with the flow direction, but with a sinusoidal imprint in the spanwise direction. Cafiero and Iuso [1] experimentally demonstrated that sinusoidal riblets cater for values of drag reduction as large as 10%, with an enhancement of about 3% when compared with straight riblets.

In the present investigation, the focus is on determining the effect of the riblet geometry on the wall-normal displacement of the coherent structures in the flow field. Through particle image velocimetry experiments in the streamwise-wall normal plane, and using vortex detection methods, the statistical location of the near-wall structures is described and compared with the reference cases of the Smooth wall and longitudinal grooves.

2 Experimental Setup

The experiments were performed in an open circuit wind tunnel at Politecnico di Torino, and the experimental setup is similar to the one used in [1, 15]. The tunnel has a settling chamber containing honeycomb and mesh screens, followed by a contraction with an area ratio of 3:1, leading into a test Sect. 5300 mm long and with a cross-section area of 500 mm × 700 mm. The flat plate where the boundary layer develops is located at mid-height of the test section and extends from its side to side, thus effectively reducing the cross-section area to 500 mm × 350 mm. This reduction is obtained through a contoured contraction. The purpose of this is twofold:

first, it avoids any sharp area variation, which would disrupt the flow quality. Second, it allows for the insulation of the bottom portion of the test section used for the plate positioning system. A trip wire is placed at the start of the flat plate to trigger an early transition to a turbulent boundary layer. A portion of the flat plate of size 260 mm × 260 mm is hollow and allows for the positioning of different plates. The correct alignment of the tested plate with respect to the remaining portion of the flat plate is ensured with micrometric accuracy through the possibility of controlling the height and inclination of the tested plate. In particular, the height is controlled using a laser optical displacement measurement optoNCDT 1320 manufactured by Micro-Epsilon, whilst the inclination is checked through a high-precision inertial unit. The alignment of the mounted plate to the surrounding flat plate is performed through the combination of the high precision laser and a calibrated flat plate. The vertical positioning of the riblet plate is changed with micrometric accuracy using a stepper motor. The plate is considered aligned when it touches the calibrated glass plate. The reading of the laser displacement measurement is considered as the datum. Four different plates were investigated: a flat plate, a riblet plate with longitudinal grooves (RLONG), and two riblet plates with sinusoidal grooves (RS1 and RS2). Each one of the tested plates is a square having a side of 258 mm and a thickness of 10 mm. This leads to a small gap between the edge of the tested plate and the flat plate. A labyrinth seal was mounted underneath the plate to avoid any leakage through the gap between the tested plate and the flat plate (Fig. 1).

The three riblet plates are characterized by the same groove profile, having parabolic shape. The height of the groove is $h = 210\,\mu$m whilst the spacing between two grooves is $s = 300\,\mu$m, thus resulting in a $h/s = 0.7$. The two sinusoidal riblets are characterized by the same wavelength $\lambda = 19.2$ mm, thus leading to at least 13 wavelengths on the tested plate; conversely, two different values of the amplitude a are considered (see Fig. 2b), with $a = 0.15$ mm in the RS1 and $a = 0.6$ mm in the RS2 case, respectively.

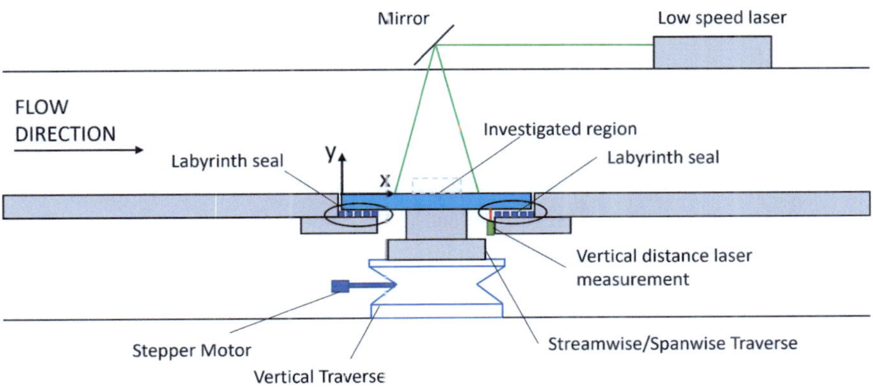

Fig. 1 Schematic representation of the test section with detail of the removable plate. The track of the laser sheet, indicated in green, is not to scale

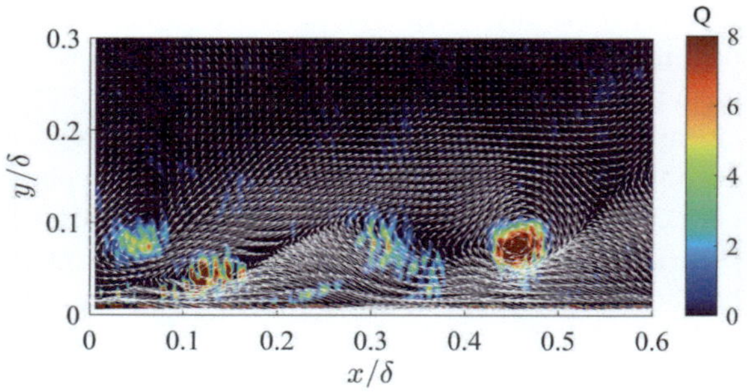

Fig. 2 Colormap of the Q values with overlaid fluctuating velocity field measured in the smooth case

3 Results and Discussion

Starting from the PIV images acquired in the x-y plane, the identification of the vortical structures can be performed using several approaches [16, 17]. In this case, it is carried out applying the Q-criterion, where Q corresponds to the second invariant of the velocity gradient tensor, defined as $Q = \frac{1}{2}(||\Omega||^2 - ||S||^2)$, and $||S||$ and $||\Omega||$ represent the traces of the symmetric and anti-symmetric components of the velocity gradient tensor.

A representative snapshot of the smooth case is reported in Fig. 2 as colormap of the Q values with overlaid velocity vectors. As expected, the majority of the turbulence activity occurs near the wall, and in particular at values of $y/\delta < 0.1$. The regions of the flow that have values of $Q > Q_{thr}$ are extracted from each snapshot using a built-in function of the Matlab suite, regionprops. This feature allows the definition of several geometrical properties of a region of an image. From each detected vortex, the location of the centroid is stored and then used to determine the vorticity magnitude at that specific location. The procedure is applied to each of the acquired images for all the investigated cases. For the present case, a value of $Q_{thr} = 1.5$ (normalised with respect to the freestream speed U_∞ and the thickness of the boundary layer δ) is considered.

The probability density function of the wall-normal location of the centroid of the identified vortices is reported in Fig. 3a. As expected, the vast majority of the vortical structures are located for $y/\delta < 0.15$, i.e. within the inner layer. The data show that the riblet geometry has an effect on the displacement of the near-wall structures. In particular, the RLong case shows a displacement of the peak towards larger values of y/δ. Nevertheless, the greatest differences are evidenced by the RS1 case. While the Smooth and RLong cases are characterized by three peaks of the pdf, located at $y/\delta \approx 0.025$, 0.075 and 0.125, the RS1 shows a more significant turbulence activity in the range of y/δ between 0.025 and 0.065.

It is also interesting to notice that the RS2 case is instead characterized by a behaviour that is similar to the Smooth flow. The quite different behaviour between the RS2 and the RS1 cases might suggest an effect of the amplitude of the riblet geometry on the way the structure of the near-wall turbulence is modified. Furthermore,

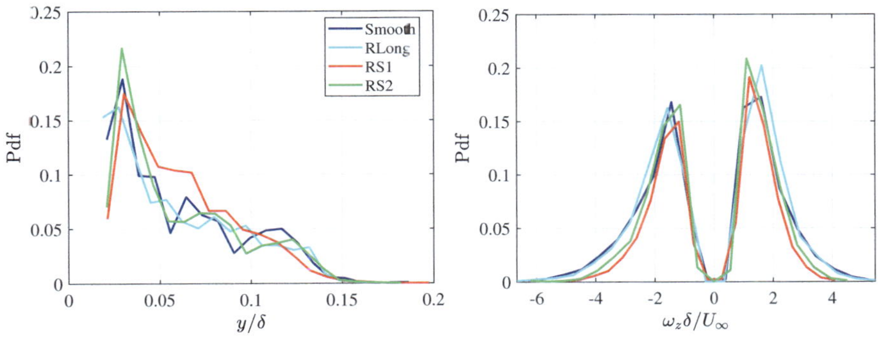

(a) (b)

Fig. 3 Pdf of the wall-normal distance of the vortices identified with the Q-criterion (**a**) and pdf of the spanwise vorticity (ω_z) calculated for the identified vortices (**b**)

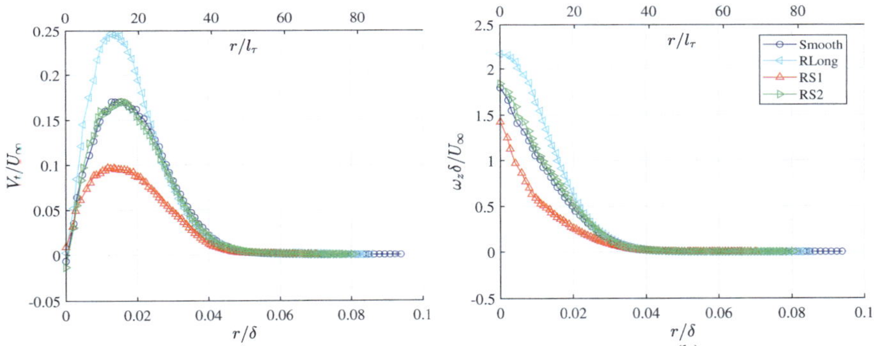

(a) (b)

Fig. 4 **a** Tangential velocity (V_t/U_∞) and **b** spanwise vorticity ($\omega_z\delta/U_\infty$) across the identified vortices

it might suggest the existence of an optimal value of the riblet amplitude such that the structures are effectively uplifted. A larger parametric space is however needed to investigate this aspect further.

The pdf of the spanwise vorticity ($\omega_z\delta/U_\infty$) calculated across the identified vortices is reported in Fig. 3b. Even though the location of the peaks of the pdf is not significantly altered by the flow manipulations, the riblet cases feature a lower probability of occurrence of structures with larger values of the modulus of the spanwise vorticity. Focusing the attention only on the sinusoidal cases, the RS1 yields the most significant attenuation. This is in agreement with the uplift of the structures towards regions of the flow characterized by a reduced mean shear.

Figure 4 shows the profiles of tangential velocity and spanwise vorticity calculated across each of the identified vortices and then averaged. It is interesting to notice how the RLong and the RS1 have different effects on the topology of the near-wall structures. In particular, the RLong case features increased values of vorticity and tangential velocity, while an opposite trend is observed in the RS1 case. The results indeed suggest that the vortices are weakened by the sinusoidal manipulation, at least in the case characterized by the smaller amplitude. The RS2 case, on the other hand, follows closely the behaviour of the Smooth plate. This result hints towards a different mechanism of drag reduction occurring for the RS2 case. Indeed, in [1], the authors showed that the RS2 was even more effective in reducing drag than the RS1 case. This aspect will be further investigated in future studies.

Acknowledgements This study was carried out within the MOST—Sustainable Mobility National Research Center and received funding from the and within the GREENER project—funded by the European Union—Next Generation EU within the PRIN 2022 program (D.D. 104 - 02/02/2022 Ministero dell'Universitá e della Ricerca). This paper reflects only the authors' views and opinions, and the Ministry cannot be considered responsible for them.

References

1. G. Cafiero, G. Iuso, Exp. Thermal Fluid Sci. **139**, 110723 (2022)
2. Y. Peet, P. Sagaut, Phys. Fluids **21**(10) (2009). https://doi.org/10.1063/1.3241993
3. M. Cannata, G. Cafiero, G. Iuso, AIAA J. **58**(5), 2042 (2020). https://doi.org/10.2514/1.J059047
4. J. Serpieri, M.T. Hehner, J. Kriegseis, *Flow, Turbulence and Combustion*, pp. 1–17 (2023)
5. G.M. Di Cicca, G. Iuso, P.G. Spazzini, M. Onorato, J. Fluid Mech. **467**, 41 (2002). https://doi.org/10.1017/S002211200200157X
6. M. Walsh, AIAA J. **82**(9), 0169 (1982). https://doi.org/10.2514/6.1982-169
7. R. Garcia-Mayoral, J. Jimenez, Phys. Fluids **24**(10) (2012)
8. S. Endrikat, R. Newton, D. Modesti, R. García-Mayoral, N. Hutchins, D. Chung, J. Fluid Mech. **952**, A27 (2022)
9. D.W. Bechert, M. Bruse, W. Hage, J.G. Van Der Hoeven, G. Hoppe, J. Fluid Mech. **338**, 59 (1997). https://doi.org/10.1017/S0022112096004673
10. K.S. Choi, J. Fluid Mech. **208**, 417 (1989). https://doi.org/10.1017/S0022112089002892
11. B. Mele, R. Tognaccini, P. Catalano, D. De Rosa, Phys. Rev. Fluids **5** (2020). https://doi.org/10.1103/PhysRevFluids.5.124609
12. Y. Peet, P. Sagaut, Y. Charron, in *38th AIAA Fluid Dynamics Conference and Exhibit* (2008). https://doi.org/10.2514/6.2008-3745
13. M. Sasamori, H. Mamori, K. Iwamoto, A. Murata, Exp. Fluids **55**(10) (2014). https://doi.org/10.1007/s00348-014-1828-z
14. M. Sasamori, O. Iihama, H. Mamori, K. Iwamoto, A. Murata, Flow Turbul. Combu. **99**(1), 47 (2017). https://doi.org/10.1007/s10494-017-9805-2
15. G. Cafiero, E. Amico, G. Iuso, Manipulation of a turbulent boundary layer using sinusoidal riblets. J. Fluid Mech. **984**, A59 (2024). https://doi.org/10.1017/jfm.2024.256.
16. J. Jeong, F. Hussain, J. Fluid Mech. **285** (1995)
17. F.D. Gregorio, A. Visingardi, G. Iuso, Energies **14** (2021)

Flow Control of a Turbulent Boundary Layer Using Uniform Blowing

G. Hasanuzzaman and C. Egbers

Abstract A brief overview on a series of flow control experiments within a Zero-Pressure-Gradient Turbulent-Boundary-Layer (ZPGTBL) is presented here. Flow control is achieved applying uniform blowing (UB) at different velocity from the wall, subsequently, increases coherence of the turbulent structures. Stereo Particle Image Velocimetry (SPIV) data shows that 'ejection' and 'sweep' events are augmented with the increased blowing velocity and they eventually contribute in the turbulence statistics.

1 Introduction

Applying efficient control on TBL flow has an wide engineering and financial benefits on modern subsonic aircraft to the high speed trains and cars. Estimations show that, nearly 50% of the total drag of an aircraft and 30% of automobiles are constituted from skin friction drag [1]. Only 1% of reduction of such skin friction drag can save 400,000 litres of fuel yearly based on subsonic Airbus A320 data. Most importantly, a major portion of the total drag for these subsonic aircraft is dependent on geometry, Reynolds number and skin friction. Therefore, a large contribution of fuel consumption and operating cost are direct outcome of this quantities. Out of these quantities, we can simulate the flow Reynolds number by performing experiments at high Reynolds numbers. In addition, streamlining the geometry can also reduce the form drag while friction drag contributes almost other half of the total drag which comes from the surface roughness and a fractional reduction of it can be interpreted as a considerable cost savings [1].

Flow control technologies (FCT) can be classified into two wide groups such as passive and active method in terms of their use of external energy. On the other hands,

G. Hasanuzzaman (✉) · C. Egbers
Department of Aerodynamics and Fluid Mechnaics, BTU Cottbus-Senftenberg, Siemens-Halske-Ring-15a, 03046 Cottbus, Germany
e-mail: hasangaz@b-tu.de

C. Egbers
e-mail: egbers@b-tu.de

R. Örlü et al. (eds.), *Progress in Turbulence X*, Springer Proceedings in Physics 404, https://doi.org/10.1007/978-3-031-55924-2_36

they are divided as open or closed-loop in terms of their control feedback. The active FCTs which uses external energy as input exhibits better drag reduction performance compared to the passive ones. However, uniform blowing (UB) particularly has shown superior performance on friction drag reduction by influencing the turbulence statistics. Here remains the question still open that whether blowing influences the coherence and the characteristics of the structures, if so, where and how?

FCTs are dependant on the turbulence mechanism. TBL is populated with turbulent structures with different scales. Where near-wall viscous sub-layer (VSL) is mostly dominated with worm-like streaks. When energized, they contribute strongly to the Reynolds stresses through 'ejection' events which are responsible for the outward vertical transport. On the other hand, outer region is mostly populated with the coherent LSMs. When they are retracted towards the wall, transport momentum from the potential flow region. This is known as 'sweeps' which brings additional impetus on the friction drag.

Recent studies show that the application of the UB at ZPGTBL flows can strongly influence the turbulence statistics at the outer region [2] and instantaneous flow patterns [3], together, they can effectively modify the statistical outer peak of the streamwise velocity Root-Mean-Square (RMS) and reduce friction drag [4], increased Reynolds stresses and energy spectra for low wave number vortices. Moreover, blowing can significantly modify the high order turbulence statistics and follows certain scaling [4, 5]. Blowing certainly modifies the characteristics of a classical TBL also known as Standard Boundary Layer (SBL, within the manuscript the quantities from a SBL is presented as subscript $(.)_{SBL}$) and imposes different wall condition. Therefore, this complex TBL requires different turbulence modelling approaches for numerical simulations. Often machine-learning or data-driven efforts are needed to complement the existing turbulence-modelling approaches [6]. The influence of blowing through perforated surface can be also extended to the thermal boundary layers and has potential outcome in terms of increasing the efficiency of heat ex-changers. In other words, one can implement such flow control to increase the efficiency of the electric aircraft's where cooling for the overheated battery plays a crucial role. At the scope of this proceeding, a ZPGTBL experiment is reported in brief at the range of wide Reynolds number $Re_\theta = U_\infty\theta/\nu$ and $Re_\tau = u_\tau\delta/\nu$, where θ, u_τ, ν and δ are the momentum thickness, friction velocity, kinematic viscosity and boundary layer thickness respectively. Figure 1a shows the range of experiments performed at two different wind tunnels namely, Department of Aerodynamics and Fluid Mechanics, Germany (BTU) and Laboratoire de Mécanique des Fluides de Lille, France respectively (LMFL). These wind tunnel facilities differed with their viscous length scales ($l^+ = \nu/u_{\tau,SBL}$), therefore, Fig. 1b shows their comparison in semi-logarithmic plot.

2 Experiment

LMFL boundary layer wind tunnel has 20.6 m long test section with 2 m^2 cross section. In order to apply blowing, a 0.55 m $\times$ 0.42 m perforated surface was placed

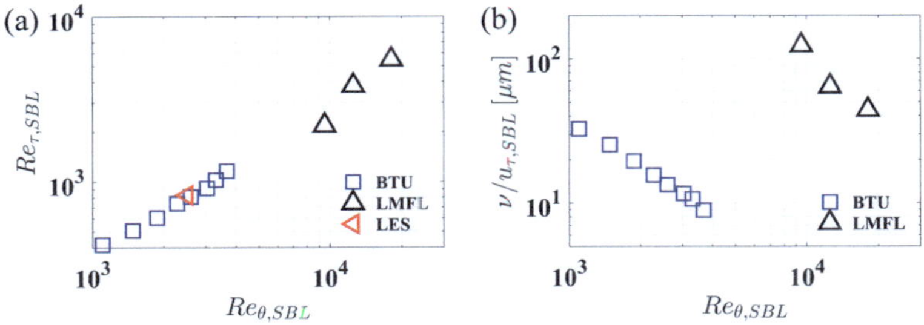

Fig. 1 **a** Shear and momentum Reynolds number space for the TBL experiment, **b** comparison of the viscous length scales at different wind tunnels [1]

at 18.42 m from the leading edge. In order to control LSMs and their influence on the wall, an active control in the form of UB through perforated wall was implemented. The location of the perforated surface is shown here with photographs presented as Fig. 2a. Multiple screening devices were mounted together in order to construct a hermetically sealed blowing assembly where inlet air was supplied through multiple tubes installed below the device as presented in Fig. 2b. The velocity of the wall normal blowing is expressed as blowing ratio ($F = V_w/U_\infty \times 100$) as a percentage of the free stream velocity (U_∞) and were varied at 0, 1, 3, 6% of the U_∞. Origin is located at the center of leading edge.

There are two principle aspect of blowing on the turbulence parameter, statistics of the velocity fields and morphology of the turbulence structure that was investigated within this experiment. Therefore, in order to capture the evolution of the turbulence structures in time and space, a finite, spanwise-wall (YZ) normal plane was measured using time resolved SPIV as shown in Fig. 2c. YZ plane was recorded at 1 kHz frequency for 3.2 seconds. Subsequently, turbulence statistics over the entire TBL was recorded using a streamwise-wall normal (XY) plane as presented in Fig. 2d [5] and was recorded at 4 Hz for 6 minutes. A total number of $10e + 3$ samples for each measurements were acquired for XY plane [5].

(a) (b) (c) (d)

Fig. 2 Photographs showing the experimental setup; **a** location of the perforated surface used for blowing at LMFL wind tunnel, **b** bottom surface of the blowing assembly with the air supply lines, **c** 2-camera SPIV setup in YZ plane, **d** 4-camera SPIV setup in XY plane

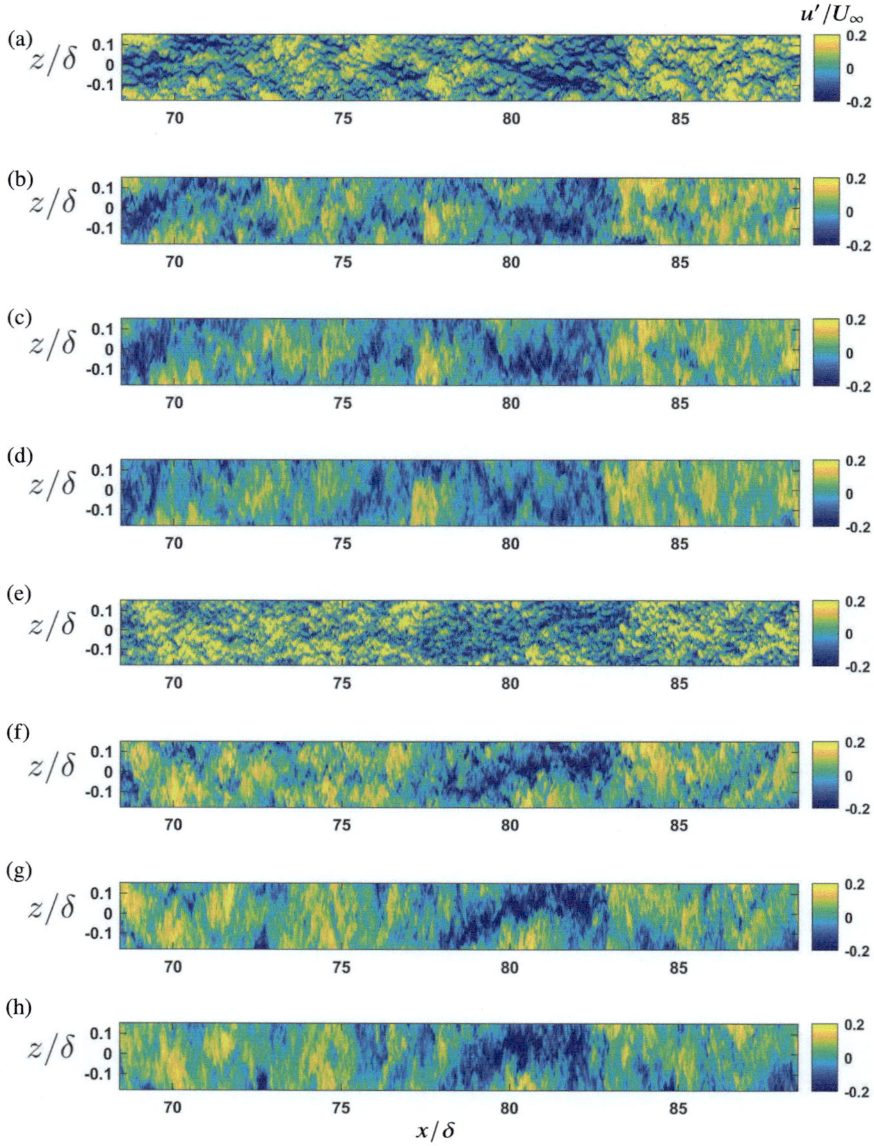

Fig. 3 At $Re_\theta = 7500$, represents the space-time diagram for outer-scaled streamwise velocity fluctuations following Taylor's frozen turbulence hypothesis in spanwise-streamwise plane at different wall normal locations: **a** and **e** at $y/\delta = 0.004$, **b** and **f** at $y/\delta = 0.06$, **c** and **g** at $y/\delta = 0.12$, **d** and **h** at $y/\delta = 0.18$, (**a**), (**b**), (**c**) and (**d**) with BR $= 0\%$, (**e**), (**f**), (**g**) and (**h**) with BR $= 6\%$. Flow is coming from left to right with the readers reference point

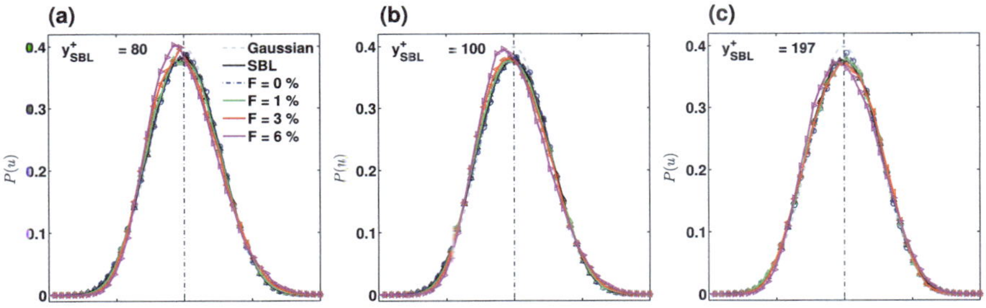

Fig. 4 PDF of streamwise velocity with different blowing rates (F) at $Re_\theta = 7500$, inner scaled wall locations are indicated in the figures

3 Results

Figure 3a to h presents the streamwise velocity fluctuations normalized with the free stream velocity and their wall normal locations are indicated in the figure description. This is presented from the YZ plane after applying Taylor' frozen turbulence hypothesis. Here, Fig. 3a–d is from a TBL with perforated plate without blowing. As reference [1] discussed that a classical TBL has very less difference compared to the TBL over perforated surface without blowing, therefore, this is presented as the indication of the coherent structures present in a classical TBL from near the wall to further away from the wall with same interval between the wall normal locations. On the other hand, Fig. 3e–h presents the measurements where blowing was applied at F = 6%. When blowing is applied, near-wall positions shows more turbulent spots and the data from furthest wall distance exhibit stronger coherence of the large-scale structures. Subsequently, Fig. 4a–c represents the probability density function ($P(u)$) of the streamwise velocity at different wall distance from near-wall to away from the wall (Figure indicate the wall distance). Here, different colors indicate the distributions of $P(u)$ from different blowing rates, F. We can see that blowing strongly changes the distribution compared to the Gaussian and $._{SBL}$ distribution at different wall locations. This is presented using the data acquired at the XY plane. Finally, the joint probability distributions (JPDF) of streamwise fluctuation presented in Fig. 5a–c at a fixed wall distance using the data from XY plane, where, $F = 1$ and $F = 6\%$ are compared. We can see that both 'ejection' and 'sweep' events are enhanced with the increased F.

4 Conclusion

We have seen that active methods such as blowing from a finite space can strongly influence the TBL turbulence statistics as well as the coherence of the large scale

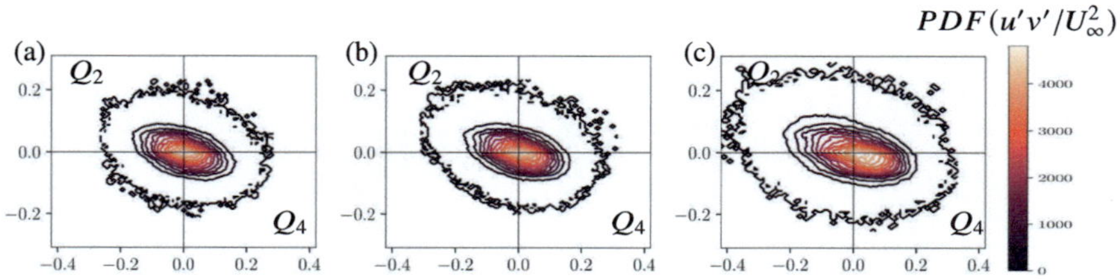

Fig. 5 Quadrant analysis of streamwise (u') and wall-normal fluctuations (v') normalized with U_∞ at $y/\delta = 0.16$ at $Re_\theta = 7500$, **a** $(.)_{SBL}$, **b** F = 1% and **c** F = 6%. Color bar represent the cross-probability density function (PDF)

structures significantly. This can be extended upto several δ in the downstream direction.

Particularly, near-wall turbulent spots and outer-layer coherence are enhanced. UB energize the small perturbations in order to enhance the transport phenomena and creates a region of high energy that prohibits the impact of sweeps on the wall.

Acknowledgements We cordially thank Dr. Vasyl Motuz and Rishindra Vallem for their support.

References

1. G. Hasanuzzaman, *Experimental Investigation of Turbulent Boundary Layer with Uniform Blowing at Moderate and High Reynolds Numbers*, 1st ed., (Cuvillier Verlag, Göttingen, Germany, 2021), pp. 1–152. ISBN: 978-3-7369-7558-3
2. G. Hassanuzzaman, S. Merbold, V. Motuz, C.H. Egbers, Experimental Investigation of Turbulent Structures and their Control in Boundary Layer Flow, GALA Fachtagung Experimentelle Strömungsmechanik, 6– 8 Sept., Dresden, Germany, pp. 64:1–6 (2016). ISBN: 978-3-9816764-2-6
3. G. Hassanuzzaman, S. Merbold, V. Motuz, C.H. Egbers, C. Cuvier, J.-M. Foucaut, Experimental investigation of active control in turbulent boundary layer using uniform blowing, ICEFM, Munich, Germany, 2–4 July 2018, pp. 1—6 (2018). ISBN: 978-3-943207-32-3
4. G. Hasanuzzaman, S. Merbold, V. Motuz, Ch. Egbers, *Enhanced Outer Peaks in Turbulent Boundary Layer using Uniform Blowing at Moderate Reynolds Number* (Turb, Jour, 2022). https://doi.org/10.1080/14685248.2021.2014058
5. G. Hasanuzzaman, S. Merbold, C. Cuvier, V. Motuz, J.-M. Foucaut, Ch. Egbers, Experimental investigation of turbulent boundary layers at high Reynolds number with uniform blowing, part I: statistics. J. Turb. **21**(3), 129–165 (2020). https://doi.org/10.1080/14685248.2020.1740239
6. G. Hasanuzzaman, H. Eivazi, S. Merbold, C. Egbers, R. Vinuesa, Enhancement of PIV measurements via physics-informed neural networks. Meas. Sci. Tech. **34**(4), 044002 (2023). https://doi.org/10.1088/1361-6501/aca9eb

On the Intermittency of the Turbulent Asymptotic Suction Boundary Layer (TASBL)

E. Foschi, R. Örlü, A. Talamelli, and P. Schlatter

Abstract This paper presents a series of direct numerical simulations (DNS) performed in order to understand the discrepancy in the literature regarding turbulent asymptotic suction boundary layers (TASBLs) at low Reynolds numbers. The hypothesis to be tested is that a significant contributing reason for higher turbulence intensities observed in experiments compared to DNS is that the latter exhibits intermittent patches of laminar flow, developing both temporally and spatially. This hypothesis we confirm here by comparing simulations with and without tripping, where the former removed patches of laminar flow thereby establishing a fully developed turbulent state with higher turbulence intensities compared to its naturally developing counterpart. The DNS were performed at different suctions rates corresponding to Reynolds numbers above the critical value of 270. The statistics taken from the simulations at different streamwise positions also support the developing character of the flow with increasing intermittency further downstream. Thus, we can conclude that the actual flow state at these marginal Reynolds numbers is indeed an intermittent one, with lower fluctuation values as the experimental data would indicate.

E. Foschi (✉)
Department of Engineering Mechanics, KTH Royal Institute of Technology, 100 44 Stockholm, Sweden
e-mail: edoardofoschi98@gmail.com

R. Örlü
Department of Mechanical, Electrical and Chemical Engineering, OsloMet – Oslo Metropolitan University, 0166 Oslo, Norway

A. Talamelli
Department of Industrial Engineering, University of Bologna, 40126 Bologna, Italy

P. Schlatter
Institute of Fluid Mechanics (LSTM), Friedrich–Alexander–Universität (FAU) Erlangen–Nürnberg, 91058, Erlangen, Germany

R. Örlü et al. (eds.), *Progress in Turbulence X*, Springer Proceedings in Physics 404,
https://doi.org/10.1007/978-3-031-55924-2_37

1 Introduction

An "asymptotic" suction boundary layer refers to the state when the the boundary layer has fully developed spatially and reached a constant boundary-layer thickness. The flow becomes thus statistically independent of the streamwise location and the non-dimensional suction rate Γ acts as the only control parameter through the Reynolds number $Re = U_\infty / V_0 = 1/\Gamma$, with V_0 and U_∞ denoting the suction velocity and free-stream velocity, respectively. The mass flux condition through the surface causes an equilibrium between the momentum loss due to wall friction and the entrainment of high-momentum fluid due to suction. In the laminar regime one can derive the velocity profile as one of the few analytical solutions of the Navier–Stokes equations,

$$\frac{U}{U_\infty} = 1 - e^{-y\frac{V_0}{\nu}}, \quad V = -V_0, \quad P = P_0 . \tag{1}$$

Methods to modify boundary layers have been studied from the beginnings of boundary layer studies, and wall-normal transpiration was considered a relatively simple control technique by Prandtl in his very first paper on boundary layer theory [9]. Despite a large number of studies, control techniques in wall turbulence remain one of the subjects of physics and engineering that need improved understanding. The stabilizing effect of wall suction on boundary layers has been investigated as a technique to delay the transition between laminar and turbulent flow regimes in order to reduce friction drag. However, it is immediately clear that the actual friction under suction increases. In addition, boundary-layer suction can be used as a means to delay separation of the boundary layer. Also, stronger suction may lead to relaminarization, which may have specific applications as well. Thus, several experiments were carried out to control laminar and turbulent boundary layers with suction [1, 5]. The first direct numerical simulation (DNS) of a turbulent asymptotic suction boundary layer (TASBL) was performed by Mariani et al. [7]. Documenting the turbulence statistics they could confirm the results obtained in previous studies [1] of a lower fluctuation level when suction is present. More extensive turbulence data was provided by Schlatter and Örlü [10] and Bobke et al. [2].

The present work aims at understanding the discrepancy at lower Re between the results of the numerical simulations by Khapko et al. [6] and the experimental data obtained by Ferro et al. [4]. The hypothesis is that the difference in the statistics for the cases studied in the aforementioned publications is due to the presence of intermittent patches of laminar flow in simulations, developing in both the temporal and spatial sense, thereby leading to a lower turbulence level. The presence of spatio-temporal intermittency in TASBL has a clear importance since laminar–turbulent coexistence still raises important questions and it has lead recently to an increase in interest, a review of which is provided in Ref. [12].

2 Numerical Setup

A series of DNS of TASBLs is performed in a box-shaped domain using lateral periodic boundary conditions, on which constant suction is applied over an otherwise flat plate with no-slip conditions. The upper boundary is in the laminar region, and allows vertical transpiration. The simulations are performed using SIMSON [3], a pseudo-spectral code developed at KTH. The solver is parallelized with shared-memory (OpenMP) and Message Passing Interface (MPI) techniques. Different *Re* and configurations are chosen to elucidate the reasons of the discrepancy between the simulation [6] and experiment [4] and are summarised in Table 1. As *Re* is varied, the value of the displacement thickness δ^* changes as well. Therefore, the initial displacement thickness δ_0^* is used as the reference length in the system. Turbulence is triggered in spatial simulations at the streamwise position $x/\delta_0^* = 40$ by numerical tripping and the suction is applied after the position $x/\delta_0^* = 500$ giving sufficient distance for the turbulent boundary layer to develop before the suction. A fringe region is implemented in the domain at position $x/\delta_0^* = 2700$ in order to guarantee the periodic boundary conditions.

The tripping is implemented as a weak random volume force in the wall normal direction, acting as a numerical "sand-paper" triggering turbulence in the domain. The main inputs are: amplitude, spanwise length scale and temporal frequency, as described in detail in Ref. [11]. Although it is impossible to associate directly the numerical implementation with a specific shape of a physical tripping device, the flow characteristics downstream of the application and the corresponding transition mechanism are expected to simulate what a trip would have induced in an experimental set-up [8]. For instance, the amplitude of the trip can be associated with the vertical shape of a roughness element and the spanwise length scale is associated with the physical scales of the trip. The time scale is related to the shedding characteristics of a physical device. The statistics of the turbulent quantities are taken at position $x/\delta_0^* = 180$, further downstream from the application, in order to be compared to the cases without tripping.

Table 1 Numerical parameters and the tripping configurations including amplitude and respective symbols used for identification in the figures

DNS	Domain size	Resolution	$\Gamma \times 10^3$	Identification	Symbol code
Temporal	$200 \times 100 \times 100$	$128 \times 257 \times 128$	3.6, 3.63, 3.66, 3.7	No tripping	Blue
				Trip amp. = 0.04	Green
				Trip amp. = 0.08	Red
Spatial	$3000 \times 50 \times 50$	$960 \times 121 \times 64$	3.56, 3.6, 3.66, 3.7	$x/\delta_0^* = 450$	Light blue
				$x/\delta_0^* = 1000$	Blue
				$x/\delta_0^* = 2000$	Green
				$x/\delta_0^* = 2500$	Red

3 Results and Discussion

In order to study the reason of the discrepancy present between the simulation [6] and experiment [4], the DNS are performed at different Reynolds numbers slightly above the critical Reynolds number of self-sustained turbulence $Re_{sst} = 270$ ($\Gamma_{sst} = 3.7 \times 10^3$) [6] below which turbulence is not sustainable and complete relaminarization is observed. The flow tripping is applied in the temporal simulations with the idea to depict the effect of the presence of laminar regions simply by removing them, making the turbulence fully developed as can be appreciated via the λ_2 visualization in Fig. 1.

The mean streamwise velocity profile of the spatial simulations in Fig. 2 shows how the turbulent profile collapses from a general BL where no suction is applied yet (light blue) to the typical one observed for ASBLs. The viscous sublayer agrees perfectly with the modified law of the wall for TASBL, and the overlap region is described with the empirical log law with modified parameters similar as in Refs. [2, 10]. A change of behavior, for different streamwise positions, is observed as far as the velocity fluctuations are concerned. This is better appreciated in Fig. 3 where the entire set of data is presented for the range of suction rates Γ specified in Table 1.

Comparing the results of the present work to the DNS [6] and the experiments [4] in Fig. 3, a very good agreement is observed between the temporal simulation performed without the application of tripping and the DNS. When the numerical trip is applied, the maximum of the velocity variance in inner units increases and becomes close to the experiments. Thus, the presence of temporal developing intermittency is a valid reason for the discrepancy since it is shown that the trip removes patches of laminar flow otherwise present.

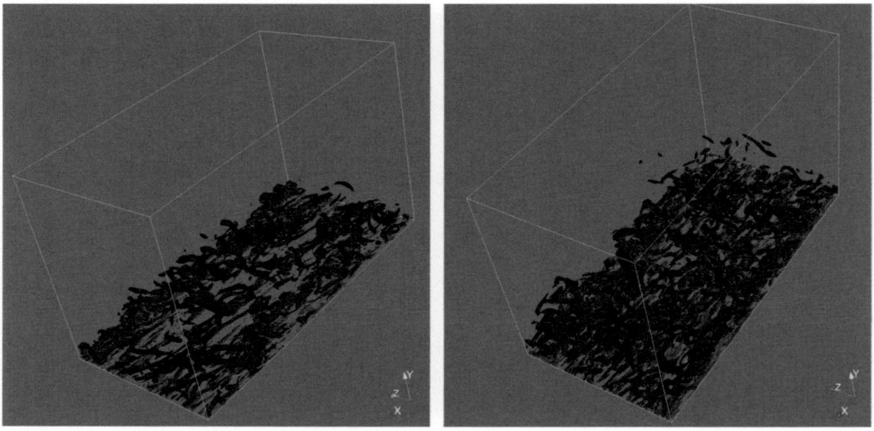

Fig. 1 λ_2 visualization in the domain of temporal simulations for $\Gamma \times 10^3 = 3.6$. Blue and red represent minimum and maximum quantities in the field respectively. *(left)* no tripping *(right)* tripping

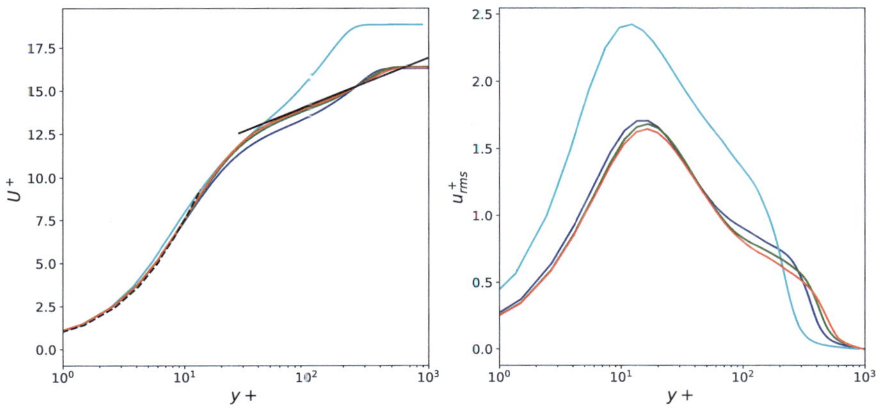

Fig. 2 Inner scaled streamwise velocity for $\Gamma \times 10^3 = 3.7$ *(left)* mean profiles, law of the wall (dashed black) and (solid black) empirical log law with $\kappa = 0.82$, $B = 8.5$ and *(right)* variance profiles. See Table 1 for symbol reference.eps

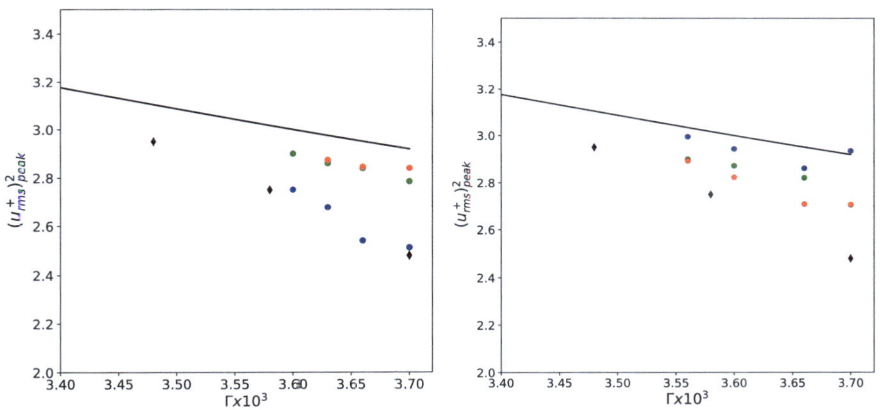

Fig. 3 Maximum of velocity variance in inner scaling of *(left)* temporal and *(right)* spatial simulations compared with Refs. [4, 6]. See Table 1 for symbol reference. Experimental fit [4] (solid black), DNS [6] (black diamonds)

Moreover, as shown in Fig. 3, only at the beginning of the asymptotic state the statistics exhibit a lower turbulence level almost collapsing on the trend observed in the experiments of Ferro et al. [4]. As the statistics are taken further downstream, the trend of the results gets closer to Khapko et al. [6], supporting the presence of spatially developing intermittency. Note that in the experiments a correction for the hotwire length was used, potentially explaining the generally higher fluctuations, but not the differing trends at low *Re*.

4 Conclusions

A series of DNS of the TASBL is performed in a periodic domain with the flow developing temporally, and a spatially developing long domain. For Reynolds numbers slightly above the critical Re_{sst} [6] it is found that the turbulence intensities increase to the values found in experiments [4] by the application of flow tripping, which removes laminar patches, making the turbulence fully extended. In the spatial setting, the statistics, taken at different streamwise positions, show the presence of increasing intermittency further downstream, confirming the developing character of the flow. Thus, we can conclude that the actual flow state, at these marginal Reynolds numbers, is indeed an intermittent one, exhibiting lower turbulence level as the experimental data would predict. Thus, the issue of long development lengths discussed by Bobke et al. [2] is still relevant. For higher $Re > 300$, intermittency seems to disappear, and the qualitative discrepancy between experiments and simulation vanishes.

Acknowledgements The simulations were performed on resources provided by the Swedish National Infrastructure for Computing (SNIC).

References

1. R. Antonia, L. Fulachier, L. Krishnamoorthy, T. Benabid, F. Anselmet, Influence of wall suction on the organized motion in a turbulent boundary layer. J. Fluid Mech. **190**, 217–240 (1988)
2. A. Bobke, R. Örlü, P. Schlatter, Simulations of turbulent asymptotic suction boundary layers. J. Turbul. **17**, 157–180 (2016)
3. M. Chevalier, P. Schlatter, A. Lundbladh, D. Henningson, SIMSON–A pseudo-spectral solver for incompressible boundary layer flow. Technical report. TRITA-MEK 2007:07, Royal Institute of Technology, Stockholm, Sweden (2012)
4. M. Ferro, B.E. Fallenius, J.H. Fransson, Experimental study on turbulent asymptotic suction boundary layers. J. Fluid Mech. **915**, A80 (2021)
5. J.M. Kay, Boundary-layer flow along a flat plate with uniform suction. Technical report, Aero. Res. Council (1948)
6. T. Khapko, P. Schlatter, Y. Duguet, D.S. Henningson, Turbulence collapse in a suction boundary layer. J. Fluid Mech. **795**, 356–379 (2016)
7. P. Mariani, P. Spalart, W. Kollmann, Direct simulation of a turbulent boundary layer with suction. Near-wall Turbul. Flows 347–356 (1993)
8. R. Örlü, P. Schlatter, Comparison of experiments and simulations for zero pressure gradient turbulent boundary layers at moderate Reynolds numbers. Exp. Fluids **54**, 1547 (2013)
9. Prandtl, L.: Über Flüssigkeitsbewegung bei sehr kleiner Reibung. Verhandl. 3rd Int. Math. Kongr. Heidelberg (1904)
10. P. Schlatter, R. Örlü, Turbulent asymptotic suction boundary layers studied by simulation. J. Phys.: Conf. Ser. **318**(2), 022020 (2011)
11. P. Schlatter, R. Örlü, Turbulent boundary layers at moderate Reynolds numbers: inflow length and tripping effects. J. Fluid Mech. **710**, 5–34 (2012)
12. L.S. Tuckerman, M. Chantry, D. Barkley, Patterns in wall-bounded shear flows. Ann. Rev. Fluid Mech. **52**, 343–367 (2020)

Flow Sensitivity Analysis for the Feedback Loop Phenomenon of Subsonic Jet Noise Generation

S. Morita, A. Yakeno, C. Bogey, and S. Obayashi

Abstract In this study, the feedback loop phenomena caused by upstream propagating waves (guided jet waves) in a Mach 0.9 free jet are investigated. A point source is placed at the end of the potential core of a three-dimensional free jet as a noise source in the jet and the sensitivity of the flow field is measured. The results show that the case $St_D = 0.47$ has the highest sensitivity, which was consistent with the allowable frequency range of the guided jet wave. This result supports the idea that guided jet waves cause free shear layer instability waves.

1 Introduction

The relationship between jet stream instability waves and the resulting acoustic noise generation is one of the important research questions in science and engineering. Jet engine exhaust noise in commercial aircraft has been a significant problem for over 70 years. This noise is caused by the motion of the vortices generated by the jet, and in this case, it is called "vortex noise". Especially in supersonic jets [1] and impinging jets [2], it has been suggested that jet noise causes flow instability waves, generates sound waves of specific frequencies, and repeats this interaction. This is called the acoustic feedback mechanism. According to a recent review paper [3], the process of the feedback loop is divided into four stages: (a) hydrodynamic perturbation is generated and it propagates to the downstream (downstream process), (b) this

S. Morita (✉) · A. Yakeno · S. Obayashi
Institute of Fluid Science, Tohoku university, 9808579, 2-1-1 Katahira, Aoba-ku, Sendai, Miyagi, Japan
e-mail: shota.morita.t7@dc.tohoku.ac.jp

A. Yakeno
e-mail: aiko.yakeno@tohoku.ac.jp

S. Obayashi
e-mail: s.obayashi@tohoku.ac.jp

C. Bogey
Laboratoire de Mécanique des Fluides et d'Acoustique, UMR 5509, F-69134 Ecully, France
e-mail: christophe.bogey@ec-lyon.fr

R. Örlü et al. (eds.), *Progress in Turbulence X*, Springer Proceedings in Physics 404,
https://doi.org/10.1007/978-3-031-55924-2_38

perturbation is converted and it propagates to the upstream direction typically in the form of an acoustic wave (generation process), (c) the acoustic wave reaches to the sensitivity point of the flow (upstream process), and (d) a new downstream instability hydrodynamic wave is reproduced by the upstream wave (receptivity process).

Recently, it has been suggested that the generation of sound in high-speed subsonic jets is also explained by a feedback loop mechanism [4–6]. The story of a subsonic jet feedback loop phenomenon follows: (a) flow instability waves propagate to the downstream direction, (b) this wave forms a vortex and generates the acoustic waves, (c) this wave propagates to the upstream direction, and (d) regenerate the flow instability waves. Particularly, the first stage of vortex coupling caused by Kelvin-Helmholtz instability which propagates to the upstream direction, and causes the repetition of flow instability. Moreover, the upstream propagating wave exists on the jet axis for an $\mathrm{M} = 0.9$ jet [7]. It is also considered as the cause of the feedback loop, However, there remain questions about how the upstream-guided jet waves occur and if they affect the flow instability waves or not. Therefore, the purpose of the study is to investigate the relationship between the guided jet waves and Mach 0.9 jet flow instability waves. However, although linear stability analysis can look at the amplification rate of the flow field over time, it cannot analyze the effects of local disturbances on the entire flow field. In this study, we measure the flow sensitivity to sound waves in a flow by placing a virtual point source in the flow field and measuring the turbulence in the flow field using flow simulation.

This paper is organized as follows. Sect. 2 describes the flow sensitivity measurement and the computational method used in this study. Section 3 presents the results of the analysis when applied to a three-dimensional jet. Finally, Sect. 4 shows the conclusion of this paper.

2 Method: Flow Sensitivity Measurement

In this study, we measure the sensitivity of flow to acoustic waves by the following method. Specifically, we computed the time-averaged flow of the free jet and introduced a single-frequency acoustic source at the end of the potential core on the jet axis. We computed the time evolution of the turbulent kinetic energy k near the jet exit as the sensitivity criteria, which formulation is shown in (1). This allows us to consider the relationship between sound waves and flow more directly. In this study, a point source emitting acoustic waves at a single frequency at the end of the potential core of the jet is reproduced by representing a small object that appears and disappears in the flow by the Volume Penalization method (VP method) [8, 9]. Here, the end of the potential core is considered the origin of the jet noise [10].

$$k = \frac{1}{2}\rho(u'^2 + v'^2 + w'^2) \tag{1}$$

We describe the governing equations used in flow analysis and the VP method used to reproduce point sources in the flow sensitivity measurement. The VP method is a type of virtual boundary method that represents an object in a flow. Specifically, an object is represented by adding an external force term called a Penalization term, which acts only in the object region, to the governing equations of the fluid. When the object does not have any porosity and movement speed, the governing equations of a VP method for compressible flow are expressed by the following (2)–(4).

$$\frac{\partial \rho}{\partial t} + \frac{\partial \rho u_j}{\partial x_j} = 0, \tag{2}$$

$$\frac{\partial \rho u_i}{\partial t} + \frac{\partial \rho u_i u_j}{\partial x_j} = -\frac{\partial p}{\partial x_i} + \frac{\partial \tau_{ij}}{\partial x_j} - \frac{\chi}{\eta} u_i, \tag{3}$$

$$\frac{\partial e}{\partial t} + \frac{\partial (e + p) u_j}{\partial x_j} = \frac{\partial u_i \tau_{ij}}{\partial x_i} + \frac{\partial}{\partial x_j}\left(\kappa \frac{\partial T}{\partial x_j}\right), \tag{4}$$

Here, ρ is the density, u_j is the flow velocity, p is the pressure, τ_{ij} is the viscous stress tensor, e is the internal energy, T is the temperature, and κ is the specific heat ratio. In addition, χ is the mask function that takes 1 in the object region and 0 in the fluid region and η is the permeability that simulates the permeability of the object.

In this study, three-dimensional jet simulations were performed using an in-house solver. This is developed for simulating the three-dimensional compressible flow. For spatial discretization, the 6th-order accurate compact scheme and the 10th-order accurate compact filter [11] were used. The free parameter α_f is set to 0.495, referring to Kawai et al. (2010) [12]. For the time integration, we used the third-order accurate TVD Runge-Kutta method [13]. The Reynolds number $Re = u_0 D/\nu$ is set to 3125 and the jet Mach number $M = u_0/c_0$ is set to 0.9. Here, D is the jet diameter, u_0 is the initial velocity of the jet and c_0 is the sonic speed of ambient pressure. The computation region used in this study is described in Fig. 1. The sponge region and low-pass filter were applied far from the boundaries to prevent acoustic reflection.

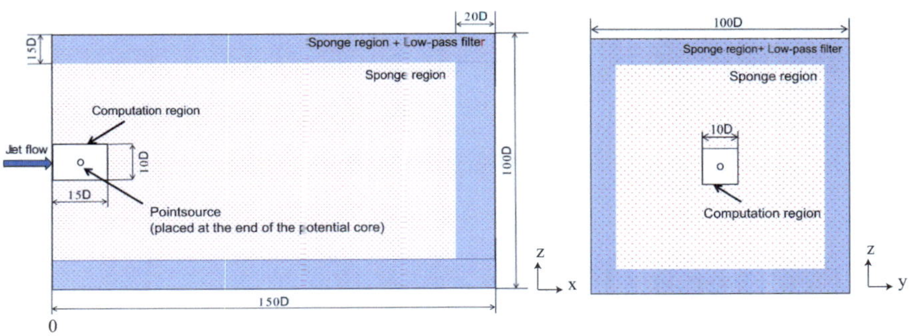

Fig. 1 The computation region in this study

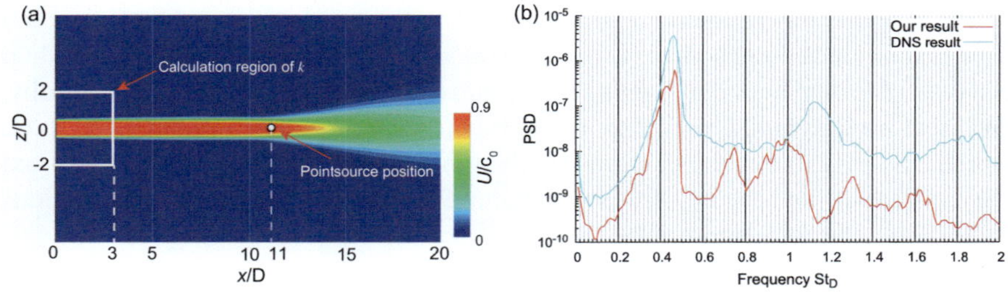

Fig. 2 Configuration of the flow-field of the jet. **a**: Time-averaged velocity field of axial velocity. The contour ranges 0 to 0.9, from blue to red. The point source is placed at the end of the potential core, where the white circle is put in this figure. **b**: Frequency spectra near the jet exit at $(x/D, y/D, z/D) = (1.0, 0.0, 0.0)$

3 Results and Discussion

3.1 Simulation Results

We obtained the time-averaged field of the 3D free jet flow using the computational conditions described in Sect. 2 to obtain the base flow of the jet for use in the sensitivity computations. The cross-section of the time-averaged field on visualization result is shown in Fig. 2a. From the contour line of this figure, it is observed that the end of the potential core is located at $x/D = 11$, which is shown in the white circle at Fig. 2a. Since jet noise is generated at the end of the potential core as mentioned in Sect. 2, we set the point source at the end of the potential core.

To estimate the frequency of the upstream propagating guided jet wave, we applied Fourier analysis near the jet exit to the time-series data of pressure fluctuation. The spectrum from the pressure at $(x/D, y/D, z/D)$=(1.0, 0.0, 0.0) is shown in Fig. 2b. Here, $St_D = fD/u_0$ is the Strauhal number. For verification of our simulation result, the spectrum from the DNS result for $M = 0.9$ and $Re = 3125$ by Bogey [10] is also shown. Three frequency peaks appear at $St_D = 0.47$, $St_D = 1.17$, and $St_D = 1.9$. These are the frequencies of the guided jet waves. However, Our result only corresponds to a $St_D = 0.47$ peak. Therefore, in this study, we focused on the frequency peak of $St_D = 0.47$, which is the first tone of the guided jet wave. From these results, sensitivity measurements were performed for 20 cases from $St_D = 0.1$ to $St_D = 2.0$ and 5 cases of $St_D = 0.42, 0.44, 0.46, 0.47, 0.48$.

3.2 Flow Sensitivity Measurement Result

The results of the flow sensitivity measurement are shown in Fig. 3a, b. Figure 3a plots the time variation of the fluctuation kinetic energy k of each 25 frequency cases. In this study, the computation region of k is set to $0 < x/D < 3.0, -2.0 < y/D < 2.0$

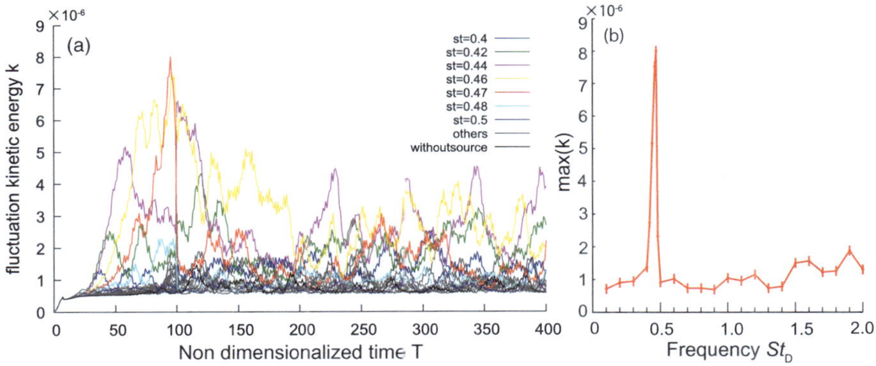

Fig. 3 Flow sensitivity measurement result. **a**: Time variation result of the fluctuation kinetic energy k. **b**: Relationship between max(k) for 3d jet and frequency of the pointsource

and $-2.0 < z/D < 2.0$. Here, T is nondimensionalized by c_0 and D. The colored lines represent the case around the guided jet wave frequency $St = 0.4$ to $St = 0.5$, a black line shows the case that no point source is introduced, and gray lines show other cases. Figure 3b plots the maximum sensitivity values between $T = 0$ and $T = 400$ for each frequency case. From Fig. 3b, the highest sensitivity is found at $St_D = 0.47$, which is consistent with the frequency of the guided-jet wave, and no other results with large sensitivity are found. Moreover, from Fig. 3a, k in the case of $St_D = 0.47$ maximizes and decays around $T = 90$. If the acoustic wave generated at the point source at the end of the potential core is considered to travel in the jet core as a sound wave, it should propagate in the jet at $u/c_0 = -0.1$ due to the relative speed between the advection velocity of the jet flow $u/c_0 = 0.9$ and the speed of sound $u/c_0 = 1$. Since the point source position in this study is $x/D = 11$, the acoustic waves reach $x/D = 2.0$ at $T = 90$. Therefore, it is suggested that the acoustic waves in a jet core that have the guided jet wave frequency affect the flow instability near the jet exit from the relation between the arrival time and kinetic energy profile. From these results, the guided jet waves induce flow instability near the jet exit, which is considered a feedback loop phenomenon.

4 Conclusion

In order to investigate the relationship between the guided jet waves propagating upstream on the jet axis and the shear layer instability waves, which play a role in the feedback loop phenomena, flow sensitivity was measured three-dimensional free jet to the acoustic waves. As the flow sensitivity measurement method, a point source oscillating at a single frequency is placed in the jet and the kinetic energy is measured near the jet exit. In this study, a point source placed in the jet is realized by the volume penalization method. 25 cases of point sources oscillating at $St_D = 0.1$ to $St_D = 2.0$ were measured. The results show that the point source with $St_D = 0.47$ has the highest sensitivity. The time variation of the sensitivity when a point source was placed corresponded to the time when the acoustic waves inside the jet reached the jet exit. This indicates that the acoustic waves generated at the end of the potential core affect the jet instability of the axisymmetric shear layer that occurs near the jet exit. This supports a feedback loop mechanism in subsonic-free jets.

Acknowledgements This work was supported by JSPS Core-to-Core Program (grant number: JPJSCCA20210005), JST SPRING (grant Number: JPMJSP2114) and ELyT collaboration research between École Centrale de Lyon (France) and Tohoku University (Japan). Numerical simulations were performed on the Supercomputer system "AFI-NITY" at the Advanced Fluid Information Research Center, Institute of Fluid Science, Tohoku University.

References

1. A. Powell, The noise of choked jets. J. Acoust. Soc. Am. **25**(3), 385–389 (1953)
2. C.-M. Ho, N.S. Nosseir, Dynamics of an impinging jet. part 1. The feedback phenomenon. J. Fluid Mech. **105**, 119–142 (1981)
3. D. Edgington-Mitchell, Aeroacoustic resonance and self-excitation in screeching and impinging supersonic jets-a review. Int. J. Aeroacoust. **18**(2–3), 118–188 (2019)
4. C. Bogey, Acoustic tones in the near-nozzle region of jets: characteristics and variations between mach numbers 0.5 and 2. J. Fluid Mech. **921**, A3 (2021)
5. J. Laufer, P. Monkewitz, On turbulent jet flows—a new perspective, in *6th Aeroacoustics Conference* (AIAA paper 1980–962) (1980)
6. C. Bogey, Interactions between upstream-propagating guided jet waves and shear-layer instability waves near the nozzle of subsonic and nearly ideally expanded supersonic free jets with laminar boundary layers. J. Fluid Mech. **949**, A41 (2022)
7. A. Towne, A.V.G. Cavalieri, P. Jordan, T. Colonius, O. Schmidt, V. Jaunet, G.A. Brès, Acoustic resonance in the potential core of subsonic jets. J. Fluid Mech. **825**, 1113–1152 (2017)
8. Q. Liu, O.V. Vasilyev, A brinkman penalization method for compressible flows in complex geometries. J. Comput. Phys. **227**(2), 946–966 (Dec2007)
9. R. Komatsu, W. Iwakami, Y. Hattori, Direct numerical simulation of aeroacoustic sound by volume penalization method. Comput. Fluids **130**, 24–36 (2016)
10. C. Bogey, Two-dimensional features of correlations in the flow and near pressure fields of mach number 0.9 jets, in *AIAA Scitech 2019 Forum* (AIAA paper 2019–0806) (2019)
11. D.V. Gaitonde, M.R. Visbal, Pade-plusmn-type higher-order boundary filters for the navier-stokes equations. AIAA J. **38**(11), 2103–2112 (2000)
12. S. Kawai, S.K. Shankar, S.K. Lele, Assessment of localized artificial diffusivity scheme for large-eddy simulation of compressible turbulent flows. J. Comput. Phys. **229**(5), 1739–1762 (Mar2010)
13. S. Gottlieb, C.-W. Shu, Total variation diminishing runge-kutta schemes. Math. Comput. **67**, 08 (1996)

Modelling and Simulations

Evaluation of Turbulence Models for Unsteady Separation

C. Y. MacDougall, F. Ambrogi, and U. Piomelli

Abstract Unsteady separation is an undesirable phenomenon that often occurs in fluid flows. Turbulence models for the Reynolds-Averaged Navier-Stokes equations are commonly used in industry to predict such flows, but they give unsatisfactory results, even for steady separation scenarios. To assess their precision in unsteady cases, we will use high-fidelity Large-Eddy Simulation (LES) results: by employing an identical numerical scheme and boundary conditions as the LES, along with a grid that achieves convergence, we pinpoint modelling errors. All turbulence models capture the fundamental characteristics of this unsteady flow, often more accurately than in analogous steady cases. The influence of the periodic nature of the unsteadiness plays a pivotal role in this result.

1 Introduction

Separation is a common phenomenon that significantly impacts flow dynamics. It can lead to heightened drag, reduced lift, increased noise, undesired vibrations, and substantial stress [11]. The complexity of the problem is further compounded by the presence of large-scale unsteadiness in numerous applications, distinct from the inherent unsteadiness of turbulent flows. The prevalent approach used in the industry to study this problem involves solving the Reynolds-Averaged Navier-Stokes (RANS) equations with turbulence models. This paper aims to assess the accuracy of these models in predicting flow with unsteady separation. Studies of this kind have been carried out for oscillating airfoils [3], components in turbo-machinery [7, 10], and rounded steps [4]. A common finding is the challenge of predicting the size of separation bubbles.

C. Y. MacDougall · F. Ambrogi · U. Piomelli (✉)
Queen's University, Kingston ON, Canada
e-mail: ugo@queensu.ca

F. Ambrogi
e-mail: f.ambrogi@queensu.ca

R. Örlü et al. (eds.), *Progress in Turbulence X*, Springer Proceedings in Physics 404,
https://doi.org/10.1007/978-3-031-55924-2_39

In particular, Park et al. [9] examined a turbulent flat-plate boundary layer exposed to unsteady Adverse Pressure Gradients (APGs) to assess the performance of the $\mathcal{K}-\omega$ and Spalart-Allmaras (SA) turbulence models. The APG was generated by a periodically varying blowing/suction velocity profile at the domain's top boundary, inducing dynamic separation and reattachment. Their findings indicate that turbulence models for the Unsteady Reynolds-Averaged Navier-Stokes (URANS) equations effectively predict the formation of separation bubble and, qualitatively, the phase response of the shear layer compared to their Direct Numerical Simulation (DNS) results. However, URANS falls short in accurately predicting the points of separation and reattachment.

Assessing turbulence models for non-equilibrium flows challenging because of the challenge of isolating various sources of error. It is difficult, and sometimes impossibe, to distinguish modelling errors from numerical ones caused by the grid resolution or the numerical method itself. Uncertainties stemming from boundary conditions and experimental errors further compound this issue. To address this challenge, the current study leverages the outcomes of a high-fidelity Large-Eddy Simulation (LES) [1]. Employing a grid that attains convergence, identical numerical scheme and boundary conditions as the LES, facilitates the isolation of modelling errors, and allows more precise examination of the turbulence-model performance.

2 Methodology

The Unsteady Reynolds-Averaged equations of conservation of mass and momentum for an incompressible flow are solved:

$$\frac{\partial \langle u_i \rangle}{\partial x_i} = 0, \quad \frac{\partial \langle u_i \rangle}{\partial t} + \frac{\partial}{\partial x_j} \langle u_i \rangle \langle u_j \rangle = -\frac{1}{p} \frac{\partial \langle p \rangle}{\partial x_i} + \frac{1}{Re} \frac{\partial^2 \langle u_i \rangle}{\partial x_j \partial x_j} - \frac{\partial}{\partial x_j} \langle u'_i u'_j \rangle. \quad (1)$$

Here, x_i denotes the Cartesian directions; we also use x for the streamwise and y for the wall-normal directions. u_i (or u and v) are the velocity components. Angle brackets or capital letters represent phase-averaged quantities, defined as

$$F = \langle f \rangle = \lim_{N \to \infty} \frac{1}{N} \sum_{n=0}^{N} f(t + nT) \quad (2)$$

T is the period of the oscillation. In this context, a prime denotes the fluctuation around the phase-averaged value: $u' = u - U$.

To parameterize the phase-averaged Reynolds stresses $\langle u'_i u'_j \rangle$, an eddy-viscosity model is used:

$$\langle u'_i u'_j \rangle = -2\nu_T \langle S_{ij} \rangle, \quad \langle S_{ij} \rangle = \frac{1}{2} \left(\frac{\partial U_i}{\partial x_j} + \frac{\partial U_i}{\partial x_j} \right). \quad (3)$$

Here, ν_T represents the eddy viscosity, and S_{ij} the rate-of-strain tensor. Three models, in their standard formulations, are tested: the Spalart-Allmaras (SA) model [12], the SST $\mathcal{K}-\omega$ model [5], and the two-layer formulation of the $\mathcal{K}-\varepsilon$ model [2]. The equations of motion (1) are solved using a fractional step method, with second-order accurate differences in space and time. The code has undergone extensive validation and has been previously applied to similar problems [13]. The computational domain is $600\delta_o^* \times 64\delta_o^*$, with δ_o^* representing the displacement thickness at the inflow, identical to the LES domain [1]. The resolution chosen (1024×160 grid points) was based on the grid-convergence study conducted by Ambrogi et al. [1].

Periodic boundary conditions are applied in the z direction, and no-slip conditions at the wall, and a convective condition [8] at the outflow. For the inflow we phase-average the LES inflow velocity. At the top boundary we prescribe the wall-normal velocity V_∞:

$$V_\infty(x,t) = -\alpha \sin\beta(x - x_r)\exp\left[-\gamma(x - x_r)^2\right] \times \sin\frac{2\pi t}{T}, x1 \quad (4)$$

$$\alpha = -1.11U_o;\ \beta = 0.017/\delta_o^*;\ \gamma = 1.7857 \times 10^{-4}/\left(\delta_o^*\right)^2;\ x_r/\delta_o^* = 450 \quad (5)$$

where U_o is the free-stream velocity at the inflow plane. The spatial variation of V_∞ aligns with that employed by [6]. While our setup shares similarities with the one studied by Park et al. [9], there is a distinction. In their case, the distribution of V_∞ resulted in a recirculation bubble present throughout the cycle. In our scenario, an FPG follows the APG for half of the cycle, and precedes it for the other half; consequently, the recirculation bubble is present only during a portion of the cycle.

The flow conditions are significantly influenced by the reduced frequency $k = \pi f L_{PG}/U_o$, where L_{PG} represents the length over which the pressure gradient varies. Three values of k were considered: $k = 0.2$, 1, and 10. Only $k = 1$ will be shown here, due to the complexity of this case. Steady calculations were conducted with V_∞ corresponding to the $\phi = 0^o,\ 90^o,\ 180^o$ and $\phi = 270^o$ phases.

3 Results and Discussion

Figure 1 displays contours of the phase-averaged streamwise velocity $\langle u \rangle$ at two opposite phases of the cycle. At $\phi = 90^o$, the strongest Adverse Pressure Gradient (APG) follows the strongest Favorable Pressure Gradient (FPG), while at $\phi = 270^o$, the opposite occurs. In the $\phi = 90^o$ phase, the flow experiences acceleration, leading to the thinning of the boundary layer in the region where $V_\infty < 0$ ($x/\delta_0^* = 150-300$). Conversely, at $\phi = 270^o$, the APG induces separation, forming a recirculation bubble, followed by reattachment in the FPG region. Subsequently, the recirculation bubble travels downstream, keeping its shape unchanged. At $\phi = 90^o$, the region of advecting slow-moving fluid is visible at $x/\delta_0 \simeq 550$.

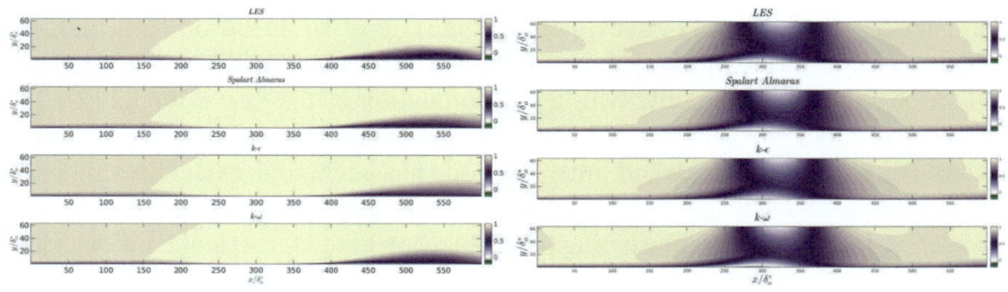

Fig. 1 Contours of the phase-averaged streamwise velocity at $\phi = 90^o$ (left column) and $\phi = 270^o$ (right column)

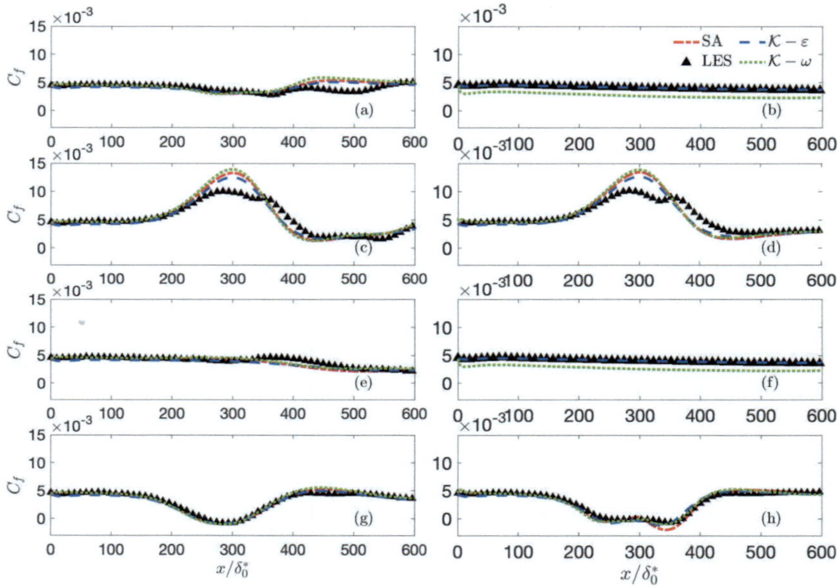

Fig. 2 Skin-friction coefficient for the steady and unsteady cases at four phases in the cycle

All turbulence models capture the general features of the flow with only minor differences with the LES data. The ability of the models to replicate the downstream shedding of the recirculation region is noteworthy. While slight variations between the models are noted in the size and shape of this region, the overall characteristics of the flow are accurately reproduced.

In Fig. 2, the skin-friction coefficient $C_f = 2\tau_w/\rho U_o^2$ (where τ_w is the phase-averaged wall stress) is depicted for both steady and unsteady cases. In the steady case, the Zero Pressure Gradient (ZPG) boundary layer is accurately predicted, as expected, given its use as a canonical case for calibrating constants. However, in the unsteady case, hysteresis effects can be observed (note the difference between the

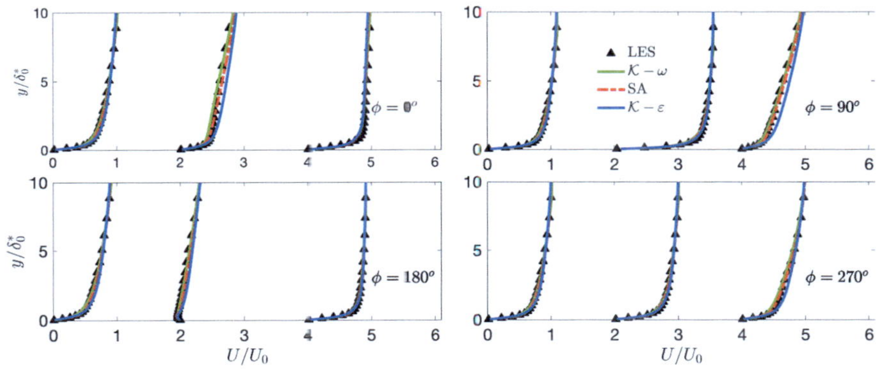

Fig. 3 Velocity profiles at $x/\delta_0^* = 200$, 300 and 450. Each profile is shifted by two units for clarity

$\phi = 0^o$ and 180° phases), as described by [1]. The most significant model error is observed in the advected recirculation region.

Panels (c) and (d) display the phase of maximum FPG. Predicting the flow under strong FPG is a known limitation of many turbulence models, as they struggle to anticipate the turbulence suppression occurring in such cases. The considerable over-prediction of C_f near the center of the domain, where the pressure gradient transitions from FPG to Adverse Pressure Gradient (APG), is notable. It is important to note that the models exhibit greater accuracy in the unsteady case. In the presence of separation, as shown in panels (g) and (h), the unsteady case is also predicted more accurately than the steady one.

In Fig. 3, velocity profiles at three locations are presented: the onset of the pressure-gradient region, the midpoint of the recirculation region, and immediately after reattachment. At the initial location, all models accurately predict the velocity, as anticipated. Some discrepancies in comparison to the LES data are noticeable in the middle of the recirculation region, reflecting the varying accuracy of the C_f prediction. Generally, the $\mathcal{K}-\varepsilon$ model exhibits the least accuracy, but the differences in the outer layer are not substantial.

The improved accuracy observed in the unsteady calculations can be attributed, at least partly, to memory effects. In the unsteady case, the pressure gradient is initially zero, and gradually increases until reaching a strong FPG-APG sequence. Subsequently, it returns to ZPG before transitioning to APG-FPG. The intermediate phases characterized by mild and zero-pressure gradients closely resemble the calibration cases and are predicted fairly accurately. These phases act as 'anchors' and prevent the solution from diverging too far from the correct one when strong pressure gradients are applied. To validate this hypothesis, we computed the difference $|C_{f,RANS} - C_{f,LES}|$, integrated over the region where the pressure gradient is applied (not shown). This measure exhibits a notably sinusoidal behavior, with the smallest values occurring in phases near FPG and larger values during the strongest

FPG or APG phases. Even at the peak pressure gradient, this quantity is significantly smaller than in the steady cases.

Park et al. [9] observed a considerably poorer agreement between Reynolds-Averaged Navier-Stokes (RANS) solutions and Direct Numerical Simulation (DNS). We propose that disparities in numerics and grid resolution may contribute to this outcome. In their work, these authors conducted the URANS simulations on a grid significantly coarser than the DNS grid, and employed a co-located method, whereas the DNS used a staggered grid. While the staggered grid conserves energy discretely, the co-located scheme does not [14]. Our conjecture is that the more substantial errors observed in [9] are not exclusively attributable to turbulence modelling inaccuracies: numerical errors arising from differences in grid resolution and grid-point arrangement, leading to excessive numerical diffusion may also contribute.

4 Conclusions

In this work we assessed the accuracy of the $\mathcal{K}-\omega$, $\mathcal{K}-\varepsilon$, and Spalart-Allmaras turbulence models in unsteady separated flows. To specifically identify modelling errors, we employed an identical numerical scheme and boundary conditions as a reference high-fidelity Large-Eddy Simulation (LES), along with a grid that achieved convergence.

These turbulence models accurately capture the overall characteristics of this flow, showing only minor differences from the LES data. The velocities in the outer layer compare well, and the most significant errors are observed in close proximity to the wall. The motion of the recirculation bubble is qualitatively captured by the models.

Notably, the turbulence models exhibit greater accuracy during phases characterized by mild and zero pressure gradients compared to phases with strong FPG-APG transitions. They reasonably predict the strong APG-FPG phases of the flow and demonstrate greater success in unsteady cases than in the corresponding steady ones. This behaviour may be due to the anchoring effects of the mild and zero pressure-gradient phases, which act as stabilizers preventing the solution from deviating too significantly from the correct one during phases with strong pressure gradients.

Acknowledgements The authors acknowledge the financial support by the Natural Sciences and Engineering Research Council of Canada (NSERC-CRNSG) under the Discovery Grant program. This research was enabled in part by computational support provided by the Digital Research Alliance of Canada.

References

1. F. Ambrogi, U. Piomelli, D.E. Rival, Characterization of unsteady separation in a turbulent boundary layer. J. Fluid Mech. **945**(A10), 1–30 (2022)
2. H. Chen, V. Patel, Near-wall turbulence models for complex flows including separation. AIAA J. **26**, 641–648 (1988)
3. J.A. Ekaterinaris, F.R. Menter, Computation of oscillating airfoil flows with one- and two-equation turbulence models. AIAA J. **32**(12), 2359–2365 (1994)
4. E. Garnier, P.Y. Pamart, J. Dandois, P. Sagaut, Evaluation of the unsteady rans capabilities for separated flows control. Comput. Fluids **61**, 39–45 (2012)
5. F.R. Menter, Two equation eddy viscosity turbulence models for engineering applications. AIAA J. **32**(8), 1598–1605 (1994)
6. Y. Na, P. Moin, Direct numerical simulation of a separated turbulent boundary layer. J. Fluid Mech. **370**, 175–201 (1998)
7. D. Nürnberger, H. Greza, Numerical investigation of unsteady transitional flows in turbomachinery components based on a rans approach. Flow. Turb. Combust. **69**, 331–353 (2002)
8. I. Orlanski, A simple boundary condition for unbounded hyperbolic flows. J. Comput. Phys. **21**, 251–269 (1976)
9. J. Park, S. Ha, D. You, On the unsteady Reynolds-Averaged Navier-Stokes capability of simulating turbulent boundary layers under unsteady adverse pressure gradients. Phys. Fluids **33**(6), 065125 (2021)
10. M. Schobeiri, *Turbomachinery Flow Physics and Dynamic Performance* (Springer, 2012)
11. R.L. Simpson, Turbulent boundary-layer separation. Annu. Rev. Fluid Mech. **21**(1), 205–232 (1989)
12. P.R. Spalart, S.R. Allmaras, A one-equation turbulence model for aerodynamic flows. La Recherche Aérospatiale **1**, 5–21 (1994)
13. W. Wu, U. Piomelli, Effects of surface roughness on a separating turbulent boundary layer. J. Fluid Mech. **841**, 552–580 (2018)
14. Y. Morinishi, T.S. Lund, O.V. Vasilyev, P. Moin, Fully conservative higher order finite difference schemes for incompressible flows. J. Comput. Phys. **143**, 90–124 (1998)

Comparison Between Linear and Hierarchical Multifidelity Models for Uncertainty Quantification in Turbulent Flows

S. Rezaeiravesh, T. Mukha, and P. Schlatter

Abstract Multifildelity models (MFMs) have received considerable attention in recent years for making outer-loop problems in computational fluid dynamics (CFD) more affordable. A recent study by Rezaeiravesh et al. [9] showed that a hierarchical MFM within a Bayesian framework can be an appropriate choice for applications with turbulent flows. In the present study, this model is compared to a linear co-kriging MFM that has been widely used in literature. The uncertainty quantification in a turbulent flow over periodic hills is considered, where two parameters affecting the geometry of the hills are assumed to be uncertain. Both MFMs are found to be capable of accurately estimating the stochastic moments and the Sobol sensitivity indices of the quantity of interest (QoI). However, for the probability distribution of the QoI, the hierarchical model showed a better performance. Future detailed examinations will reveal the conditions under which any of these MFM strategies can be used with high reliability in the results.

1 Introduction

In data analytics, an outer-loop is performed for the purpose of uncertainty quantification (UQ), sensitivity analysis, prediction, optimisation, etc. In these problems, for each of many samples taken for the uncertain/design parameters, a realisation of the quantities of interest (QoIs) should be obtained, where the number of realisations

S. Rezaeiravesh
Department of Fluids and Environment, The University of Manchester, M139PL Manchester, UK
e-mail: saleh.rezaeiravesh@manchester.ac.uk

T. Mukha
Computer, Electrical and Mathematical Sciences and Engineering Division, King Abdullah University of Science and Technology, 23955 Thuwal, Saudi Arabia
e-mail: timofey.mukha@kaust.edu.sa

P. Schlatter (✉)
Institute of Fluid Mechanics (LSTM), Friedrich–Alexander–Universität (FAU) Erlangen–Nürnberg, 91058 Erlangen, Germany
e-mail: philipp.schlatter@fau.de

R. Örlü et al. (eds.), *Progress in Turbulence X*, Springer Proceedings in Physics 404,
https://doi.org/10.1007/978-3-031-55924-2_40

have a direct impact on the accuracy and validity of the outer-loop problem. For the applications relying on turbulent flow simulations, obtaining many accurate QoIs using high-fidelity (HF) approaches such as direct numerical simulation (DNS) and large-eddy simulation (LES), can be computationally expensive and in many cases, e.g., high Reynolds numbers flows, become impractical. On the other hand, relying only on the low-fidelity (LF) data, e.g., obtained from Reynolds-Averaged Navier–Stokes (RANS) simulations, will deteriorate the predictive accuracy of a data-driven model. Therefore, multifidelity models (MFMs) [7] have been developed to combine many LF datasets with fewer HF samples in such a way that the overall computational cost is reduced while retaining a high predictive accuracy. MFMs have been increasingly used for applications relying on CFD and turbulent flow simulations, see, e.g., [9, 12] and the references therein.

Recently, Voet et al. [12] applied several MFMs to study the influence of two parameters controlling the shape of periodic hills with turbulent flow. None of the MFMs, including the co-kriging [2] model (CKM) was found capable of obtaining accurate estimation for the stochastic moments of the QoIs (the UQ problem). The same case was revisited in [9], but with using a hierarchical MFM with calibration (HCM) [4] which showed satisfactory results. In the present study, we compare the performance of the HCM and CKM for the same flow case described in Sect. 2. The two MFMs are shortly explained in Sect. 3 and their results applied to the flow problem are presented and discussed in Sect. 4.

2 Flow Simulations and Problem Statement

To test the two MFMs, we consider turbulent flow over periodic hills at Reynolds number 5600, see the right plot in Fig. 1. Following [13], a parameterisation of the geometry is introduced based on two parameters, α and γ, controlling the length of the hill and the flat portion of the bottom wall, respectively. The total length of the domain L_x is computed as $L_x/h = 3.858\alpha + 5.142\gamma$, where h is the height of the hill. The values $\alpha = 1, \gamma = 1$ give a baseline case studied previously (see, e.g., [3]).

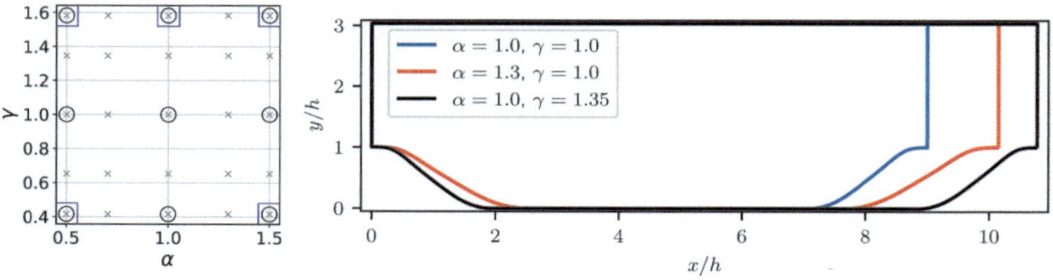

Fig. 1 (Left) Data samples associated to RANS (LF): ×, DNS (HF): □, DNS (ground truth): ○. (Right) Geometry of the periodic hill simulations associated with a number of $\alpha - \gamma$ values

As QoI for uncertainty propagation and global sensitivity analysis we use the height of the separation bubble, H_{bubble}/h, at the streamwise location $x/h = 2.5$. In this framework, α and γ act as uniformly distributed uncertain parameters.

The DNS data provided in [13] is used both to form the ground truth and the results from the HF simulator. The former consists of results for a total of 9 combinations of the parameters, obtained as the tensor product of $\alpha \in \{0.5, 1.0, 1.5\}$ and $\gamma \in \{0.4166, 1.0, 1.5834\}$. Four of these combinations are then excluded to give the HF dataset, as shown in the left plot in Fig. 1. As the LF simulator, we use the results of RANS simulations obtained with the standard k - ω model, as implemented in Ansys Fluent [1]. Quadrilateral grids were used, with the total grid size ranging from $\approx 150 \cdot 10^3$ to $\approx 500 \cdot 10^3$ cells, depending on the values of α and γ. The RANS covers the same range of α and γ as the DNS, but adds $\alpha = \{0.702, 1.297\}$ and $\gamma = \{0.653, 1.347\}$, giving 25 simulations in total.

3 Linear and Hierarchical Multifidelity Models

Consider $\mathbf{x} \in \mathbb{Q} \subset \mathbb{R}^p$ to be p uncertain/design parameters which are sampled in an outer loop problem (here, $\mathbf{x} = \{\alpha, \gamma\}$). By running a simulator for a sample $\mathbf{x}^{(i)}$, the realisation $y^{(i)}$ for the QoI is obtained. A useful method to construct a surrogate based on a dataset $\{(\mathbf{x}^{(i)}, y^{(i)})\}_{i=1}^{n}$ is to use Gaussian process regression (GPR) [8]:

$$y = \hat{f}(\mathbf{x}) + \varepsilon \,, \tag{1}$$

where the surrogate $\hat{f}(\mathbf{x})$ has a multivariate Gaussian distribution

$$\hat{f}(\mathbf{x}) \sim \mathcal{GP}\left(m(\mathbf{x}), k(\mathbf{x}, \mathbf{x}')\right) \,, \tag{2}$$

with m and k being the mean and covariance functions, respectively. When the prior distribution of the GP $\hat{f}(\mathbf{x})$ is defined, it means functions for $m(\mathbf{x})$ and $k(\mathbf{x}, \mathbf{x}')$ are defined which are dependent on a set of hyerparameters $\boldsymbol{\beta}$ which are calibrated using a training dataset $\mathcal{D}$. As a result, the posterior distributions of $\hat{f}(\mathbf{x})$ and hence y are obtained.

Here, we consider only two fidelities, low (L) and high (H). In the co-kriging model (CKM), the GPs modelling the fidelities are assumed to be linearly correlated through a parameter ρ:

$$\hat{f}_H(\mathbf{x}) = \rho \hat{f}_L(\mathbf{x}) + \hat{\delta}(\mathbf{x}) \,, \tag{3}$$

where $\hat{\delta}(\mathbf{x})$ accounts for the discrepancy (bias) between the two models. This MFM was first proposed by Kennedy and O'Hagan [5] and then extended to optimisation by Forrester et al. [2]. The correlation ρ is inferred along with the hyperparameters of the GPs using the training data. In another landmark work, Kennedy and O'Hagan [6] proposed a framework for calibration of computer models adopting a

Bayesian inference technique. This was further extended to an arbitrary number of fidelities by Goh et al. [4] leading to the hierarchical MFM with calibration (HCM). By construction, the GPs at each fidelity can also be dependent on a set of extra fidelity-specific parameters **t** where their posteriors $\boldsymbol{\theta}$ can be learned along with the hyperparameters using a Bayesian inference technique. As explained in [9], this model is consistent with the hierarchy of simulation approaches for turbulent flows, as for instance the numerical and modelling parameters in RANS and LES can be considered as **t**. For two fidelities, the HCM is written as,

$$\begin{cases} y_H = \hat{f}(\mathbf{x}_H, \boldsymbol{\theta}) + \hat{\delta}(\mathbf{x}_H) + \varepsilon \\ y_L = \hat{f}(\mathbf{x}_L, \mathbf{t}) \end{cases} . \tag{4}$$

Another important point discussed in [9] is that the predictive accuracy of an MFM can be dependent on the accuracy of the learned parameters and hyperparameters. In particular, a fully Bayesian approach based on a Markov Chain Monte Carlo (MCMC) technique was shown result in a more accurate MFM compared to the use a point-estimator such as maximum likelihood or maximum a-posteriori estimator.

4 Results and Discussion

The two MFMs, CKM and HCM introduced in the previous section are applied to the periodic hill case. The posterior distribution of the bubble height obtained from the HCM and CKM are shown Fig. 2. For reference, the probability density function (PDFs) obtained from the LF, HF, and ground truth are also presented. In all cases, the predictions of the models at the same set of arbitrary test points have been used. In general, CKM is surprisingly good, given it is linear and has been reported to yield a poor performance for RANS and DNS data that are basically not highly correlated, see [12]. For the CKM, a standard gradient-based optimisation method was adopted to maximize the likelihood function. On the other hand, for the HCM, a fully Bayesian

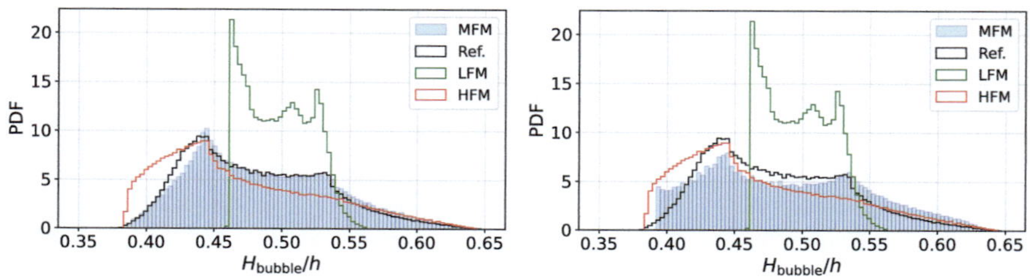

Fig. 2 PDF of the QoI H_{bubble}/h at $x/h = 2.5$ due to the variation of α and γ obtained from (left) HCM and (right) CKM. In each plot, the PDF resulting from the MFM is compared to the PDFs of the ground truth, LF data, and HF data

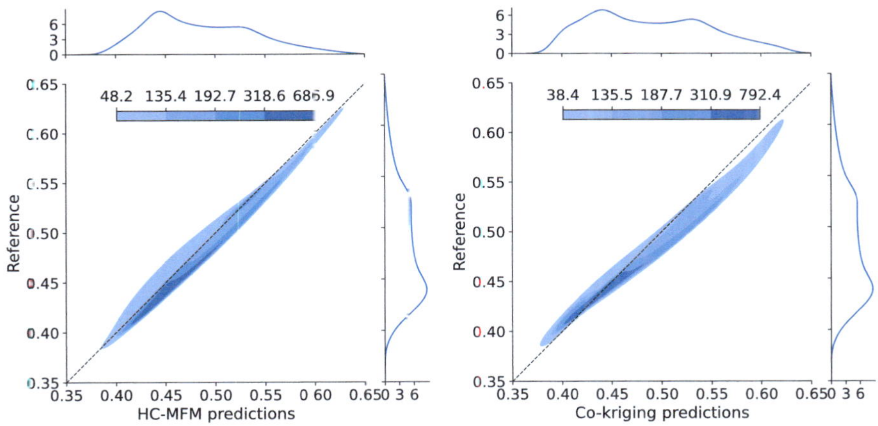

Fig. 3 Joint PDFs of H_{bubble}/h at $x/h = 2.5$ obtained from (left) HCM and (right) CKM

MCMC method is used (with implementation in PyMC3 [10], see [9]) to sample from the posterior distribution of the relevant hyperparameters. Figure 3 shows the PDF of the QoI predicted by the MFMs jointly plotted against the PDF of the ground truth. Here, the better performance of the HCM than the CKM is more visible as reflected in the better alignment of the associated joint PDF with the diagonal dashed line. Finally, the stochastic moments and Sobol sensitivity indices [11] of the QoI with respect to α and γ are computed, see Table 1. Both MFMs as well as the LF and HF models predict the stochastic mean of the QoI close to the reference value. However, the standard deviation is being predicted most accurately by the HCM. Based on the Sobol indices, all models predict higher sensitivity of the QoI to the variation of α than γ, with the MFMs showing considerable improvement in the estimated values of the indices as compared to the LF and HF models. The CKM provides a more accurate estimate of the index with respect to α, although the HCM also shows a high accuracy. Since the stochastic moments and Sobol indices are integrated quantities, see [9, 11], it is not straightforward to understand how the CKM which showed a less accurate PDF than the HC-MFM could give more accurate Sobol indices.

Table 1 Estimated mean, standard deviation, and total Sobol indices of the QoI $y = H_{\text{bubble}}/h$ at $x/h = 2.5$ due to the uncertainty in α and γ

	Moments due to α, γ		Total Sobol indices of the QoI w.r.t.	
Data set	$\mathbb{E}[y]$	$\mathbb{S}[y]$	α	γ
Reference	0.48104	0.04873	0.94551	0.08093
High-Fidelity (HF)	0.46981	0.05198	0.82327	0.18237
Low-Fidelity (LF)	0.49759	0.02362	0.99075	0.02639
HCM	0.48634	0.05266	0.91188	0.10419
CKM	0.48877	0.05854	0.94190	0.07391

5 Conclusions

A comparison between the hierarchical (HCM) and linear co-kriging models (CKM) is presented for the non-intrusive uncertainty quantification in a turbulent flow over periodic hills. Data from RANS simulations [9] and DNS [13] are used associated to the samples taken for the uncertain parameters which control the geometry in the periodic hill simulations. Overall, both MFMs are found to be accurate and clearly show improvement compared to using a single fidelity model solely based on the low or high-fidelity data. Based on the PDF of the predicted QoI, the HCM is found to be more accurate than the CKM. However, considering only the stochastic moments and Sobol indices of the QoI, the difference between the two models is small. Making a general statement about the superiority of each model to the other requires further study and different applications. Moreover, two different inference approaches were adopted in the present study to find the hyperparameters in the Gaussian processes; a task known to be sensitive. Keeping this approach the same (an ongoing research) allows for a more fair comparison between the two MFMs.

References

1. ANSYS Fluent, *Theory Guide*. Release 2019R3 (2019)
2. A.I. Forrester, A. Sóbester, A.J. Keane, Multi-fidelity optimization via surrogate modelling. Proceed. R. Soc. A. **463**(2088), 3251–3269 (2007)
3. J. Fröhlich, C.P. Mellen, W. Rodi, L. Temmerman, M.A. Leschziner, Highly resolved large-eddy simulation of separated flow in a channel with streamwise periodic constrictions. J. Fluid Mech. **526**, 19–66 (2005)
4. J. Goh, D. Bingham, J.P. Holloway, M.J. Grosskopf, C.C. Kuranz, E. Rutter, Prediction and computer model calibration using outputs from multifidelity simulators. Technometrics **55**(4), 501–512 (2013)
5. M.C. Kennedy, A. O'Hagan, Predicting the output from a complex computer code when fast approximations are available. Biometrika **87**(1), 1–13 (2000)
6. M.C. Kennedy, A. O'Hagan, Bayesian calibration of computer models. J. R. Stat. Soc. Ser. B. (Stat. Methodol.) **63**(3), 425–464 (2001)
7. B. Peherstorfer, K. Willcox, M. Gunzburger, Survey of multifidelity methods in uncertainty propagation, inference, and optimization. SIAM Rev. **60**(3), 550–591 (2018)
8. Rasmussen, C.E., Williams, C.K.I.: *Gaussian Processes for Machine Learning (Adaptive Computation and Machine Learning)* (The MIT Press, 2005)
9. S. Rezaeiravesh, T. Mukha, P. Schlatter, Efficient prediction of turbulent flow quantities using a Bayesian hierarchical multifidelity model. J. Fluid Mech. **964**, A13 (2023)
10. J. Salvatier, T.V. Wiecki, C. Fonnesbeck, Probabilistic programming in Python using PyMC3. PeerJ Comput. Sci. **2**, e55 (2016)
11. I. Sobol, Global sensitivity indices for nonlinear mathematical models and their Monte Carlo estimates. Math. Comput. Simul. **55**(1), 271–280 (2001)
12. L.J. Voet, R. Ahlfeld, A. Gaymann, S. Laizet, F. Montomoli, A hybrid approach combining DNS and RANS simulations to quantify uncertainties in turbulence modelling. Appl. Math. Model. **89**, 885–906 (2021)
13. H. Xiao, J.L. Wi, S. Laizet, L. Duan, Flows over periodic hills of parameterized geometries: a dataset for data-driven turbulence modeling from direct simulations. Comput. Fluids **200**, 104431 (2020)

Physics-Informed Minimal Error Simulation Methods for Turbulent Flow Predictions

S. Heinz

Abstract Simulations of high Reynolds number (Re) separated turbulent flows have faced significant problems for decades as large eddy simulation (LES) is computationally too expensive while Reynolds-averaged Navier-Stokes (RANS) methods and hybrid RANS-LES methods often provide unreliable results. This situation causes serious consequences, we are currently unable to reliably predict very high Re regimes, which hampers applications and our understanding of turbulence structures. The paper reports the advantages of using a strict mathematical approach to derive partially resolving turbulence models. In contrast to popular hybrid RANS-LES, this approach includes a dynamic modification of the turbulence model in response to the actual flow resolution: the model can increase (decrease) its contribution to the simulation in dependence of a low (high) flow resolution. The model is physics-informed via the involvement of information about the resolved motion. Applications of such methods, referred to as continuous eddy simulation (CES), reveal significant advantages, e.g., for the simulation of asymptotic Re regimes. It is worth mentioning that CES methods can be used as resolving LES by avoiding the problem of involving the filter width as artificial length scale, which can become unphysical on coarse grids.

1 Introduction

Turbulent flows of practical relevance are often characterized by a high Reynolds number (Re) and (depending on the surface geometry) flow separation. It is well-known that reliable and efficient simulations of separated high Re turbulent flows face significant problems [1]. Reynolds-averaged Navier-Stokes (RANS) methods are known to be often unreliable since the exclusion of instantaneous turbulence implies unphysical simulation results. Large eddy simulation (LES) focuses on a maximum of flow resolution, it is aiming at a realistic simulation of instantaneous turbulence. However, LES is limited to not too high Re flow simulations because of

S. Heinz (✉)
Department of Mathematics and Statistics, University of Wyoming, 1000 E. Univ. Ave., Laramie, WY 82071, USA
e-mail: heinz@uwyo.edu

R. Örlü et al. (eds.), *Progress in Turbulence X*, Springer Proceedings in Physics 404,
https://doi.org/10.1007/978-3-031-55924-2_41

their significant computational cost requirements. An illustration of consequences in regard to the prediction of very high *Re* flows is given in Sect. 3.

The most appropriate balance between RANS and LES approaches is the design of hybrid RANS-LES methods [1–3]. Many different hybrid RANS-LES were introduced so far, for example, wall-modeled LES (WMLES) [4], detached eddy simulation (DES) concepts used in a variety of ways [2, 5], unified RANS-LES (UNI-LES) [6, 7], partially averaged Navier-Stokes (PANS) [8], partially integrated transport modeling (PITM) [2], and scale adaptive simulation (SAS) methods [3]. The problem with these hybrid RANS-LES is their limited reliability: there is a lack of proven predictive power, all such hybrid RANS-LES results need validation, which is often hardly possible. For example, a common problem with popular hybrid RANS-LES, WMLES and DES methods, is the significant uncertainty of predictions depending on adjustable model settings. Such predictions depend on different (equilibrium, nonequilibrium) wall models applied, definitions of regions where different models and grids are applied, different mesh distributions, and set-up options to manage the data transfer between such different flow regions [1].

An explanation for the problems faced by popular hybrid RANS-LES provides the concept of these methods to generate conditions which enable the model to produce as much as possible resolved motion. This idea does not ensure at all a physically meaningful simulation. The best example is LES performed on coarse grids. Although there is a lot of resolved motion, such simulations are known to usually provide unphysical results. It is plausible that the uncontrolled balance between modeled motion represented by the model and produced resolved motion implies the sensitive dependence of model results on changes of simulation settings (see the discussion in the preceding paragraph). The most relevant problem is that usually applied hybrid RANS-LES do not enable a seamless transition between RANS and LES regimes. For example, simulations of the effect of increasing *Re* using the same grid face a decrease of the amount of resolved motion. The model needs to compensate this loss of total kinetic energy by increasing its contribution to the simulation. In general, a functional RANS-LES swing requires that the model contribution to the simulation is relatively low (high) if the flow resolution is high (low). The latter requires that the model receives information about the amount of resolved motion, which is not the case in regard to usually applied hybrid RANS-LES.

These issues lead to the question of which more appropriate simulation concepts can be applied to overcome these serious problems. The most promising approach is the design of a computational method that minimizes the hybridization error: the expectation is that a minimal error requires an appropriate balance of resolved and modeled motion. Such methods were introduced recently: they are referred to as continuous eddy simulation (CES) methods [9–17]. The motivation for presenting this paper is to describe the main characteristics of CES methods in conjunction with their ability to overcome the issues of popular hybrid RANS-LES. The mathematical approach will be described next followed by a discussion of implications.

2 Minimal Error Simulation Methods

Incompressible flow is considered for simplicity, corresponding compressible formulations can be found elsewhere [15]. The incompressible continuity equation $\partial \widetilde{U}_i/\partial x_i = 0$ and momentum equation are considered,

$$\frac{D\widetilde{U}_i}{Dt} = -\frac{\partial(\widetilde{p}/\rho + 2k/3)}{\partial x_i} + 2\frac{\partial(\nu + \nu_t)\widetilde{S}_{ik}}{\partial x_k}. \tag{1}$$

Here, $D/Dt = \partial/\partial t + \widetilde{U}_k \partial/\partial x_k$ denotes the filtered Lagrangian time derivative and the sum convention is used throughout this paper. $\widetilde{U}_i$ refers to the ith component of the spatially filtered velocity. We have here the filtered pressure $\widetilde{p}$, ρ is the constant mass density, k is the modeled energy, ν is the constant kinematic viscosity, and $\widetilde{S}_{ij} = (\partial\widetilde{U}_i/\partial x_j + \partial\widetilde{U}_j/\partial x_i)/2$ is the rate-of-strain tensor. The modeled viscosity is given by $\nu_t = C_\mu k^{1/2} L$. Here, C_μ is a model parameter with standard value $C_\mu = 0.09$, and L is a characteristic length scale. L can be calculated in different ways using $L = k^{3/2}/\varepsilon = k^{1/2}\tau = k^{1/2}/\omega$, where the dissipation rate $\varepsilon = k/\tau$ of modeled kinetic energy, the dissipation time scale τ, and the turbulence frequency $\omega = 1/\tau$ are involved.

The minimal error approach can be applied in conjunction with a variety of turbulence models [16]. Here, the $k - \varepsilon$ model is applied to determine L via $L = k^{3/2}/\varepsilon$,

$$\frac{Dk}{Dt} = P - \varepsilon + D_k, \qquad \frac{D\varepsilon}{Dt} = C_{\varepsilon_1}\frac{\varepsilon^2}{k}\Big(\frac{P}{\varepsilon} - \alpha^*\Big) + D_\varepsilon. \tag{2}$$

$P = \nu_t S^2$ is the production of k, where $S = (2\widetilde{S}_{mn}\widetilde{S}_{nm})^{1/2}$ is the characteristic shear rate. C_{ε_1} is a constant with standard value $C_{\varepsilon_1} = 1.44$, and $\sigma_\varepsilon = 1.3$. The diffusion terms are given by $D_k = \partial[\nu_t \, \partial k/\partial x_j]/\partial x_j$, $D_\varepsilon = \partial[(\nu_t/\sigma_\varepsilon)\, \partial\varepsilon/\partial x_j]/\partial x_j$. In RANS, α^* is considered to be constant: $\alpha^* = \alpha = C_{\varepsilon_2}/C_{\varepsilon_1}$ where $C_{\varepsilon_2} = 1.92$ implies $\alpha = 1.33$. Here, $\alpha^*$is considered to be an undetermined parameter that needs to be chosen to minimize the hybridization error.

Corresponding implications of minimal error methods are shown in Table 1 including key steps of the derivation. The abbreviations KES and KEK refer to the consideration of the $k - \varepsilon$ model where the hybridization is accomplished in the scale equation (involving α^* that needs to be determined) or k-equation (involving ψ_α that needs to be determined), respectively. Both approaches can provide equivalent results as long as the coefficient relation $\alpha^* = 1 + \alpha - \psi_\alpha$ is taken into account. In regard to both KES and KEK models there are two analysis options. Option O_1 is an exact hybridization where total viscosities need to be applied in turbulent transport terms. Option O_2 is an hybridization where the usual model viscosities are involved in turbulent transport terms; an approximation is involved by the neglect of substantial derivatives (Dk/Dt, $D\varepsilon/Dt$) in regard to the model coefficient calculation. The derivation of implications is based on the consideration of the residual of the ε equation depending on the hybridization applied. By involving the bracket relations and setting the first variation of the residual equal to zero we obtain, e.g., in the Option

Table 1 Minimal error $k-\varepsilon$ models: both KES and KEK hybridizations are considered in analysis options O_1, O_2 depending on ν_t^*. Model errors λ, first variations, and resulting mode controls α^* are provided. Variations applied are given in brackets

$$\frac{Dk}{Dt} = P - \varepsilon + D_k, \quad \frac{D\varepsilon}{Dt} = C_{\varepsilon_1}\frac{\varepsilon^2}{k}\Big(\frac{P}{\varepsilon} - \alpha^*\Big) + D_\varepsilon, \quad D_k = \frac{\partial}{\partial x_j}\nu_t^*\frac{\partial k}{\partial x_j}, \quad D_\varepsilon = \frac{\partial}{\partial x_j}\frac{\nu_t^*}{\sigma_\varepsilon}\frac{\partial \varepsilon}{\partial x_j} \qquad \text{(KES)}$$

$$\frac{Dk}{Dt} = P - \psi_\alpha\varepsilon + D_k, \quad \frac{D\varepsilon}{Dt} = C_{\varepsilon_1}\frac{\varepsilon^2}{k}\Big(\frac{P}{\varepsilon} - \alpha\Big) + D_\varepsilon, \quad \alpha^* = 1 + \alpha - \psi_\alpha \qquad \text{(KEK)}$$

• Analysis option O_1 [with Dk/Dt, $D\varepsilon/Dt$, $\nu_t^* = \nu_{t,tot}$]: $\quad \Big[\frac{\delta(Dk/Dt)}{Dk/Dt} = \frac{\delta D_k}{D_k} = \frac{\delta k}{k}, \ \frac{\delta(D\varepsilon/Dt)}{D\varepsilon/Dt} = \frac{\delta D_\varepsilon}{D_\varepsilon} = \frac{\delta\varepsilon}{\varepsilon}\Big]$

$$\lambda_1 = C_{\varepsilon_1}\frac{\varepsilon^2}{k}\Big(\frac{P}{\varepsilon} - \alpha_1^*\Big) + D_\varepsilon - \frac{D\varepsilon}{Dt}, \quad \delta\Big(\frac{\lambda_1}{\varepsilon}\Big) = \frac{C_{\varepsilon_1}}{\tau}(\alpha_1^* - 1)\Big[\frac{\delta\tau}{\tau} - \frac{\delta\alpha_1^*}{\alpha_1^*-1}\Big], \quad \int_\alpha^{\alpha_1^*}\frac{dx}{x-1} = \int_{\tau_{tot}}^{\tau}\frac{dy}{y}, \quad \frac{\alpha_1^*-1}{\alpha-1} = \tau_+$$

• Analysis option O_2 [without Dk/Dt, $D\varepsilon/Dt$, $\nu_t^* = \nu_t$]: $\quad \big[\delta D_k/D_k = 3\delta k/k - \delta\varepsilon/\varepsilon, \ \delta D_\varepsilon/D_\varepsilon = 2\delta k/k\big]$

$$\lambda_2 = C_{\varepsilon_1}\frac{\varepsilon^2}{k}\Big(\frac{P}{\varepsilon} - \alpha_2^*\Big) + D_\varepsilon, \quad \delta\Big(\frac{\lambda_2}{k^2}\Big) = \frac{C_{\varepsilon_1}}{L^2}(\alpha_2^* - 1)\Big[\frac{\delta L^2}{L^2} - \frac{\delta\alpha_2^*}{\alpha_2^*-1}\Big], \quad \int_\alpha^{\alpha_2^*}\frac{dx}{x-1} = \int_{L_{tot}^2}^{L^2}\frac{dy}{y}, \quad \frac{\alpha_2^*-1}{\alpha-1} = L_+^2$$

O_2 the relation $\delta\alpha_2^*/[\alpha_2^* - 1] = \delta L^2/L^2$. This relation can be integrated from the RANS state to a state with a certain level of resolved motion (see Table 1) leading to $\alpha^* = 1 + L_+^2(\alpha - 1)$, which is dynamically updated during the simulation.

The structure of CES methods differs significantly from popular RANS-LES methods due to the involvement of resolution indicators like $0 \leq L_+ = L/L_{tot} \leq 1$ in model equations [$L = k^{3/2}/\varepsilon$ is the model length scale and L_{tot} is the total (RANS-type) length scale]. L_+ measures the degree of flow resolution: $L_+ \approx 1$ represents an almost completely modeled (RANS) regime, whereas $L_+ \approx 0$ represents an almost completely resolved (LES) regime. Most importantly, in contrast to usually applied hybrid RANS-LES the model is informed about the actual resolution in this way, and the model is able to respond to resolution variations implied by changes of the model coefficient α^*. For example, a higher resolution (L_+ becomes smaller) decreases α^*. Thus, there is less dissipation of ε, ε increases which decreases the model viscosity $\nu_t = C_\mu k^2/\varepsilon$. To minimize the error of a hybrid method means, therefore, to minimize the uncontrolled coexistence of resolved and modeled motion seen in popular hybrid RANS-LES. In addition to features of popular hybrid RANS-LES (see above), it is also worth noting that this approach only depends on a minimum of model settings given by the model parameters involved (C_μ, C_{ε_1}, C_{ε_2}, σ_ε).

3 Applications of Minimal Error Simulation Methods

One of the CES applications is the simulation of periodic hill flows as illustrated in Fig. 1 [10]. This flow is a channel flow involving periodic restrictions. This flow, which is used a lot for the evaluation of turbulence models [1], involves features such as separation, recirculation, and natural reattachment [18, 19]. A thorough evaluation of the performance of CES methods in regard to simulating periodic hill flows at the highest $Re = 37,000$ (Re is based on the hill height and the bulk velocity) for which experimental data for model evaluation are still available can be found elsewhere [10]. A representative example of advantages is given in Fig. 2. This figure shows reattachment point predictions of experiments and CES results depending on

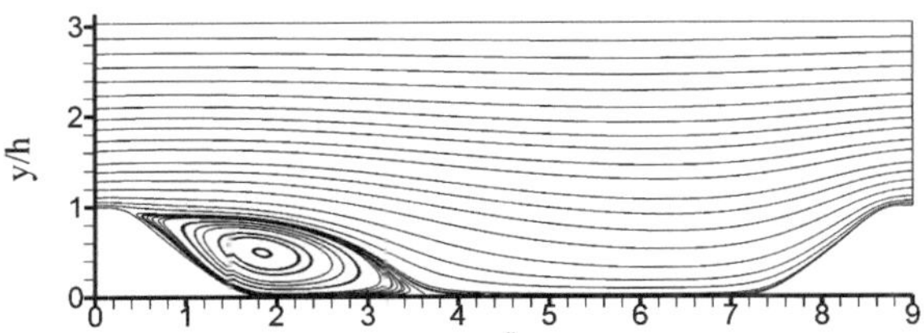

Fig. 1 Velocity streamlines seen in periodic hill flows: results obtained by continuous eddy simulation at $Re = 37,000$. Reprinted with permission from [10]. Copyright 2020 AIP publishing

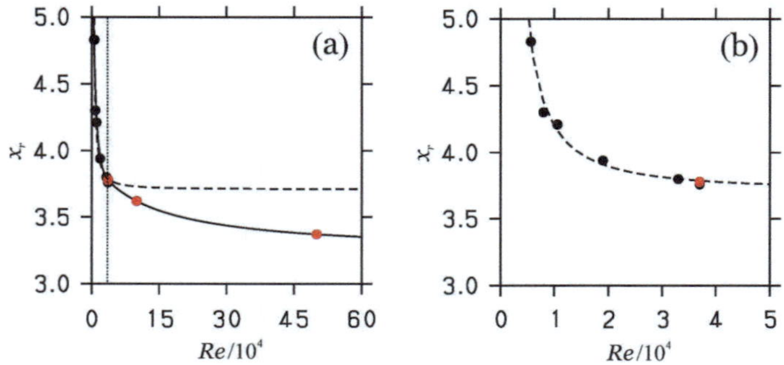

Fig. 2 Reattachment point predictions x_r versus Re: experimental (black dots) [18, 19] and CES results (red dots). The black line in **a** shows $x_r = 3.23\big(1 + 0.5e^{-1.5\times10^{-4}Re}\big)[1 + 15.1 \times 10^4/Re]/[1 + 12.4 \times 10^4/Re]$ obtained from CES predictions, the dashed line shows $0.49(Re/10^4)^{-1.4} + 3.71$ of Kähler et al. [19]. The vertical dotted line shows the range of previously available results. Figure **b** shows the zoomed-in curve fit of [19] compared to previous experimental and $Re = 37K$ CES results. Reprinted with permission from [10]. Copyright 2020 AIP publishing

Re. Available experimental results support the curve fit $0.49(Re/10^4)^{-1.4} + 3.71$ of Kähler et al. [19]. It may be seen, however, that this curve fit can hardly predict reattachment point changes above Re considered before in experiments. On the other hand, in consistency with studies of extreme Re regimes of canonical wall-bounded turbulent flows, the curve fit derived from CES predictions (see Fig. 2) plausibly explains how the asymptotic Re regime is reached [10, 20, 21].

4 Summary

The paper reports the advantages of partially resolving simulation methods based on strict mathematics. Corresponding minimal error simulation methods include an essential mechanism that is missing in popular hybrid RANS-LES methods: the model integrates physics, it can dynamically respond to the actual amount of flow resolution. The implied stability of the model in regard to redistributions of resolved and modeled motion opens the way to reliably predict very high Re regimes in absence of validation data: see Fig. 2 in regard to periodic hill flow applications.

Acknowledgements I would like to acknowledge support from the National Science Foundation (AGS, Grant No. 2137351, with Dr. N. Anderson as Technical Officer) and support from the Hanse-Wissenschaftskolleg (Delmenhorst, Germany).

References

1. S. Heinz, A review of hybrid RANS-LES methods for turbulent flows: Concepts and applications. Prog. Aerosp. Sci. **114**, 100597/1–100597/25 (2020)
2. B. Chaouat, The state of the art of hybrid RANS/LES modeling for the simulation of turbulent flows. Flow Turbul. Combust. **99**(2), 279–327 (2017)
3. F. Menter, A. Hüppe, A. Matyushenko, D. Kolmogorov, An overview of hybrid RANS-LES models developed for industrial CFD. Appl. Sci. **11**(6), 2459 (2021)
4. J. Larsson, S. Kawai, J. Bodart, I. Bermejo-Moreno, Large eddy simulation with modeled wall-stress: recent progress and future directions. Mech. Eng. Rev. **3**(1), 15–00418/1–15–00418/23 (2016)
5. C. Mockett, M. Fuchs, F. Thiele, Progress in DES for wall-modelled LES of complex internal flows. Comput. Fluids **65**, 44–55 (2012)
6. R. Mokhtarpoor, S. Heinz, M. Stoellinger, Dynamic unified RANS-LES simulations of high Reynolds number separated flows. Phys. Fluids **28**(9), 095101/1–095101/36 (2016)
7. R. Mokhtarpoor, S. Heinz, Dynamic large eddy simulation: Stability via realizability. Phys. Fluids **29**(10), 105104/1–105104/22 (2017)
8. S. Girimaji, Partially-averaged Navier-Stokes method for turbulence: A Reynolds-averaged Navier-Stokes to direct numerical simulation bridging method. ASME J. Appl. Mech. **73**(3), 413–421 (2006)
9. S. Heinz, The large eddy simulation capability of reynolds-averaged navier-stokes equations: Analytical results. Phys. Fluids **31**(2), 021702/1–021702/6 (2019)
10. S. Heinz, R. Mokhtarpoor, M.K. Stoellinger, Theory-based reynolds-averaged navier-stokes equations with large eddy simulation capability for separated turbulent flow simulations. Phys. Fluids **32**(6), 065102/1–065102/20 (2020)
11. S. Heinz, The continuous eddy simulation capability of velocity and scalar probability density function equations for turbulent flows. Phys. Fluids **33**(2), 025107/1–025107/13 (2021)
12. S. Heinz, J. Peinke, B. Stoevesandt, Cutting-edge turbulence simulation methods for wind energy and aerospace problems. Fluids **6**, 288/1–288/15 (2021)
13. S. Heinz, Theory-based mesoscale to microscale coupling for wind energy applications. Appl. Math. Model. **98**, 563–575 (2021)
14. S. Heinz, Remarks on energy partitioning control in the PITM hybrid RANS/LES method for the simulation of turbulent flows. Flow. Turb. Combust. **108**(4), 927–933 (2022)
15. S. Heinz, From two-equation turbulence models to minimal error resolving simulation methods for complex turbulent flows. Fluids **7**(12), 368 (2022)
16. S. Heinz, Minimal error partially resolving simulation methods for turbulent flows: A dynamic machine learning approach. Phys. Fluids **34**(5) 051705/1–051705/7 (2022)
17. S. Heinz, A mathematical solution to the computational fluid dynamics (CFD) dilemma. Mathematics **11**, 3199/1–3199/17 (2023)
18. C. Rapp, M. Manhart, Flow over periodic hills-an experimental study. Exp. Fluids **51**, 247–269 (2011)
19. C.J. Kähler, S. Scharnowski, C. Cierpka, Highly resolved experimental results of the separated flow in a channel with streamwise periodic constrictions. J. Fluid Mech. **796**, 257–284 (2016)
20. S. Heinz, On mean flow universality of turbulent wall flows. I. High Reynolds number flow analysis. J. Turbul. **19**(11–12), 929–958 (2018)
21. S. Heinz, On mean flow universality of turbulent wall flows. II. Asymptotic flow analysis. J. Turbul. **20**(2), 174–193 (2019)

Direct Numerical Simulation of Turbulent Open Channel Flow: Streamwise Turbulence Intensity Scaling and Its Relation to Large-Scale Coherent Motions

C. Bauer, Y. Sakai, and M. Uhlmann

Abstract We conducted direct numerical simulations of turbulent open channel flow (OCF) and closed channel flow (CCF) of friction Reynolds numbers up to $\mathrm{Re}_\tau \approx 900$ in large computational domains up to $L_x \times L_z = 12\pi h \times 4\pi h$ to analyse the Reynolds number scaling of turbulence intensities. Unlike CCF, our data suggests that the streamwise turbulence intensity in OCF scales with the bulk velocity for $\mathrm{Re}_\tau \gtrsim 400$. The additional streamwise kinetic energy in OCF with respect to CCF is provided by larger and more intense very-large-scale motions in the former type of flow. Therefore, compared to CCF, larger computational domains of $L_x \times L_z = 12\pi h \times 4\pi h$ are required to faithfully capture very-large-scale motions in OCF—and observe the reported scaling. OCF and CCF turbulence statistics data sets are available at https://doi.org/10.4121/88678f02-2a34-4452-8534-6361fc34d06b.

1 Introduction

Plane Poiseuille flow, also known as closed channel flow (CCF), is one of the most studied canonical flows by means of direct numerical simulations (DNSs, [2–4] [5, 6]). The numerical domain is defined by doubly-periodic boundary condition in the stream- and spanwise directions, and impermeable no-slip walls at the bottom and the top. Conversely, less attention has been paid to open channel flow (OCF), where one of the no-slip walls is replaced by a free-slip plane, despite its even more direct relevance in environmental flows. Most of the early OCF DNSs were restricted to small friction

C. Bauer (✉)
Institute of Aerodynamics and Flow Technology, German Aerospace Center, Göttingen, Germany
e-mail: christian.bauer@dlr.de

Y. Sakai
TUM School of Engineering and Design, Technical University of Munich, Munich, Germany

M. Uhlmann
Institute for Water and Environment, Karlsruhe Institute of Technology, Karlsruhe, Germany

R. Örlü et al. (eds.), *Progress in Turbulence X*, Springer Proceedings in Physics 404,
https://doi.org/10.1007/978-3-031-55924-2_42

Reynolds numbers $Re_\tau = u_\tau h/\nu$, based on friction velocity u_τ, channel height h and kinematic viscosity ν, and/or small computational domains. Recently, Yao et al. [7] compared CCF and OCF by means of DNS data in computational domains of $L_x \times L_z = 8\pi h \times 4\pi h$ for Reynolds numbers up to $Re_\tau = 2000$. In agreement with experiments [8, 9] and DNSs [10], they reported that so-called very-large-scale motions (VLSMs)—which are a common feature of wall-bounded turbulent flows—are more energetic and larger in OCF than in CCF. Generally, VLSMs become more energetic with increasing Reynolds number [7, 9, 10], leading to the failure of wall scaling of the streamwise intensity in CCF [4]. While for CCF the streamwise turbulence intensity scales neither with the friction velocity nor with the bulk velocity, Afzal et al. [11] suggested that the streamwise turbulence intensity in OCF scales with the bulk velocity based on an experimental data set. Other experimental studies, such as Duan et al. [8, 9], on the contrary, show the—relatively scattered—streamwise turbulence intensity scaled in wall units together with universal scaling functions proposed by Nezu and Rodi [12]. Thus, additional highly accurate OCF data sets are required to answer the question whether the streamwise turbulence intensity in OCF scales in bulk units.

Since the streamwise length of VLSMs is an order of magnitude larger than the channel half width, exceptionally large domains are required to faithfully capture VLSMs in DNS [13, 14]. Moreover, Pinelli et al. [15] found that the small scales—responsible for the mass transfer across the free surface in OCF—are spatially organised by VLSMs, which requires an additional grid refinement towards the free surface. In the present study, we investigate the scaling of the streamwise turbulence intensity in OCF by means of DNSs. For this purpose, we generate new OCF and CCF DNS datasets at comparable Reynolds numbers up to $\mathrm{Re}_\tau = 900$ in identical computational domains of $L_x \times L_z = 12\pi h \times 4\pi h$, with grid refinements towards no-slip and free-slip walls. By removing Reynolds number or domain size discrepancies between OCF and CCF, we are able to quantify the impact of the free-slip boundary condition on turbulence one-point statistics in the cleanest possible manner.

2 Methodology

We solve the incompressible Navier-Stokes equations in their wall-normal velocity/vorticity formulation with a pseudo-spectral representation of the flow variables [2]. We apply periodic boundary conditions in the stream- and the spanwise directions, the smooth no-slip wall at the bottom and the free-slip boundary condition at the top of the channel. Although the numerical method has shown its validity in numerous CCF simulations [2, 3, 16], the introduction of a free-slip surface has to be regarded separately. A Chebyshev-Gauss-Lobatto grid, which is adopted in the surface-normal direction provides a grid refinement both towards the wall and the free surface. We set up six OCF and four CCF cases, where the Reynolds number and the computational domain length are varied (cf. Table 1). Hereinafter, the veloc-

Table 1 Turbulent channel flow simulation cases. N_x, N_y and N_z are the number of grid points in the streamwise, wall-normal and spanwise direction; Δx^+, the streamwise grid spacing; Δz^+, the spanwise grid spacing; Δy^+_{max}, the maximum grid spacing in the wall-normal direction. ΔT is the temporal averaging interval for turbulent statistics. Prefix "O" indicates OCF cases, "C" CCF cases

Case	O200	O400	O600	O900L4	O900L8	O900L12	C200	C400	C600	C900
Re_τ	200	399	596	899	895	894	200	397	593	890
Re_b	3170	6969	11047	17512	17512	17512	3170	6969	11047	17512
L_x/h	12π	12π	12π	4π	8π	12π	12π	12π	12π	12π
L_z/h	4π	4π	4π	2π	4π	4π	4π	4π	4π	4π
N_x	768	1536	1536	1536	2048	3072	768	1536	2048	3072
N_y	129	193	257	385	385	385	129	193	257	385
N_z	512	1024	1536	1024	2048	2048	512	1024	1536	2048
Δx^+	9.8	9.8	14.6	7.4	11.0	11.0	9.8	9.8	14.6	11.0
Δz^+	4.9	4.9	4.8	5.5	5.5	5.5	4.9	4.9	4.8	5.5
Δy^+_{max}	2.5	3.3	3.7	3.7	3.7	3.7	4.9	6.5	7.3	7.3
$\Delta T u_b/h$	8660	1925	1460	417	570	1054	8600	3260	1757	1013
$\Delta T u_\tau^2/\nu$	109700	43900	46980	19080	26100	44700	108200	73730	55960	45820

ity fluctuation in i-direction is defined as $u_i' = u_i - \langle u_i \rangle$, where angular brackets denote averaging over both homogeneous directions and time. Normalisation in viscous or "wall units" is indicated by the $+$ superscript. Corresponding scales are u_τ and the viscous length scale $\delta_\nu = \nu/u_\tau$. Characteristic scales in "bulk units", on the other hand, are the bulk velocity u_b and the channel height h, which forms the bulk Reynolds number $Re_b = u_b h/\nu$.

3 Computational Domain Size

In the following, the effect of the constraints upon large-scale coherent motions in the vicinity of the free surface due to the finite size of the computational domain is investigated. Figure 1 displays instantaneous realisations of the streamwise fluctuating velocity for the three open channel simulation cases with different computational domain size O900L4, O900L8, and O900L12 in the vicinity of the free-slip boundary ($y/h \approx 0.95$), together with closed channel data for comparison. A coherent structure with a streamwise extent comparable to the domain length $l_x \approx 12\pi h$ and a spanwise extent of $l_z \approx 2h$ is clearly visible in Fig. 1c for O900L12 at $z/h \approx 9$, but not for the smaller boxes (a, b), where the flow domain interferes with the natural extensions of such a structure. Moreover, the structures in the vicinity of the free-slip boundary in OCF appear to be more intense and larger than the structures near the centreline of CCF (Fig. 1d). Thus, larger computational domains than in CCF are required to fully capture these scales. Figure 2 displays one-dimensional pre-multiplied energy spectra of the streamwise velocity fluctuations for $\mathrm{Re}_\tau \approx 900$

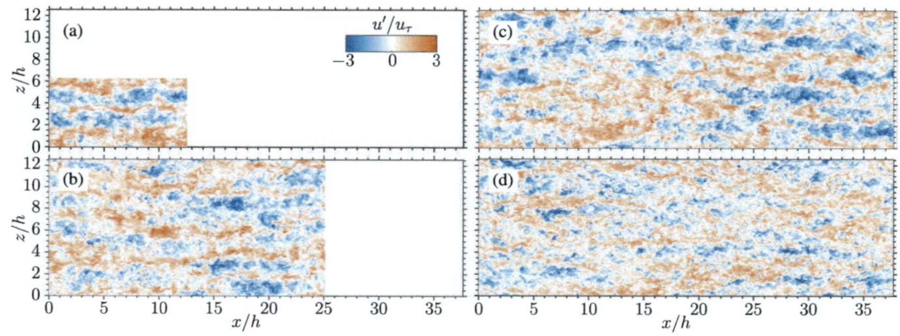

Fig. 1 Instantaneous streamwise velocity fluctuations u'^+ in a wall-parallel plane directly below the free surface in OCF (**a**, **b**, **c**) and the vicinity of the channel centreline in CCF (**d**, $y/h \approx 0.95$). **a** O900L4; **b** O900L8; **c** O900L12; **d** C900

and different computational domain length as a function of the wavelength and the wall distance. Although at first glance both the streamwise and the spanwise premultiplied energy spectra decay at large wavelengths for O900L4 (Fig. 2a, b), in fact larger box sizes are required to physically capture the largest scales that only appear in the larger domains. This can be seen in the streamwise spectra for O900L8 (Fig. 2c), where a spectral pile-up at large wavelength occurs. For O900L12 (Fig. 2e, f), on the other hand, both the streamwise and the spanwise spectra decay at large wavelengths again. Thus, a domain size of $L_x \times L_z = 12\pi h \times 4\pi h$ appears to be sufficiently large to capture a relevant amount of VLSMs. Moreover, a comparison of the spanwise spectra (Fig. 2b, d, f) indicates an artificially large accumulation of energy at spanwise wavelengths around $\lambda_z/h \approx 2$ in the smallest box ($L_x = 4\pi h$, $L_z = 2\pi h$). In summary, as in CCF, VLSMs appear in the bulk flow, penetrating into the buffer layer and therefore through these motions the whole flow domain is influenced by the free-surface layer. Due to their large spatial extent, larger computational domains than in CCF are needed to fully capture these scales. A domain size of $L_x = 12\pi h$ was found to be sufficient at $\mathrm{Re}_\tau = 900$.

4 Streamwise Turbulence Intensity Scaling

Let us now focus on the variation with Reynolds number of the streamwise turbulence intensity normalised with the bulk velocity u_b. Both open and closed channel turbulence intensities obtained from DNSs with the largest computational domain ($L_x \times L_z = 12\pi h \times 4\pi h$) are presented in Fig. 3. While the spanwise and wall-normal contributions to the turbulence intensity scale well with the friction velocity u_τ in the bulk region of the flow for $Re_\tau \geq 400$ for both flow configurations (figure omitted), the streamwise contribution differs. As Fig. 3a indicates, the streamwise turbulence intensity in OCF appears to scale with the bulk velocity u_b for $Re_\tau \geq 400$

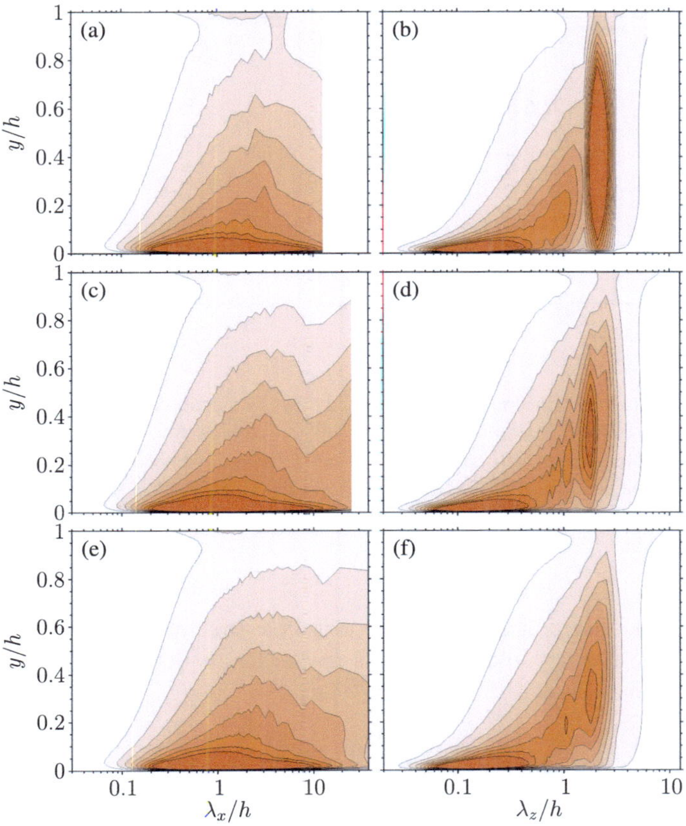

Fig. 2 Contour lines of the one-dimensional pre-multiplied energy spectra of the streamwise velocity fluctuations as a function of the wavelength and the wall distance. **a**, **c**, **e** Streamwise spectra $\kappa_x \phi_{uu}^+(\lambda_x)$; **b**, **d**, **f** spanwise spectra $\kappa_z \phi_{uu}^+(\lambda_z)$. **a**, **b** O900L4; **c**, **d** O900L8; **e**, **f** O900L12. Spectra are normalised by their peak values. Values range from 0 (white) to 0.5 (orange) with an increment of 0.05

and $y/h \gtrsim 0.3$, whereas in the closed channel case it neither scales with u_τ (figure omitted) nor with u_b (Fig. 3b). This scaling property of OCF is in qualitative agreement with the experimental study of OCF by [11], who observed the scaling of the outer flow streamwise turbulence intensity with u_b for $1260 \leq Re_\tau \leq 5280$. Besides, the streamwise turbulence intensity profile obtained from recent OCF measurements by [8, 9] at $\mathrm{Re}_\tau = 2407$ normalised with u_b collapses well with our data (Fig. 3a inset). The above observation differs significantly from CCF, where the spectral signature of VLSMs first appears at $\mathrm{Re}_\tau \approx 600$ (figure omitted) and the contribution of VLSMs to the streamwise turbulence intensity is much lower than in OCF [8]. Even at $\mathrm{Re}_\tau = 10049$, which is currently the highest Reynolds number ever been achieved by a CCF DNS [6], the bulk scaling of the streamwise turbulence intensity has not

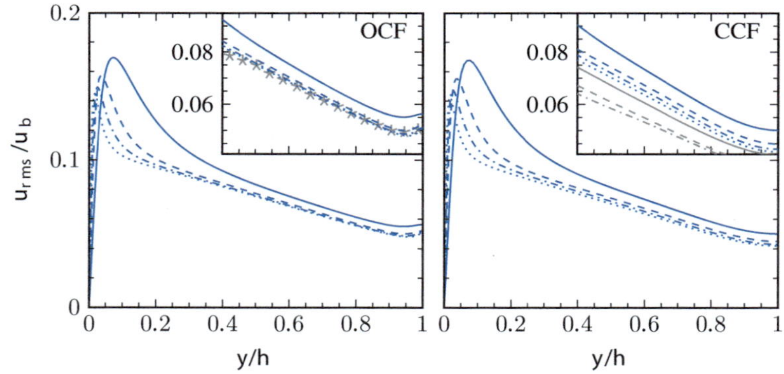

Fig. 3 Streamwise turbulence intensity normalised by u_b as function of the distance from the wall y/h for OCF (left) and CCF (right): —, $\mathrm{Re}_\tau \approx 200$; - - -, $\mathrm{Re}_\tau \approx 400$; -·-, $\mathrm{Re}_\tau \approx 600$; ..., $\mathrm{Re}_\tau \approx 900$. The insets show a zoom for the upper channel half. Left: The symbols (∗) indicate a profile from OCF measurements at $\mathrm{Re}_\tau = 2407$ [8]. Right: The gray lines indicate CCF DNS data at $\mathrm{Re}_\tau = 2003$ (—, [4]), $\mathrm{Re}_\tau = 5186$ (- - -, [5]), $\mathrm{Re}_\tau = 10049$ (-·-, [6])

been observed. Hence, it is remarkable that OCF streamwise turbulence intensity scales with u_b at a Reynolds number as low as $\mathrm{Re}_\tau = 400$, if a sufficiently large numerical domain is used.

5 Conclusion

We performed DNSs of OCF and CCF in computational domains up to $L_x \times L_z = 12\pi h \times 4\pi h$ and Reynolds numbers up to $\mathrm{Re}_\tau = 900$. The enhancement of VLSMs in OCF leads to more stringent requirements on the domain size than in the closed channel counterpart. Since VLSMs appear at lower Re in OCF, a domain size $L_x \geq 12\pi h$ is already required at the moderate value $Re_\tau = 400$. Unlike CCF, the streamwise turbulence intensity in OCF scales with the bulk velocity u_b for $Re_\tau \geq 400$ and $y/h \gtrsim 0.3$, which is attributed to the larger and more intense VLSMs in OCF. Although it was suggested by the experiment of Afzal et al. [11], this is the first time the bulk scaling of the streamwise turbulence intensity has been observed in DNS data. In addition, only with $L_x = 12\pi h$ the bulk-scaling profile appears in the $Re_\tau = 400$ case (figure omitted), which indicates that VLSMs in OCF already play a significant role at this Reynolds number. Thus, it is of critical importance to use sufficiently large numerical domains. Finally, both OCF and CCF turbulence statistics data sets are available online [1].

Acknowledgements Many fruitful discussions with Genta Kawahara throughout this work are gratefully acknowledged. The simulations were carried out at UC2 of SCC Karlsruhe and at DLR's CARA cluster. The computer resources and assistance provided by these centres are gratefully acknowledged. YS and MU acknowledge funding by DFG through grant UH 242/3-1.

References

1. C. Bauer, Y. Sakai, M. Uhlmann, *Data underlying the publication: Direct numerical simulation of turbulent open channel flow*, 4TU.ResearchData, https://doi.org/10.4121/88678F02-2A34-4452-8534-6361FC34D06B.V1 (2023)
2. J. Kim, P. Moin, R.D. Moser, J. Fluid Mech. **177**, 133–166 (1987)
3. R.D. Moser, J. Kim, N.N. Mansour, Phys. Fluids **11**, 943–945 (1999)
4. S. Hoyas, J. Jiménez, Phys. Fluids **18**, 11702 (2006)
5. M. Lee, R.D. Moser, J. Fluid Mech. **774**, 395–415 (2015)
6. M. Oberlack, S. Hoyas, S.V. Kraheberger, F. Alcántara-Ávila, J. Laux, Phys. Rev. Lett. **128**, 024502 (2022)
7. J. Yao, X. Chen, F. Hussain, J. Fluid Mech. **953**, A19 (2022)
8. Y. Duan, Q. Chen, D. Li, Q. Zhong, J. Fluid Mech. **892**, A3 (2020)
9. Y. Duan, Q. Zhong, G. Wang, P. Zhang, D. Li, J. Fluid Mech. **918**, A40 (2021)
10. C. Bauer, *Master's Thesis* (KIT, Karlsruhe, 2015)
11. B. Afzal, M.A. Faruque, R. Balachandar, J. Hydraul. Res. **47**(1), 66–81 (2009)
12. I. Nezu, W. Rodi, J. Hydraul. Eng. **112**, 335–355 (1986)
13. A. Lozano-Durán, J. Jiménez, Phys. Fluids **26**, 011702 (2014)
14. D. Feldmann, C. Bauer, C. Wagner, J. Turbul. **19**(3), 274–295 (2018)
15. M. Pinelli, H. Herlina, J.G. Wissink, M. Uhlmann, J. Fluid Mech. **933**, A49 (2022)
16. J.C. del Álamo, J. Jiménez, Phys. Fluids **15**(6), L41–L44 (2003)

Temperature Assimilation for Convective Flows by Means of Convolutional Neural Networks

M. Mommert, C. Bauer, and C. Wagner

Abstract A convolutional encoder-decoder network trained on instantaneous velocity fields is used to assimilate corresponding temperature fields of convective flows. In particular, synthetic data of Rayleigh-Bénard convection in a cubic sample at $Ra = 10^8$, $Pr = 0.7$ and $Ra = 10^{10}$, $Pr = 6.9$ is studied and the shape and size of the windowed input and output of the network are varied to determine a favorable domain. Additionally, the amount of training data is also varied to determine it's extrapolation potential. This approach proves to predict the temperature fields well for all parameter variations considered. Particularly good correlations between the predictions and the ground truth are achieved for horizontal planar domains and large amounts of training data.

1 Motivation

Flows that transport heat by convection play an important role in nature and in many engineering applications. Accurate predictions of such convective flows can be made using direct numerical simulations (DNS), but these are computationally expensive. Therefore, DNS can only be used for predictions over comparatively short time periods and for flow problems that, due to their dimensions, can usually be studied in the laboratory [1]. On the other hand, measurements can capture the velocity field over long periods of time. However, in order to be able to determine the convective heat transport of flows, additional spatial temperature measurements are required. Such combined measurements [2] are also laborious and therefore rarely applied.

To facilitate the acquisition of heat flux fields over long time periods, we pursue the assimilation of temperature fields (T) based on existing velocity fields ($\boldsymbol{u}$). Besides

M. Mommert (✉) · C. Bauer · C. Wagner
Department Ground Vehicles, Institute of Aerodynamics and Flow Technology, German Aerospace Center, Bunsenstraße 10, 37073 Göttingen, Germany
e-mail: michael.mommert@dlr.de

C. Wagner
Institute of Thermodynamics and Fluid Mechanics, Technische Universität Ilmenau, 98684 Ilmenau, Germany

R. Örlü et al. (eds.), *Progress in Turbulence X*, Springer Proceedings in Physics 404,
https://doi.org/10.1007/978-3-031-55924-2_43

the numerical approach [3], several machine learning approaches have been found suitable for different assimilation tasks in convective flows [4–7]. The present study has two main objectives regarding the assimilation using purely data-driven machine learning techniques: First, a favorable domain shape–three- or two-dimensional (including vertical or horizontal)–is to be determined. Second, the extrapolation capability of this approach is investigated.

2 Methodology

This study was performed on two DNS datasets of Rayleigh-Bénard convection in a cubic sample, allowing access to temperature fields as ground truth data. Both data sets cover a period of 10 free-fall time units (t_{ff}) and have been downsampled to 64^3 grid points. The datasets represent flows with the following combinations of Rayleigh (Ra) and Prandtl (Pr) numbers: $Ra = 10^8$, $Pr = 0.7$ and $Ra = 10^{10}$, $Pr = 6.9$. There were 201 and 667 flow snapshots in each dataset, respectively. For both datasets, the first 5 snapshots were reserved for calculating the validation metrics, followed by the training snapshots, which were $N_t \in \{1; 4; 16; 64\}$. Finally, the last 30 snapshots were used to test the performance of the models.

An encoder-decoder architecture based on multiple convolutional layers is used to assimilate the temperature fields. It exploits the structural information of the velocity fields and provides an order reduction that is advantageous for potentially noisy inputs. Figure 1 illustrates that the inputs and outputs of the network resemble windows clipped from the respective fields. This reduces the size of the neural networks and augments the training data, since the window views were generated with a step size of 2 grid points in each direction. In addition, a windowed approach is not affected by the large-scale flow state for extrapolative inference tasks because the smaller flow structures are presented to the models independently of the large-scale flow state. Since this requires that the windows of size $N_x \times N_y \times N_z$ is large enough to contain structures, we tested different dimensions of 5, 7, 9, or 11. The 3D domain shape was tested for cubic windows ($N_x = N_y = N_z$), while for the 2D shapes, $N_x = 1$ for a vertical window and $N_z = 1$ for a horizontal window.

The sizes of the convolution kernels were adjusted based on the window size, which can be taken from the annotations of Fig. 1. There, N_* denotes the dimensions of the input and output windows not equal to 1. ReLU functions were used to activate all layers except the last one, which was activated with a sigmoid function.

In this way, the number of trainable model variables ranged from about 11,000 for a $5 \times 5 \times 1$ window to 180,000 for an $11 \times 11 \times 11$ window.

For inference, the temperature fields were reconstructed by averaging over all available window views. The overlapping windows result in averaging over a maximum of $N_x \times N_y \times N_z$ different views for grid points more than half the window size away from the walls, which also smooths the results.

To account for the random weight initialization and the stochastic learning process, 3 training runs of 500 epochs were performed (mean squared error loss function with

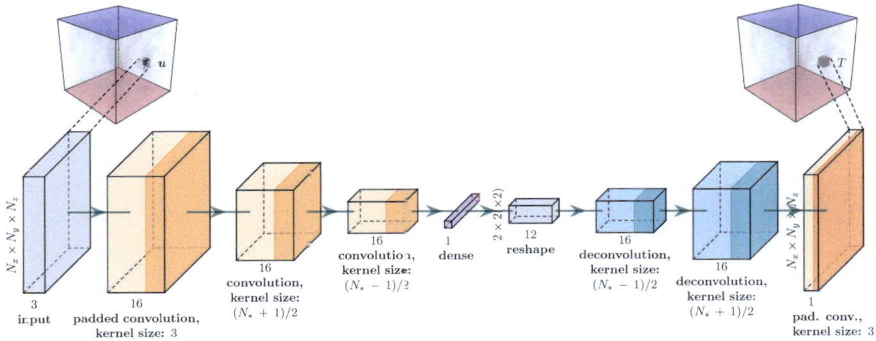

Fig. 1 Architecture of the encoder-decoder neural networks visualized with [8]

Adam optimizer–learning rate 0.0025, batch size 512), from which the run with the best final validation loss was used for further analysis.

3 Analysis

To illustrate the performance of selected models, sample temperature fields for selected window configurations are shown in Fig. 2. The top row of the figure resembles the downsampled DNS data for the last snapshot of the test interval, while the bottom rows show the corresponding reconstructions for models trained with 64 or 1 snapshot. The cubic domains are cropped to display the diagonal plane of the large-scale circulation and half of the top planes.

The visualizations show that the main temperature structures are recovered by all models. Nevertheless, the models are not able to reconstruct the full extent of the temperature amplitudes and small structures due to the smoothing effect, which is particularly evident in the central part of the domain. Increasing the number of training snapshots seems to mitigate these shortcomings.

However, even the shortest training period produces robust results.

3.1 Influence of the Window Size and Shape

To measure the performance of the different window sizes and shapes, the correlation coefficients $\rho(T_{\text{gt}}, T_{\text{pd}})$ between the ground truth (gt) and predicted (pd) temperatures were calculated using the snapshots of the test interval. The results of all cases for the training interval $N_t = 64$ are shown in Fig. 3.

The largest correlation coefficients between reconstructed and ground truth temperature fields for $Ra = 10^8$ and $Pr = 0.7$, left side of Fig. 3, were obtained with the

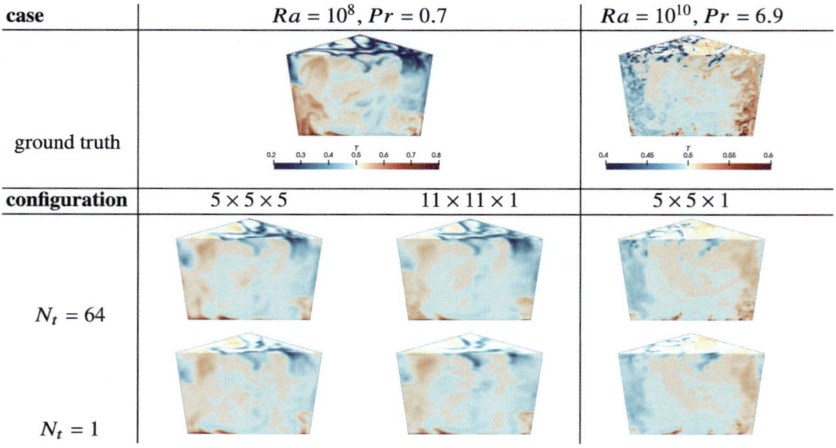

Fig. 2 Sample reconstructed temperature fields of selected window configurations compared to the ground truth of each case

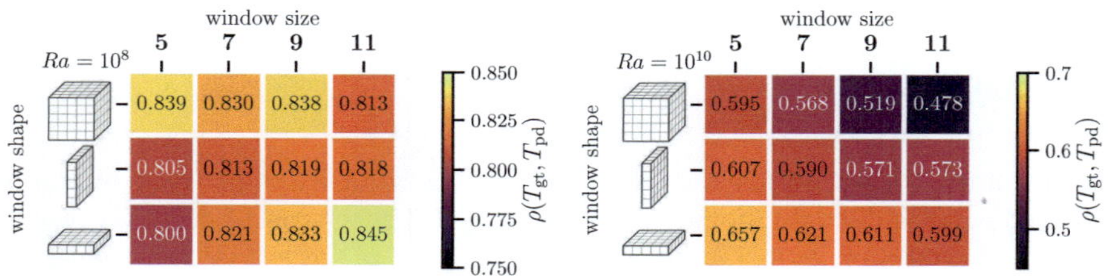

Fig. 3 Correlation results of the predicted temperatures with respect to the ground truth within the test interval for the different cases and window configurations. Left: Case of $Ra = 10^8$, $Pr = 0.7$. Right: Case of $Ra = 10^{10}$, $Pr = 6.9$

largest horizontal window. This indicates, that the flow structures within a horizontal slice are more valuable for temperature assimilation than those present in vertical slices. The advantages of having larger windows, which contain more information about the flow and allow for more complex models, become apparent only for the 2D windows. However, reconstructions based on larger windows are also associated with a stronger smoothing effect, which can be detrimental to the correlation result. This causes the results of the 3D windows to be almost independent of the window size, with the smallest size showing the best performance.

The detrimental smoothing effect of large windows is even more pronounced in the case of $Ra = 10^{10}$ and $Pr = 6.9$ displayed on the right side of Fig. 3, which contains finer flow structures, especially with respect to temperatures. Thus, the correlation values are lower overall, with the best results being obtained for the smallest window size for each shape. As for the lower Rayleigh number, the horizontal shape yields the best results.

3.2 Extrapolation Capability

For the favorable window configurations, we examine the effect of the number of snapshots provided for training by plotting the correlation coefficient $\rho(T_{\mathrm{gt}}(t), T_{\mathrm{pd}}(t))$ for individual snapshots in Fig. 4. The left side of the figure shows the time evolution of the correlations after the last training snapshot at $t_{\mathrm{l\,tr}}$. In all cases, increasing the number of training snapshots improves the assimilation performance. However, the differences between 1 and 4 snapshots are negligible. Since we expect a similar effect for high N_t, increasing it should further improve the results. Furthermore, the primary correlation decay occurs within the first $0.75\,t_{\mathrm{ff}}$ after the last training snap-

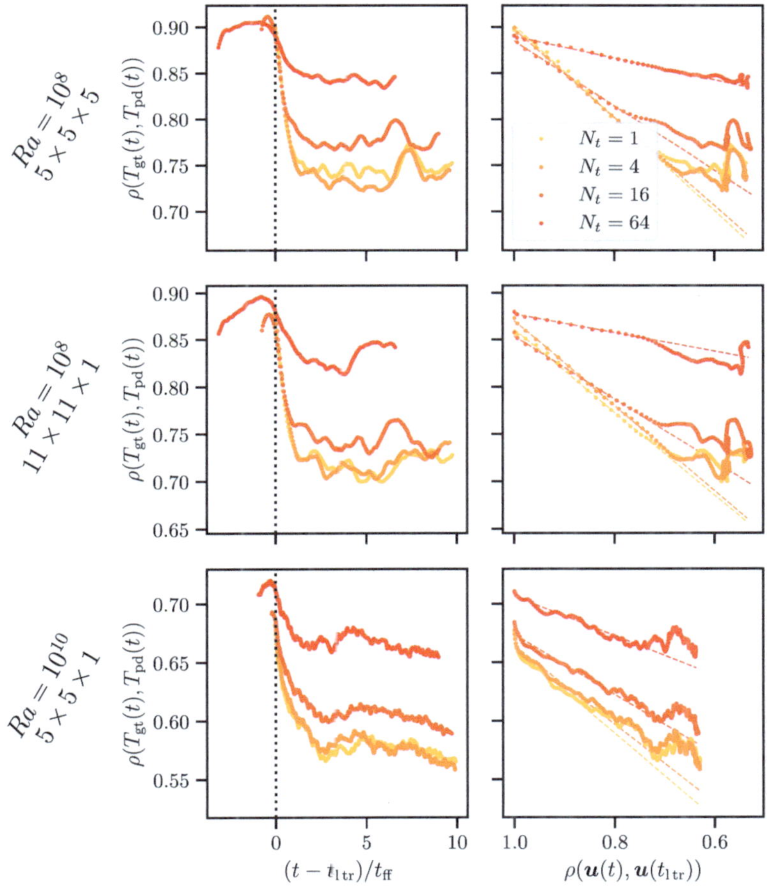

Fig. 4 Snapshot-wise correlation coefficient of the reconstructed temperature fields and the ground truth for selected window configurations. Left: Correlation over time after the last training snapshot at $t_{\mathrm{l\,tr}}$. Right: Correlation in relation to the correlation coefficients of the corresponding velocity fields and the one of the last training snapshot

shot. After that, the existing fluctuations make it difficult to conclude whether the correlation values stabilize or continue to decay slowly.

To gain further insight, we relate the decay to the evolution of the velocity field by plotting it against the correlation coefficient $\rho(\boldsymbol{u}(t), \boldsymbol{u}(t_{\text{l tr}}))$ of the corresponding velocity fields $(\boldsymbol{u}(t))$ to that of the last training snapshot $(\boldsymbol{u}(t_{\text{l tr}}))$ on the right side of Fig. 4. This shows a linear relationship for the initial decay (dashed line fitted for the first $0.75\,t_{\text{ff}}$ after training). But after that, the prediction achieves higher correlations than the linear projection. This suggests that the reconstruction performance loses its dependence on the ongoing evolution of the velocity fields.

4 Conclusion and Outlook

Robust temperature predictions based on instantaneous velocity fields were produced by training the presented convolutional encoder-decoder model architecture. Correlation coefficient analysis revealed the benefits of horizontal planar domains and extended training data. Single snapshot correlation results were also encouraging in terms of stable reconstruction performance over time.

However, for a full proof of the extrapolation capability, the presented approach should be further tested on larger datasets or by applying a DNS-trained model to measured data. Another open question is the precision with which the dimensionless numbers of the training and inference cases need to matched to achieve sufficient results, as preliminary tests showed a noticeable reduction of the correlation performance when models trained on one case with a certain Ra number are applied to a case with another Ra number.

Acknowledgements The authors are thankful for the computing resources provided by the CARA cluster.

References

1. M. Kaczorowski, C. Wagner, *Study on the resolution requirements for DNS in turbulent Rayleigh-Bénard convection, in Notes on Numerical Fluid Mechanics and Multidisciplinary Design* (Springer, Berlin Heidelberg, 2010), pp.199–205
2. M. Mommert, K. Niehaus, D. Schiepel, D. Schmeling, C. Wagner, Measurement of the turbulent heat fluxes in mixed convection using combined stereoscopic PIV and PIT. Exp. Fluids **64**(6) (2023)
3. C. Bauer, D. Schiepel, C. Wagner, Assimilation and extension of particle image velocimetry data of turbulent Rayleigh-Bénard convection using direct numerical simulations. Exp. Fluids **63**(1) (2022)
4. J. Kim, C. Lee, Prediction of turbulent heat transfer using convolutional neural networks. J. Fluid Mech. **882** (2019)

5. S. Cai, Z. Wang, F. Fuest, Y.J. Jeon, C. Gray, G.E. Karniadakis, Flow over an espresso cup: inferring 3-D velocity and pressure fields from tomographic background oriented Schlieren via physics-informed neural networks. J. Fluid Mech. **915** (2021)
6. P. Clark Di Leoni, L. Agasthya, M. Buzzicotti, L. Biferale, Reconstructing Rayleigh–Bénard flows out of temperature-only measurements using physics-informed neural networks. Eur. Phys. J. E **46**(3) (2023)
7. P. Teutsch, T. Käufer, P. Mäder, C. Cierpka, Data-driven estimation of scalar quantities from planar velocity measurements by deep learning applied to temperature in thermal convection. Exp. Fluids **64**, 191 (2023). https://doi.org/10.1007/s00348-023-03736-2
8. H. Iqbal, Harisiqbal88/plotneuralnet v1.0.0 (2018)

Mixed Convection in a Particle-Laden Channel Flow: One and Two-Way Coupling Regimes

D. Zaza and M. Iovieno

Abstract In the present work, by means of Eulerian-Lagrangian DNSs within the point-particle approach, we have studied the modulation of the fluid and particle temperature fields in the channel flow operated by the momentum and energy transfer between the two phases, when also buoyancy forces acting on the fluid are taken into account. We performed simulations at a single friction Reynolds number, $\mathrm{Re}_\tau = 180$, and Prandtl number, $\mathrm{Pr} = 0.71$, while we varied the Richardson number from 0.272 to 27.2, and the Stokes number between 0.6 and 120. Both the one and the two-way coupling regimes were explored in order to evaluate the influence of particles on the fluid. When buoyancy effects are weak, particle feedback significantly influences fluid statistics, whereas its impact seems to diminish and become negligible as natural convection becomes dominant. Particle statistics are showcased across various Stokes numbers, emphasizing how buoyancy-induced enhancement of vertical motions reduces particle concentration near the wall.

1 Introduction

Particle-laden flows occur in many engineering applications and natural phenomena. The most representative examples are the evolution of droplet clouds, sandstorms, soil erosion processes, sediment transport in watercourses, combustion in solid propellant endoreactors, and fluidized bed reactors. The importance of detailed investigations of such flows for modeling purposes is therefore indisputable, and Direct Numerical Simulation (DNS), more than any other approach, has proven to be essential in providing new physical insights into them, particularly when the particle feedback on the fluid phase cannot be neglected, i.e. in the so-called two-way coupling regime [1], as opposed to the one-way coupling, where only the actions of the fluid on the

D. Zaza (✉) · M. Iovieno
DIMEAS, Politecnico di Torino, Corso Duca degli Abruzzi 24, 10129 Torino, Italy
e-mail: domenico.zaza@polito.it

M. Iovieno
e-mail: michele.iovieno@polito.it

R. Örlü et al. (eds.), *Progress in Turbulence X*, Springer Proceedings in Physics 404,
https://doi.org/10.1007/978-3-031-55924-2_44

particles are taken into account. For isotropic turbulence, particle volume fractions, φ, larger than 10^{-6} are in fact effective in modulating turbulence, enhancing turbulent kinetic energy production, or conversely dissipation, depending on the particle Stokes number [2]. The thermal modulation operated by the two-way coupling has been widely investigated as well in both homogeneous isotropic turbulence [3–5] and wall-bounded turbulence [6–8]. In the channel flow, fluid-particle exchanges were found to significantly enhance heat transfer as the specific heat of the particles increased [7], while [6] showed that particle feedback affects the wall heat flux depending on their size. Therefore, the aim of this work is to explore the influence of particle feedback on both fluid and particle temperature fields in the presence of combined natural and forced convection mechanisms.

2 Physical Model

We investigate the non-isothermal particle-laden flow in the plane channel in which also heat transfer by natural convection is taken into account. The channel is $2h$ high in the y coordinate, the computational domain is $2\pi h$ in the streamwise (x) and spanwise (z) directions. The flow is driven by a uniform pressure gradient, $\partial\langle p\rangle/\partial x$, and is seeded with small, spherical, heavy and rigid particles. A temperature difference, $T_1 - T_2$, is imposed between the two isothermal walls, with the lower wall warmer than the upper, to produce an unstable stratification, as in [9]. Periodic boundary conditions are set for both velocity and temperature fields in the homogeneous directions x and z, while no-slip boundary conditions are enforced at the wall. By rescaling the problem using the mean density of the fluid, the channel half height, h, the friction velocity, u_τ, and the half temperature difference between the walls, $(T_1 - T_2)/2$, as mass, length, velocity, and temperature variations reference scales respectively, the flow is governed by the friction Reynolds number, $\mathrm{Re}_\tau = u_\tau h/\nu$, the Prandtl number, $\mathrm{Pr} = \nu/\kappa$, and the friction Richardson number, $\mathrm{Ri}_\tau = h\alpha g\Delta T_{\mathrm{ref}}/u_\tau^2$, where ν, κ and α are respectively the kinematic viscosity, the thermal diffusivity, and the thermal expansion coefficient of the fluid, whereas g is the gravitational acceleration. Since particles are assumed to be smaller than the smallest dynamically significant flow scale, the point-particle approach can be adopted [1]. Therefore, the dispersed phase is followed in a Lagrangian way by solving the well-known Maxey-Riley equations [10], in which only the Stokes drag force contribution is retained. The gravitational force is neglected to avoid deposition, and inter-particle exchanges of momentum and heat are neglected, having minimal effects at low volume fractions [2]. Thus, for the j-th particle one can write the following non-dimensional equations:

$$\frac{\mathrm{d}\mathbf{x}_j}{\mathrm{d}t} = \mathbf{v}_j(t), \qquad \frac{\mathrm{d}\mathbf{v}_j}{\mathrm{d}t} = \frac{\mathbf{u}(t,\mathbf{x}_j) - \mathbf{v}_j(t)}{\tau_u}\left(1 + 0.15\,\mathrm{Re}_j^{0.687}\right), \tag{1}$$

$$\frac{\mathrm{d}\vartheta_j}{\mathrm{d}t} = \frac{\vartheta\left(t,\mathbf{x}_j\right) - \vartheta_j(t)}{\tau_\vartheta}\frac{\mathrm{Nu}_j}{2}, \qquad \mathrm{Nu}_j = 2 + 0.6\,\mathrm{Re}_j^{1/2}\mathrm{Pr}^{1/3}. \tag{2}$$

Here, $\mathbf{x}_j(t)$, $\mathbf{v}_j(t)$ and $\vartheta_j(t)$ are the position, velocity and temperature of the j-th particle, respectively, while $\mathbf{u}(t, \mathbf{x}_j)$ and $\vartheta\left(t, \mathbf{x}_j\right)$ are the fluid non-dimensional velocity and temperature evaluated at the particle location. In (1–2) the empirical corrections already employed in [6] for finite particle Reynolds numbers Re_j have been introduced. The quantities τ_u and τ_ϑ represent the momentum and thermal relaxation times of the particles, and have non-dimensional form: $\tau_u = 2/9\, \rho_p\, r^2\, \mathrm{Re}_\tau$ and $\tau_\vartheta = 1/3\, \rho_p c_p\, r^2 \mathrm{Re}_\tau \mathrm{Pr}$. In these expressions, ρ_p and c_p refer respectively to the dimensionless density and specific heat of the particles. The ratio of the relaxation times to the viscous time scale of the fluid ($\tau_\eta = 1/\mathrm{Re}_\tau$) gives the Stokes number, $\mathrm{St} = \tau_u/\tau_\eta$, and the thermal Stokes number, $\mathrm{St}_\vartheta = \tau_\vartheta/\tau_\eta$, which govern particle dynamics. The fluid phase is instead described by the incompressible Navier-Stokes equations, coupled with the advection-diffusion equation for the fluid temperature

$$\frac{\partial u_j}{\partial x_j} = 0, \tag{3}$$

$$\frac{\partial u_i}{\partial t} + \frac{\partial (u_i u_j)}{\partial x_j} = -\frac{\partial p}{\partial x_i} + \frac{1}{\mathrm{Re}_\tau} \frac{\partial^2 u_i}{\partial x_j^2} + \mathrm{Ri}_\tau \vartheta \delta_{i,2} + \delta_{i,1} - C_{u,i}, \tag{4}$$

$$\frac{\partial \vartheta}{\partial t} + \frac{\partial \left(\vartheta u_j\right)}{\partial x_j} = \frac{1}{\mathrm{Re}_\tau \mathrm{Pr}} \frac{\partial^2 \vartheta}{\partial x_j^2} - C_\vartheta, \tag{5}$$

where $u_i\,(t, \mathbf{x})$ is the i-th component of the fluid velocity field, $p\,(t, \mathbf{x})$ stands for the fluctuating kinematic pressure, and $\vartheta\,(t, \mathbf{x})$ is the temperature field. The term $\mathrm{Ri}_\tau \vartheta \delta_{i,2}$ in (4) is the net buoyancy force within the Boussinesq approximation, while $C_{u,i}$ and C_ϑ in (4) and (5) are the momentum and thermal particle feedback on the carrier phase, i.e. the momentum and heat exchanged per unit volume by all the N_p particles with the fluid. Present only in the two-way coupling, they are given by

$$\mathbf{C}_u = \frac{4}{3}\pi \rho_p r^3 \sum_{j=1}^{N_p} \frac{\mathrm{d}\mathbf{v}_j}{\mathrm{d}t} \delta(\mathbf{x} - \mathbf{x}_j), \qquad C_\vartheta = \frac{4}{3}\pi \rho_p c_p r^3 \sum_{j=1}^{N_p} \frac{\mathrm{d}\vartheta_j}{\mathrm{d}t} \delta(\mathbf{x} - \mathbf{x}_j).$$

The carrier flow equations (3–5) are reformulated and numerically solved in terms of the wall-normal component of the vorticity vector and the wall-normal component of the velocity field as in [11]. Pseudo-spectral methods are employed for the spatial discretization of the fluid equations: the Fourier-Galërkin approach in the two homogeneous directions x and z, and the Chebyshev-tau method in the wall-normal direction y. The aliasing errors are removed via the 2/3-rule [12]. The equations are advanced in time through the implicit Crank-Nicolson scheme for the diffusive terms, and the explicit two-step Adams-Bashforth scheme for the remaining ones. Particle equations (1–2) are integrated in time using the same Adams-Bashforth scheme. Tri-linear interpolation and tri-linear projection are employed for the evaluation of the fluid variables at the particle location and the feedback terms, respectively.

3 Results and Discussion

Twelve simulations were performed at $\mathrm{Re}_\tau = 180$, $\mathrm{Pr} = 0.71$, particle volume fraction $\varphi = 2.5 \times 10^{-4}$, and non-dimensional particle radius $r = 6 \times 10^{-4}$. The resulting number of particles, N_p, was 22 million. We considered two friction Richardson numbers, $\mathrm{Ri}_\tau = 0.272$ and $\mathrm{Ri}_\tau = 27.2$, which exemplify two different limiting cases in which forced convection and natural convection prevail, and three different Stokes numbers, $\mathrm{St} = 0.6$, 60, and 120, in both the one-way and two-way coupling regimes. The particle specific heat value, c_p, was set to 4.17 in each simulation, while the particle density, ρ_p, was adjusted to obtain the desired Stokes numbers. The fluid mean temperature and temperature variance profiles (Fig. 1) show the existence of two different thermal regimes. As the Richardson number grows, and buoyancy forces become more important, the temperature profile in the central region of the channel flattens out, and temperature gradients are confined to a narrow layer near the wall (Fig. 1a). Indeed, an unstable stratification promotes turbulent exchanges, redistributes temperature fluctuations, particularly in the core region, and drives gradients towards the walls [9]. Particle feedback slightly modulates the temperature profile at $\mathrm{Ri}_\tau = 0.272$, producing a small deviation from the the one-way coupling regime, especially for $\mathrm{St} = 0.6$, in agreement with the results obtained by [6] for particles with relatively small inertia (Fig. 1a), while particles do not seem to have any influence on the mean temperature of the fluid at $\mathrm{Ri}_\tau = 27.2$. An abrupt change of the fluid temperature variance is observed as the Richardson number increases (Fig. 1b): the peaks near the wall become stronger and move further towards the wall, while the maxima in the centerline fade to a plateau, confirming the redistribution effect of stratification observed in [9]. Particle feedback significantly modifies the fluid temperature variance $\langle \vartheta' \vartheta' \rangle$ when $\mathrm{Ri}_\tau = 0.272$: the two-way coupling noticeably amplifies the near-wall peak compared to the reference unladen case, and this effect appears more pronounced at lower inertia. Meanwhile, within the core region, particle inertia filters fluid temperature fluctuations in a non-monotonic way. Particle feedback effect is reduced at higher Ri_τ.

Figures 2a, b, which show particle mean temperature in wall units for $\mathrm{Ri}_\tau = 0.272$ and $\mathrm{Ri}_\tau = 27.2$ respectively, indicate that thermal inertia strongly reduces particle temperature in both flow regimes, especially for $y^+ < 30$. This is mainly due to the deposition/resuspension mechanism which promotes particle exchanges between the core and near-wall regions, where temperatures are considerably different [6]. This mechanism is also fostered by the enhanced vertical motions due to buoyancy, which also reduces the preferential clustering of particles in the near-wall region, as can be noticed by comparing Fig. 3a, b. In fact, buoyancy tends to redistribute the particles along the height of the channel, smoothing the near-wall concentration peaks, and driving the dispersed phase towards the logarithmic layer. Particle temperature variance is consistently larger than that of the fluid in each case, and monotonically increases with the Stokes number when $\mathrm{Ri}_\tau = 0.272$ (Fig. 2c). When natural convection becomes dominant (Fig. 2d), the profile of the particle temperature variance mimics that of the fluid, but reaches much higher values: it shows a maximum near the

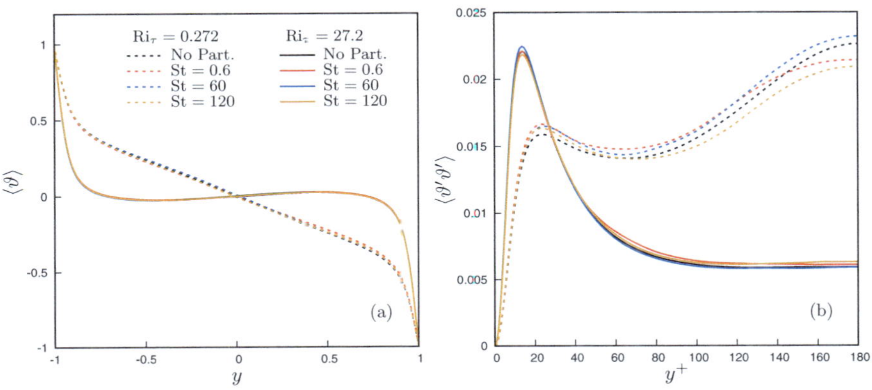

Fig. 1 Fluid mean temperature **a** and temperature variance **b** in the one-way and two-way coupling regimes for $\mathrm{Ri}_\tau = 0.272$, 27.2 and St = 0.6, 60, 120

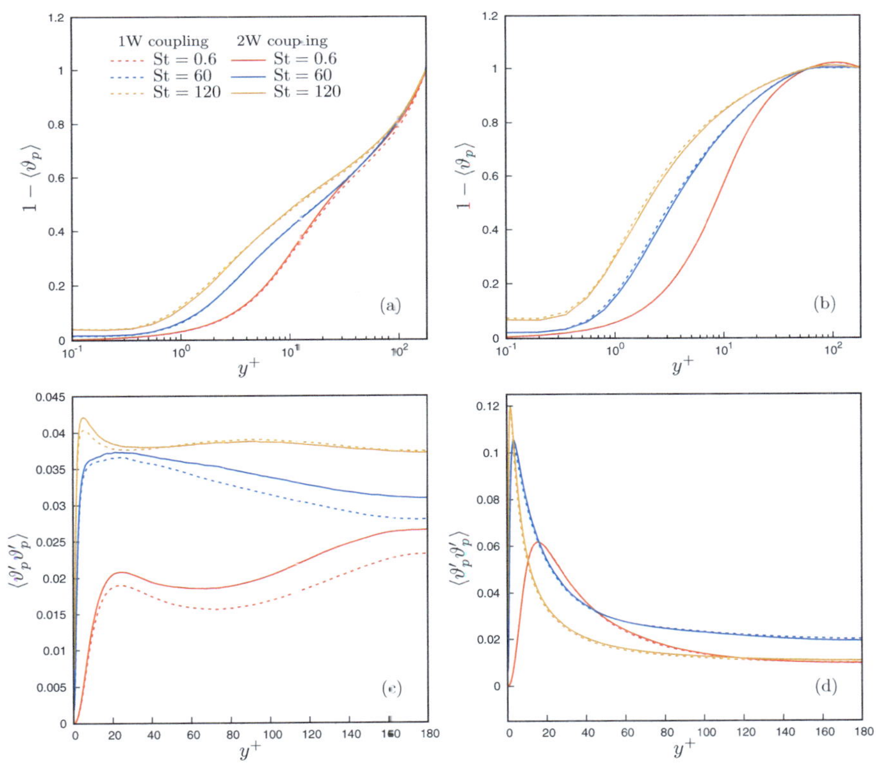

Fig. 2 Particle mean temperature (first row) and temperature variance (second row) in the one-way and two-way coupling regimes for $\mathrm{Ri}_\tau = 0.272$ (left column) and $\mathrm{Ri}_\tau = 27.2$ (right column)

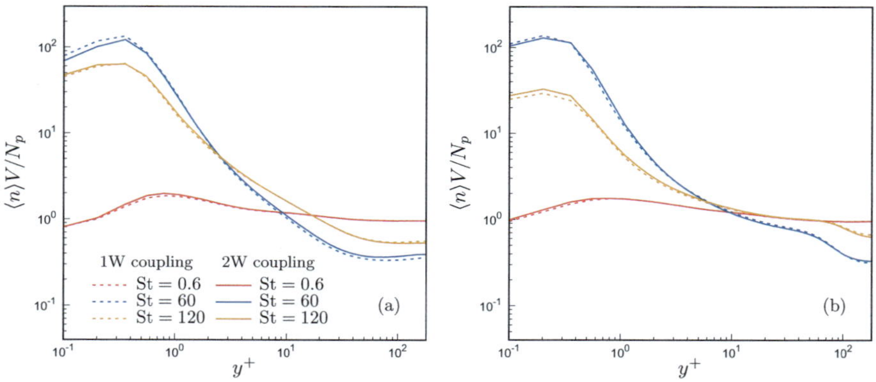

Fig. 3 Normalized particle concentration in the one and two-way coupling regimes: effect of the Stokes number, St, for $\mathrm{Ri}_\tau = 0.272$ **a** and $\mathrm{Ri}_\tau = 27.2$ **b**

wall that increases significantly with Stokes number, and a central plateau that reaches maximum values for St $= 60$. Finally, in the two-way coupling, particle feedback tends to considerably increase particle temperature variance for $\mathrm{Ri}_\tau = 0.272$, especially at lower inertia, while particle modulation appears negligible at $\mathrm{Ri}_\tau = 27.2$.

Acknowledgements The authors acknowledge the computer resources provided by CINECA under the ISCRA initiative, project HP10CIUF9S. Additional resources were provided by HPC@POLITO (http://www.hpc.polito.it).

References

1. J.G.M. Kuerten, Point-particle dns and les of particle-laden turbulent flow-a state-of-the-art review. Flow Turbul. Combust. **97**(3), 689–713 (2016)
2. S. Elghobashi, On predicting particle-laden turbulent flows. Appl. Sci. Res. **52**(4), 309–329 (1994)
3. Y. Sato, E. Deutsch, O. Simonin, Direct numerical simulations of heat transfer by solid particles suspended in homogeneous isotropic turbulence. Int. J. Heat Fluid Flow **19**(2), 187–192 (1998)
4. H.R. Zandi Pour, M. Iovieno, Heat transfer enhancement by suspended particles in a turbulent shearless flow, in *33rd Congress of the International Council of the Aeronautical Sciences, ICAS 2022*, vol. 4 (2022), pp. 2452–2463
5. H.R. Zandi Pour, M. Iovieno, Heat transfer in a non-isothermal collisionless turbulent particle-laden flow. Fluids **7**(345), 11 (2022)
6. F. Zonta, C. Marchioli, A. Soldati, Direct numerical simulation of turbulent heat transfer modulation in micro-dispersed channel flow. Acta. Mechanica. **195**, 305–326 (2008)
7. J.G.M. Kuerten, C.W.M. van der Geld, B.J. Geurts, Turbulence modification and heat transfer enhancement by inertial particles in turbulent channel flow. Phys. Fluids **23**(12), 123301 (2011)
8. F. Rousta, B. Lessani, Near-wall heat transfer of solid particles in particle-laden turbulent flows. Int. Commun. Heat Mass Transf. **112**, 104475 (2020)
9. S. Pirozzoli, M. Bernardini, R. Verzicco, P. Orlandi, Mixed convection in turbulent channels with unstable stratification. J. Fluid Mech. **821**, 482–516 (2017)

10. M.R. Maxey, J.J. Riley, Equation of motion for a small rigid sphere in a nonuniform flow. Phys. Fluids **26**(4), 883–889 (1983)
11. J. Kim, P. Moin, R. Moser, Turbulence statistics in fully developed channel flow at low Reynolds number. J. Fluid Mech. **177**, 133–166 (1987)
12. C. Canuto, M.Y. Hussaini, A. Quarteroni, T.A. Zang, *Spectral Methods in Fluid Dynamics* (Springer, Heidelberg, 1988)

Effect of Coherent Fluctuation in Stellar Convection Viewed From Non-equilibrium Turbulence

N. Yokoi, Y. Masada, and T. Takiwaki

Abstract The non-equilibrium effect associated with plume motions in convective turbulent flows is investigated. Plume motions are treated as coherent fluctuations in a time–space double averaging framework. Non-equilibrium turbulence effect is incorporated into turbulence modelling through temporal variations of turbulence along plume motions.

1 Entropy-Gradient Driven and Surface-Cooling Driven Convections

The surface cooling and resultant down-flowing plumes play a dominant role in the turbulent mixing in the stellar convection [1]. The properties of turbulent transports as well as spatiotemporal field configurations are fairly different between the surface cooling non-locally driven case and the adiabatically locally driven case (Fig. 1) [2].

The spatial distribution of the turbulent fluxes (turbulent mass flux, Reynolds stress, turbulent heat flux, etc.) in the surface-cooling or non-locally driven case are fairly different from the counterparts in the entropy-gradient or locally driven case. The results for the turbulent internal-energy fluxes $\langle e'\mathbf{u}'\rangle$ (e' : internal-energy fluctuation, $\mathbf{u}'$: velocity fluctuation) obtained by direct numerical simulations (DNSs)

N. Yokoi (✉)
Institute of Industrial Science (IIS), University of Tokyo, 4-6-1, Komaba, Meguro, Tokyo 153-8505, Japan
e-mail: nobyokoi@iis.u-tokyo.ac.jp

Y. Masada
Faculty of Science, Department of Applied Physics, Fukuoka University, Fukuoka 814-0180, Japan
e-mail: ymasada@fukuoka-u.ac.jp

T. Takiwaki
Division of Science, National Astronomical Observatory of Japan (NAOJ), Mitaka, Tokyo 181-8588, Japan
e-mail: takiwaki.tomoya.astro@gmail.com

R. Örlü et al. (eds.), *Progress in Turbulence X*, Springer Proceedings in Physics 404,
https://doi.org/10.1007/978-3-031-55924-2_45

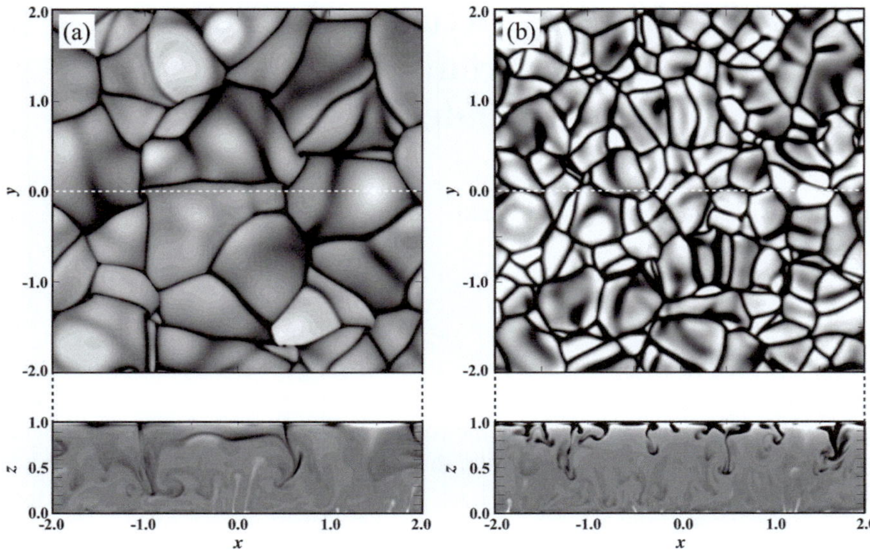

Fig. 1 Entropy distributions in our direct numerical simulations (DNSs) for the locally driven case **a** and the non-locally driven case **b**. The horizontal cross-section at the top surface (upper) and the vertical cross-section of the entropy fluctuation

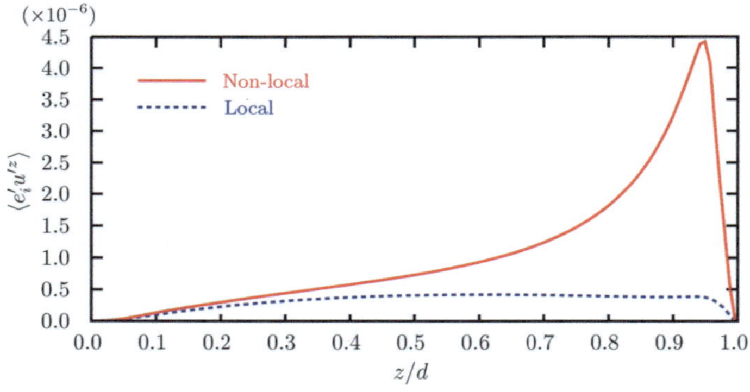

Fig. 2 Spatial profiles of the turbulent internal-energy flux $\langle e' \mathbf{u}' \rangle$ in our DNSs for the locally driven case (dotted line) and non-locally driven case (solid line)

are shown in Fig. 2. In the non-locally driven case, the total amount of flux is much larger than that in the locally-driven case. The spatial distribution of the flux is much more localised in the near surface region in the non-locally driven case, and the peak value is one order higher than the locally-driven counterpart. Such behaviour in the non-locally driven convection cannot be explained by the standard gradient-diffusion type model with the mixing-length theory [3].

2 Response Function Formulation and Non-Equilibrium Effect

From the multiple-scale direct-interaction approximation (MSDIA), the first-order velocity field in the wave number space, $\mathbf{u}'_1(\mathbf{k};\tau)$, is expressed in terms of the zeroth-order velocity $\mathbf{u}'_0(\mathbf{k};\tau_1)$ and the response function $G(\mathbf{k};\tau,\tau_1)$ as

$$
\begin{aligned}
u_1'^{a}(\mathbf{k};\tau) = & -\frac{\partial U^b}{\partial X^a}\int_{-\infty}^{\tau} d\tau_1 G^{\alpha b}(\mathbf{k};\tau,\tau_1) u_0'^{a}(\mathbf{k};\tau_1) - \int_{-\infty}^{\tau} d\tau_1 G^{\alpha a}(\mathbf{k};\tau,\tau_1)\frac{Du_0'^{a}(\mathbf{k};\tau_1)}{DT} \\
& +2M^{dab}(\mathbf{k})\iint \delta(\mathbf{k}-\mathbf{p}-\mathbf{q})d\mathbf{p}d\mathbf{q}\int_{-\infty}^{\tau} d\tau_1 G^{\alpha d}(\mathbf{k};\tau,\tau_1)\frac{q^b}{q^2}u_0'^{a}(\mathbf{p};\tau_1)\frac{\partial u_0'^{c}(\mathbf{q};\tau_1)}{\partial X^c} \\
& -M^{abcd}(\mathbf{k})\iint \delta(\mathbf{k}-\mathbf{p}-\mathbf{q})d\mathbf{p}d\mathbf{q}\int_{-\infty}^{\tau} d\tau_1 G^{\alpha d}(\mathbf{k};\tau,\tau_1)\frac{\partial}{\partial X^c}u_0'^{a}(\mathbf{p};\tau_1)u_0'^{b}(\mathbf{q};\tau_1),
\end{aligned}
\tag{1}
$$

where $\mathbf{U}$ is the mean velocity, $\mathbf{X}$ and T are slow space and time variables, respectively, and $\mathbf{M}$s are combined projection operators in wavenumber space. (For the formulation of the MSDIA and the detailed definitions of the quantities, the reader is referred to, for example, a recent review textbook [4].) Here, the second term, expressed by the Lagrangian derivative $D/DT(=\partial/\partial t + \mathbf{U}\cdot\nabla)$ of $\mathbf{u}'_0$ coupled with the Green's function $\mathbf{G}$, arises from the variation of velocity fluctuation along the mean velocity. This represents the non-equilibrium property of turbulence [5].

Using (1), the turbulent energy K is calculated by

$$
K \equiv \langle \mathbf{u}'^2\rangle/2 = \frac{1}{2}\int d\mathbf{k}\, \langle u'^{\ell}(\mathbf{k};\tau)u'^{\ell}(\mathbf{k}';\tau)\rangle/\delta(\mathbf{k}+\mathbf{k}') \tag{2}
$$

with expansion

$$
\langle u'^{\ell}u'^{\ell}\rangle = \langle u_0'^{\ell}u_0'^{\ell}\rangle + \delta\langle u_1'^{\ell}u_0'^{\ell}\rangle + \delta\langle u_0'^{\ell}u_1'^{\ell}\rangle + \cdots . \tag{3}
$$

The turbulent energy is expressed as

$$
K = C_{K1}\varepsilon^{2/3}\ell_C^{2/3} - C_{K2}\varepsilon^{-3/2}\ell_C^{4/3}\frac{D\varepsilon}{Dt} - C_{K3}\varepsilon^{1/3}\ell_C^{1/3}\frac{D\ell_C}{Dt}, \tag{4}
$$

where ε is the energy dissipation rate, ℓ_C is the characteristic length of energy-containing eddy (C_{K1}, C_{K2} and C_{K3}: model constants). This can be solved with respect to ℓ_C by iteration as

$$
\ell_C = C_{\ell 1}K^{3/2}\varepsilon^{-1} + C_{\ell 2}K^{3/2}\varepsilon^{-2}\frac{DK}{Dt} - C_{\ell 3}K^{5/2}\varepsilon^{-3}\frac{D\varepsilon}{Dt} \tag{5}
$$

($C_{\ell 1}$, $C_{\ell 2}$ and $C_{\ell 3}$: model constants). The first term corresponds to the Kolmogorov scaling based on the local equilibrium of turbulence. The second and third terms arise from the non-equilibrium property of turbulence related to the Lagrangian derivatives of K and ε. Including this non-equilibrium contributions, from (5), the length scale of turbulence ℓ_{C} is approximated as

$$\ell_{\mathrm{C}} = \ell_{\mathrm{E}} \left(1 - C'_{\mathrm{N}} \frac{1}{K} \frac{D}{Dt} \frac{K^2}{\varepsilon} \right), \tag{6}$$

where $\ell_{\mathrm{E}} = K^{3/2}/\varepsilon$ is the equilibrium length scale (C'_{N}: model constant).

The eddy viscosity for non-equilibrium turbulent flow can be modelled as

$$\nu_{\mathrm{T}} = \nu_{\mathrm{E}} \left(1 \pm C_{\mathrm{N}} \frac{1}{K} \frac{D}{Dt} \frac{K^2}{\varepsilon} \right)^{\mp 1}, \tag{7}$$

where the double-signs correspond in the same order. The upper (lower) signs are for the positive (negative) $(D/Dt)(K^2/\varepsilon)$. This non-equilibrium expression for eddy viscosity has been applied to homogeneous shear flow turbulence, where the standard $K - \varepsilon$ model with the equilibrium eddy viscosity ν_{E} overestimates the turbulent energy. It was confirmed that this non-equilibrium model with $C_{\mathrm{N}} = 0.8$ could successfully reproduce the temporal behaviour of the turbulent energy observed in direct numerical simulations (DNSs) [6].

3 Time–Space Double Averaging

In the Reynolds averaging procedure with a space or ensemble averaging, the plume motions cannot be properly captured since collective motions of plumes are smeared out through the space or ensemble averaging. In order to extract the effect of plumes, a time–space double averaging procedure is adopted, where a field quantity f is decomposed as

$$f = \overbrace{\langle \overline{f} \rangle + \underbrace{\widetilde{f}}_{\overline{f} - \langle \overline{f} \rangle}}^{\overline{f}} + \underbrace{f''}_{f - \overline{f}}. \tag{8}$$

Here, $\langle \overline{f} \rangle$ is the space average of the time averaged $\overline{f}$, $\widetilde{f}$ is the dispersion part of $\overline{f}$ and is denoted as the *coherent fluctuations*, f'' is the deviation from the time average $\overline{f}$ and is denoted as *incoherent fluctuations*. In this framework, the plumes are regarded as the *coherent fluctuations* represented by $\widetilde{\mathbf{u}}$ (the fluctuation velocity $\mathbf{u}'$ is divided into $\mathbf{u}' = \widetilde{\mathbf{u}} + \mathbf{u}''$).

In the present work, we adopt the simplest possible time averaging by filter G_{F} as

$$\overline{f}(\mathbf{x}; t) = \int f(\mathbf{x}; s) G_{\mathrm{F}}(t - s) ds \tag{9}$$

with

$$G_{\rm F}(t-s)=\begin{cases}1/T & (|t-s|\le T/2),\\ 0 & \text{(otherwise)},\end{cases} \tag{10}$$

and the space averaging in the horizontal plane

$$\langle f\rangle(z;t)=\frac{1}{\Delta x\Delta y}\int_S f(x,y,z;t)dxdy. \tag{11}$$

With this time–space double averaging formulation, the evolution equations of the coherent fluctuation energy and the incoherent fluctuation energy for the Boussinesq convection flow are written as

$$\begin{aligned}\left(\frac{\partial}{\partial t}+\langle u\rangle^\ell\frac{\partial}{\partial x^\ell}\right)\left\langle\frac{1}{2}(\tilde{u}^a)^2\right\rangle &= -\langle\tilde{u}^a\tilde{u}^\ell\rangle\frac{\partial\langle u\rangle^a}{\partial x^\ell}-\frac{1}{\rho_0}\left\langle\tilde{p}\left(\frac{\partial\tilde{u}^a}{\partial x^a}\right)\right\rangle-\nu\left\langle\left(\frac{\partial\tilde{u}^a}{\partial x^\ell}\right)^2\right\rangle\\ &+\frac{g}{\Theta_0}\left\langle\widetilde{\theta\tilde{u}^\ell}\right\rangle\delta^{\ell 3}+\frac{\partial}{\partial x^\ell}\left\langle-\tilde{u}^\ell\frac{1}{2}(\tilde{u}^a)^2+\tilde{p}\tilde{u}^\ell+\nu\frac{\partial}{\partial x^\ell}\frac{1}{2}(\tilde{u}^a)^2\right\rangle\\ &+\left\langle\widetilde{u''^\ell u''^a}\frac{\partial\tilde{u}^a}{\partial x^\ell}\right\rangle-\frac{\partial}{\partial x^\ell}\left\langle\widetilde{u''^\ell u''^a}\tilde{u}^a\right\rangle,\end{aligned} \tag{12}$$

$$\begin{aligned}\left(\frac{\partial}{\partial t}+\langle u\rangle^\ell\frac{\partial}{\partial x^\ell}\right)\left\langle\frac{1}{2}(u''^a)^2\right\rangle &= -\langle u''^a u''^\ell\rangle\frac{\partial\langle u\rangle^a}{\partial x^\ell}-\frac{1}{\rho_0}\left\langle p''\frac{\partial u''^a}{\partial x^a}\right\rangle-\nu\left\langle\left(\frac{\partial u''^a}{\partial x^\ell}\right)^2\right\rangle\\ &+\frac{g}{\Theta_0}\left\langle\theta'' u''^\ell\right\rangle\delta^{\ell 3}+\frac{\partial}{\partial x^\ell}\left\langle-u''^\ell\frac{1}{2}(u''^a)^2+p''u''^\ell+\nu\frac{\partial}{\partial x^\ell}\frac{1}{2}(u''^a)^2\right\rangle\\ &-\left\langle\widetilde{u''^\ell u''^a}\frac{\partial\tilde{u}^a}{\partial x^\ell}\right\rangle-\frac{\partial}{\partial x^\ell}\left\langle\widetilde{\frac{1}{2}(u''^a)^2\tilde{u}^\ell}\right\rangle,\end{aligned} \tag{13}$$

where g/Θ_0-related terms represent the fluctuation energies due to buoyancy (g: gravity, Θ_0: reference mean temperature).

The first terms in the third lines of (12) and (13) represent the energy transfer between the coherent and incoherent fluctuations, $\widetilde{K}\equiv\langle\widetilde{\mathbf{u}}^2\rangle/2$ and $K''\equiv\langle\mathbf{u}''^2\rangle/2$. They are expressed by the production mechanisms as

$$P_{\tilde{K}}\equiv\left\langle\widetilde{u''^\ell u''^i}\frac{\partial\tilde{u}^i}{\partial x^\ell}\right\rangle=-P_{K''}. \tag{14}$$

This means that, in the presence of the coherent velocity shear, $\nabla\tilde{\mathbf{u}}$, energy transfer between the coherent and incoherent fluctuation components is occurred mediated by the dispersion part of the random fluctuation velocity correlation, $\widetilde{\mathbf{u}''\mathbf{u}''}(\equiv\overline{\mathbf{u}''\mathbf{u}''}-\langle\overline{\mathbf{u}''\mathbf{u}''}\rangle)$ (Fig. 3).

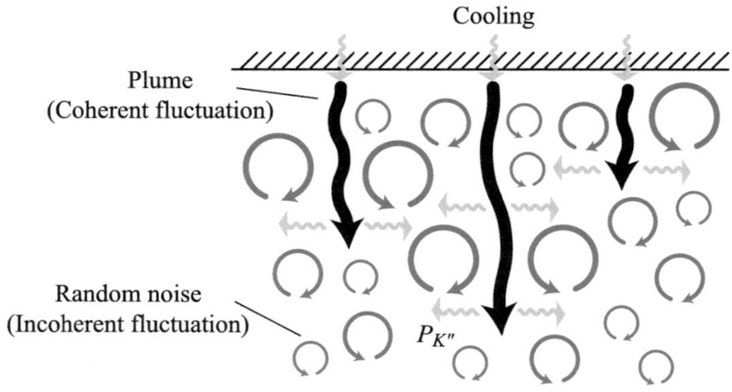

Fig. 3 Schematic picture of the interaction between the coherent and incoherent fluctuations. Plumes (coherent fluctuations, depicted by thick black curved lines) are driven by the surface cooling. The energy of plume motions is transferred to the energy of random noise (incoherent fluctuations, depicted by grey circle eddies) by the interaction $P_{K''}$ if $P_{K''} > 0$

In addition to the role of coherent fluctuations, the non-equilibrium effect along plume motion is expected to control the turbulent fluxes. On the analogy of the non-equilibrium effect in the Reynolds averaged model (7), the non-equilibrium effect in the time-space averaging procedure is expressed as [3]

$$\kappa_{\rm NE} = \begin{cases} \kappa_{\rm E} \left[1 - C_{\tilde{\tau}} \dfrac{\tilde{\tau}}{\langle \mathbf{u}'^2 \rangle} \tilde{\Gamma}_D \right] & \text{for } \tilde{\Gamma}_D < 0, \\ \kappa_{\rm E} \left[1 + C_{\tilde{\tau}} \dfrac{\tilde{\tau}}{\langle \mathbf{u}'^2 \rangle} \tilde{\Gamma}_D \right]^{-1} & \text{for } \tilde{\Gamma}_D > 0, \end{cases} \tag{15}$$

with $\widetilde{\Gamma}_D = \langle (\widetilde{\mathbf{u}} \cdot \nabla) \overline{\mathbf{u}'^2} \rangle$, where $\tilde{\tau}$ is the timescale of coherent fluctuation and $C_{\tilde{\tau}}$ is the model constant. The non-equilibrium effect along the plume motions is represented by $\widetilde{\Gamma}_D$.

The turbulent internal-energy flux $\langle e' \mathbf{u}' \rangle$ in the non-locally driven case is plotted in comparison with the DNS (Fig. 4). The spatial distribution of $\langle e' \mathbf{u}' \rangle$, which cannot be reproduced by the standard gradient-diffusion model with the mixing-length expression, is well reproduced by the non-equilibrium model.

The role of plumes in stellar convection can be incorporated into turbulence model in the framework of a time–space double averaging, as the non-equilibrium effect of turbulence through variation of turbulence along plume motions.

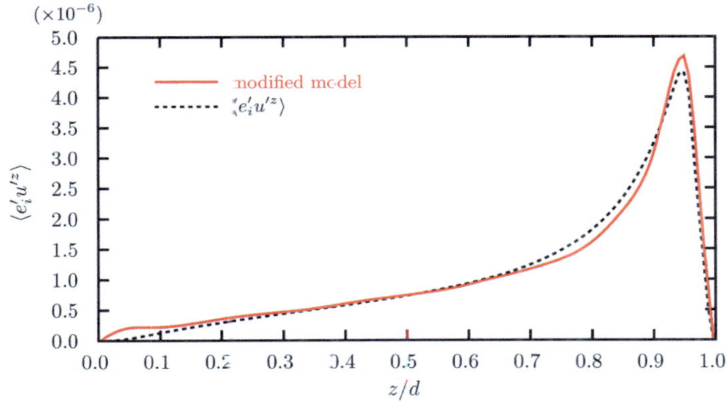

Fig. 4 Spatial profiles of the turbulent internal-energy flux $\langle e' \mathbf{u}' \rangle$ in the non-locally driven case. The results by the DNS (dotted line) and by the non-equilbrium model (solid line)

References

1. H.C. Spruit, Convection in stellar envelopes: A changing paradigm. Memorie della Società Astronomica Italiana **68**, 397–413 (1997)
2. J.-F. Cossette, M.P. Rast, Supergranuration as the largest buoyantly driven convective scale of the Sun. Astrophys. J. **829**, L17-1–5 (2016)
3. N. Yokoi, Y. Masada, T. Takiwaki, Modelling stellar convective transport with plumes–I. Non-equilibrium turbulence effect in double-averaging formulation. Mon. Not. R. Astron. Soc. **516**, 2718–2735 (2022)
4. N. Yokoi, Turbulence, transport and reconnection, in *Topics in Magnetohydrodynamic Topology, Reconnection and Stability Theory*, ed. by D. MacTaggart, A. Hillier (Springer, 2020), pp. 177–265
5. N. Yokoi, Non-equilibrium turbulent transport in convective plumes obtained from closure theory. Atmosphere **14**, 01013-1–22 (2023)
6. A. Yoshizawa, S. Nisizima, A non-equilibrium representation of the turbulent viscosity based on the two-scale turbulence theory. Phys. Fluids A **5**, 3302–3304 (1993)

Corner Circulation Scaling Laws of Turbulent Rayleigh-Bénard Convection in a Cubic Cell

R. Barta, C. Bauer, D. Schiepel, and C. Wagner

Abstract Rayleigh-Bénard convection (RBC) in a cubic cell is studied at Prandtl number $\mathrm{Pr} = 6.9$ in the hard turbulent regime with Rayleigh numbers $\mathrm{Ra} \gg \mathrm{Pr} \cdot 10^6$. The predominant flow structure is the large-scale circulation oriented along one of the cell diagonals. Corner circulations can develop in the cell corners along both cell diagonals. Here, the scaling laws for the Reynolds number of the corner circulations are calculated analyzing two direct numerical simulations performed for two Rayleigh numbers $\mathrm{Ra} \in \{3 \cdot 10^8, 1 \cdot 10^{10}\}$. Two scaling laws are found, one for each cell diagonal and the corner circulations along the same diagonal seem to obey the same scaling law. One of the scaling laws is similar to the results obtained for an two-dimensional (2D) RBC sample. The scaling laws are validated with a particle tracking velocimetry experiment of a cubic Rayleigh-Bénard convection cell filled with water ($\mathrm{Pr} \approx 6.9$) at $\mathrm{Ra} \approx 1 \cdot 10^{10}$.

1 Theoretical Background

We consider turbulent Rayleigh-Bénard convection (RBC) in a cubic cell with aspect ratio $\Gamma = 1$, see Fig. 1. For an incompressible fluid, the underlying set of equations

$$\frac{\partial \mathbf{u}}{\partial t} + \mathbf{u} \cdot \nabla \mathbf{u} = -\nabla p + \sqrt{\frac{\mathrm{Pr}}{\mathrm{Ra}}}\, \Delta \mathbf{u} + \theta \cdot \mathbf{e}_z \tag{1}$$

$$\frac{\partial \theta}{\partial t} + \mathbf{u} \cdot \nabla \theta = \frac{1}{\sqrt{\mathrm{Pr}\,\mathrm{Ra}}}\, \Delta \theta \tag{2}$$

$$\nabla \cdot \mathbf{u} = 0 \tag{3}$$

R. Barta (✉) · C. Bauer · D. Schiepel · C. Wagner
Institute of Aerodynamics and Flow Technology, DLR, Göttingen, Germany
e-mail: robin.barta@dlr.de

C. Wagner
Institute of Thermodynamics and Fluid Mechanics, TU Ilmenau, Ilmenau, Germany

R. Örlü et al. (eds.), *Progress in Turbulence X*, Springer Proceedings in Physics 404,
https://doi.org/10.1007/978-3-031-55924-2_46

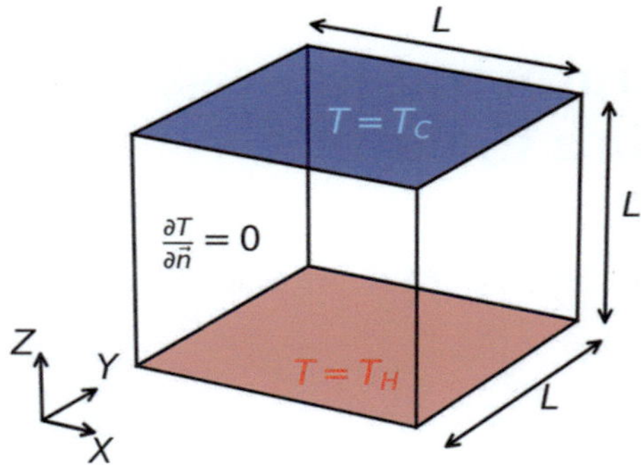

Fig. 1 A sketch of a cubic RBC cell with volume L^3 and L being the cell height. All walls are no-slip boundaries, top and bottom walls are isothermal and side walls are adiabatic

(1)–(3) in dimensionless form is given by the Navier-Stokes equation in Boussinesq approximation with the fluid temperature θ. The hard turbulent regime is defined by the relation (4) between the system parameters.

$$\mathrm{Ra} \gg \mathrm{Pr} \cdot 10^6 \tag{4}$$

In this we consider the following Rayleigh and Prandtl numbers.

$$\mathrm{Ra} = \frac{g\,\lambda\,\Delta T\,L^3}{\nu\,\kappa} \in \{3 \cdot 10^8, 1 \cdot 10^{10}\} \quad , \quad \mathrm{Pr} = \frac{\nu}{\kappa} = 6.9 \tag{5}$$

Here, g is the absolute value of the gravitational acceleration on Earth, λ is the heat expansion coefficient, $\Delta T = T_H - T_C$ the temperature difference between heating and cooling plates with temperatures T_H and T_C, respectively, ν the kinematic viscosity and κ the thermal diffusivity.

RBCs in cubic cells are predominated by a large scale circulation (LSC) along one of the cell diagonals [1]. Pandey et al. [2] derived the scaling law in equation (6) for the Reynolds number of the LSC in the turbulent regime (4) for a cubic cell.

$$\mathrm{Re}_{\mathrm{LSC}} = \frac{\bar{U}_{\mathrm{LSC}} \cdot L}{\nu} \approx \sqrt{\frac{7.5}{\mathrm{Pr}}}\,\mathrm{Ra}^{0.38} \tag{6}$$

The maximum velocity magnitude of the LSC in a time-averaged flow field is denoted by $\bar{U}_{\mathrm{LSC}}$. Secondary flow structures known as Corner Circulations (CC), arise at the corners of the enclosed domain and are coupled to the LSC. In cubic cells, CC can be found in both cell diagonals no matter how the LSC is oriented. In general, four CCs are found. Zhou et al. [3] derived the scaling law in equation (7) for the CC Reynolds number in 2D and quasi-2D domains with aspect ratio $\Gamma = 1$.

$$\mathrm{Re}_{CC} = \frac{\bar{U}_{CC} \cdot r_{CC}}{\nu} \approx \mathrm{Ra}^{0.25} \tag{7}$$

Here, r_{CC} defines the average distance of the CC core to the intersections to the surrounding walls and $\bar{U}_{CC}$ is the maximum velocity magnitude of the corresponding CC in a time-averaged flow field. In the following, we analyze the CC in a cubic 3D RBC cell and provide scaling laws for its Reynolds number.

2 Scaling Law Analysis

Two direct numerical simulations (DNS) are performed with control parameters from equation (5), see Barta et al. [4] for details on the DNS. The reference scales for estimating physical values from the DNS are $L \in \{300, 500\}$ mm for $\mathrm{Ra} \in \{3 \cdot 10^8, 1 \cdot 10^{10}\}$, $\Delta T = 6$ K, a mean temperature of the medium water of $\bar{T} = 20\,°\mathrm{C}$ and the kinematic viscosity of water $\nu = 1.003\ \frac{\mathrm{mm}^2}{s}$. Each DNS provides a collection of CCs. The corresponding CC Reynolds numbers are estimated from the DNS velocity fields which are temporally averaged over a turnover time of the LSC. Figures 2 and 3 show slices of the mean velocity field along the cell diagonal and its off-diagonal for the DNS with $\mathrm{Ra} = 1 \cdot 10^{10}$, respectively.

A convex hull $\partial\Omega$ of each CC region Ω is estimated by coherent structure detection [5] and thus isolated from the LSC. The center of each CC is estimated as the 3D position where the velocity magnitude reaches its lowest value inside the convex hull. The average distance of each CC r_{CC} is estimated by averaging the distances between the CC center and the convex hull $\partial\Omega$. $\bar{U}_{CC}$ is measured as the maximum absolute velocity in Ω. Finally, the CC Reynolds number Re_{CC} is calculated using equation (7). Table 1 gives an overview of all measured properties of the LSC and the CC of both simulations.

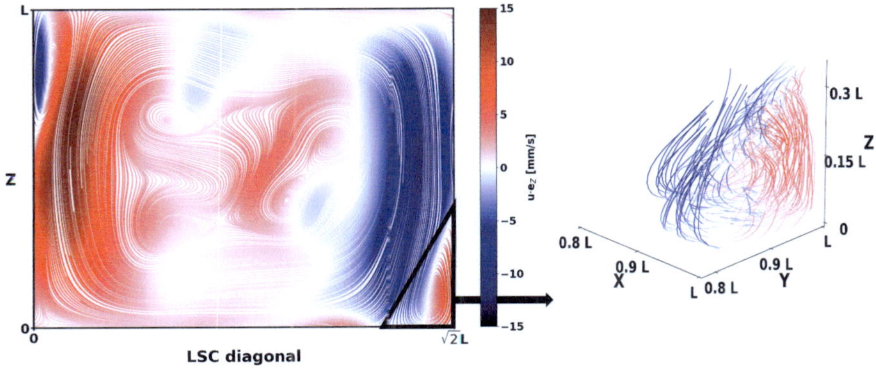

Fig. 2 The streamlines of the mean velocity field along the LSC diagonal from the DNS with $\mathrm{Ra} = 1 \cdot 10^{10}$ (left) and close-up view of the 3D streamlines of an isolated CC (right)

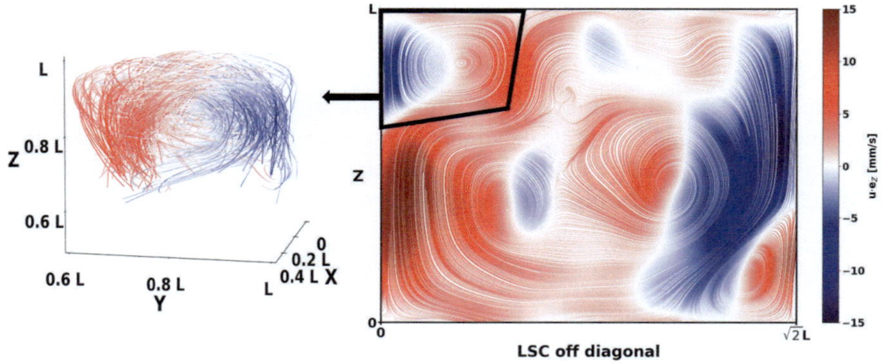

Fig. 3 The streamlines of the mean velocity field along the LSC off-diagonal from the DNS with $\mathrm{Ra} = 1 \cdot 10^{10}$ (right) and close-up view of the 3D streamlines of an isolated CC (left)

Table 1 Measured velocity and length scales of the LSC and four CC from the DNS results with $\mathrm{Ra} \in \{3 \cdot 10^8, 1 \cdot 10^{10}\}$. The Prandtl number is fixed at $\mathrm{Pr} = 6.9$. The last two lines of the table show the calculated Reynolds numbers for each circulation based on the measured properties. The kinematic viscosity of water $\nu = 1.003\ \frac{\mathrm{mm}^2}{s}$ at $\bar{T} = 20°\mathrm{C}$ is used

Ra	$\bar{U}_{\mathrm{LSC}}\left[\frac{\mathrm{mm}}{s}\right]$	$\bar{U}_{\mathrm{CC}}^{\mathrm{diag},1}\left[\frac{\mathrm{mm}}{s}\right]$	$\bar{U}_{\mathrm{CC}}^{\mathrm{diag},2}\left[\frac{\mathrm{mm}}{s}\right]$	$\bar{U}_{\mathrm{CC}}^{\mathrm{offdiag},1}\left[\frac{\mathrm{mm}}{s}\right]$	$\bar{U}_{\mathrm{CC}}^{\mathrm{offdiag},2}\left[\frac{\mathrm{mm}}{s}\right]$
$3 \cdot 10^8$	10.75	3.41	5.53	7.68	6.54
$1 \cdot 10^{10}$	13.53	6.17	6.71	6.02	6.67
Ra	L [mm]	$r_{\mathrm{CC}}^{\mathrm{diag},1}$[mm]	$r_{\mathrm{CC}}^{\mathrm{diag},2}$[mm]	$r_{\mathrm{CC}}^{\mathrm{offdiag},1}$[mm]	$r_{\mathrm{CC}}^{\mathrm{offdiag},2}$[mm]
$3 \cdot 10^8$	300	44.31	33.73	38.02	37.43
$1 \cdot 10^{10}$	500	45.61	49.96	85.26	67.36
Ra	$\mathrm{Re}_{\mathrm{LSC}}$	$\mathrm{Re}_{\mathrm{CC}}^{\mathrm{diag},1}$	$\mathrm{Re}_{\mathrm{CC}}^{\mathrm{diag},2}$	$\mathrm{Re}_{\mathrm{CC}}^{\mathrm{offdiag},1}$	$\mathrm{Re}_{\mathrm{CC}}^{\mathrm{offdiag},2}$
$3 \cdot 10^8$	3133.91	161.87	172.69	270.34	262.38
$1 \cdot 10^{10}$	6568.60	301.78	310.02	475.88	481.98

The calculated CC Reynolds numbers are used to estimate the scaling laws for the CC along each cell diagonal, see equation (8).

$$\mathrm{Re}_{\mathrm{CC}}^{\mathrm{diag}} \approx \mathrm{Ra}^{0.256} \quad , \quad \mathrm{Re}_{\mathrm{CC}}^{\mathrm{offdiag}} \approx \mathrm{Ra}^{0.271} \tag{8}$$

The 3D CCs located in the LSC diagonal obey almost the same scaling law as derived for a 2D RBC by as Zhou et al. [3], see equation (7). The scaling exponent for CCs along the LSC off-diagonal differs from this scaling by about 0.12. At the high Rayleigh numbers, the corresponding difference is not insignificant.

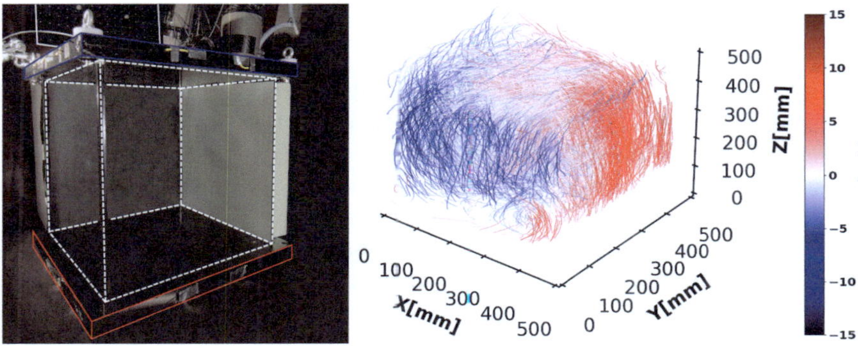

Fig. 4 **Left:** the cubic cell used in the experiment. The heating plate is indicated by a red frame and the cooling plate by a blue frame. **Right:** All particle trajectories processed with pyPTV resolving the LSC and a CC along the LSC off-diagonal

3 Experimental Verification

The estimated scaling laws in equation (8) are verified with experimental data from Schiepel et al. [6]. A cubic cell with side length $L = 500$ mm, temperature difference $\Delta T = 6$ K and water as medium is used to provide the control parameters $\mathrm{Pr} \approx 6.9$ and $\mathrm{Ra} \approx 1 \cdot 10^{10}$. The measurement is processed over 100 time steps with a recoding frequency of 6 Hz using our in-house PTV framework pyPTV, see Barta et al. [4]. A CC along the LSC off-diagonal is fully resolved and can be seen in the lower front corner in Fig. 4. The figure also shows the cubic cell used in the experiment. A mean field over a turnover time is estimated based on the particle tracks and the CC Reynolds number is calculated using the measured average CC distance and velocity scale, see equation (9).

$$\bar{U}_{\mathrm{CC}}^{\mathrm{offdiag}} = 6.71\,\frac{\mathrm{mm}}{\mathrm{s}} \quad , \quad r_{\mathrm{CC}}^{\mathrm{offdiag}} = 82.61\mathrm{mm} \quad \rightarrow \quad \mathrm{Re}_{\mathrm{CC}}^{\mathrm{offdiag}} = 552.65 \tag{9}$$

The scaling law for the CC along the LSC off-diagonal in equation (8) can be used to predict the corresponding CC Reynolds number. The resulting Reynolds number is shown in equation (10). The predicted CC Reynolds number and the measured CC Reynolds number deviates by approximately 7.2%.

$$\mathrm{Re}_{\mathrm{CC}}^{\mathrm{offdiag}} \approx \mathrm{Ra}^{0.271} \approx 512.86 \tag{10}$$

4 Conclusion

The corner circulations of turbulent Rayleigh-Bénard convection in a cubic cell are studied. The Reynolds numbers of the corner circulations along the same cell diagonal are found to obey the same scaling law. Two direct numerical simulations of RBC with system parameters (5) are used to estimate two scaling laws, one for each diagonal of the cubic cell, see equation (8). Interestingly, the scaling law for the corner circulations along the LSC diagonal is close to the result obtained by Zhou et al. [3] for a 2D RBC. This CC scaling law is probably similar in 2D and 3D since the LSC and CCs are coupled in their dynamics and the Reynolds number scaling law (6) for the LSC in 2D and 3D rectangular cells with aspect ratio $\Gamma = 1$ is similar as shown by Pandey et al. [2]. The resulting scaling law for the corner circulation along the LSC off-diagonal is verified with an experiment with similar system parameters as one of the direct numerical simulations. Particle tracking velocimetry is used to measure the flow field. A corner circulation is fully resolved by the particle tracks. The measured and predicted Reynolds numbers differ by 7.2 %. Further investigation of corner circulations in systems with different system parameters is necessary to refine the estimated scaling laws and is therefore one of our ongoing research goals. New high resolution measurement data will help us to find even more corner circulations in real experiments to test these scaling laws.

References

1. V.T. Vishnu, A.K. De, P.K. Mishra, Dynamics of large-scale circulation and energy transfer mechanism in turbulent Rayleigh-Bénard convection in a cubic cell. Phys. Fluids **32**, 095115 (2020)
2. A. Pandey, A. Kumar, A.G. Chatterjee, M.K. Verma, Dynamics of large-scale quantities in Rayleigh-Bénard convection. Phys. Rev. E **94**, 053106 (2016)
3. W.F. Zhou, J. Chen, Similarity model for corner roll in turbulent Rayleigh-Bénard convection. Phys. Fluids **30**, 111705 (2018)
4. R. Barta, C. Bauer, C. Wagner, pyPTV–a complete framework to reconstruct physical flow fields from camera images, in *Conference Proceeding International Symposium on Particle Image Velocimetry* (2023)
5. B.E. Husic, K.L. Schlueter-Kuck, J.O. Dabiri, Simultaneous coherent structure coloring facilitates interpretable clustering of scientific data by amplifying dissimilarity. PLoS ONE **14**(3), e0212442 (2019)
6. D. Schiepel, J. Bosbach, C. Wagner, Tomographic particle image velocimetry of turbulent Rayleigh-Bénard convection in a cubic sample. J. Flow Visual. Image Process. **20** (2013)

MIX
Papier aus verantwortungsvollen Quellen
Paper from responsible sources
FSC® C105338

If you have any concerns about our products,
you can contact us on
ProductSafety@springernature.com

In case Publisher is established outside the EU,
the EU authorized representative is:
Springer Nature Customer Service Center GmbH
Europaplatz 3, 69115 Heidelberg, Germany

Printed by Libri Plureos GmbH
in Hamburg, Germany